Purple Comet! Math Meet

the first ten years

Purple Comet! Math Meet

the first ten years

Titu Andreescu

Jonathan Kane

Library of Congress Control Number: 201292068

ISBN-13: 978-0-9799269-1-4 ISBN-10: 0-9799269-1-2

9 8 7 6 5 4 3 2 1

www.awesomemath.org

Cover design by Iury Ulzutuev

Prologue

The **Purple Comet! Math Meet** is a free, annual, on-line, team, mathematics competition designed for middle and high school students. Problems written for this contest take into account the contest format. Because the math meet is a team contest for teams of up to six students, some problems require analysis of many cases; something we shy away from for individual contests because of time restrictions, but something better suited to a team contest. Although contest participants are welcome to use calculators and computers to help solve problems, all problems are designed to be done without calculators.

Similar to the format of the American Invitational Mathematics Exam (AIME), all problems for the **Purple Comet! Math Meet** have non-negative integer answers. Unlike the AIME, answers are not limited to three-digit integers, and indeed, it is common for problems have answers greater than 1000. Problems are always stated in a way so that the answers are non-negative integers. The following examples demonstrate how this is done.

PROBLEM 1 Find the sum of the four smallest odd prime numbers.

ANSWER The four smallest odd prime numbers are 3, 5, 7, and 11, so their sum is $3 + 5 + 7 + 11 = 26$.

PROBLEM 2 If a fair coin is flipped two times, the probability that one flip will be HEADS and the other will be TAILS is $\frac{m}{n}$ where m and n are relatively prime positive integers. Find $m + n$.

ANSWER Using H to represent a flip of HEADS, and T to represent a flip of TAILS, there are four equally likely ways to flip a fair coin twice: HH, HT, TH, and TT. In two of these four cases, one flip is H and one flip is T. The correct probability is $\frac{2}{4}$. But 2 and 4 are not relatively prime; that is, the fraction $\frac{2}{4}$ is not in lowest terms. Reducing the fraction to lowest terms gives $\frac{1}{2}$. Now 1 and 2 are relatively prime positive integers. The requested sum is $1 + 2 = 3$.

PROBLEM 3 The diagram below shows a square with area 3 with an equilateral triangle with the same side length as the square mounted on top. The total height of this figure can be put in the form $\frac{m+\sqrt{n}}{p}$ where m and p are relatively prime positive integers. Find $m+n+p$.

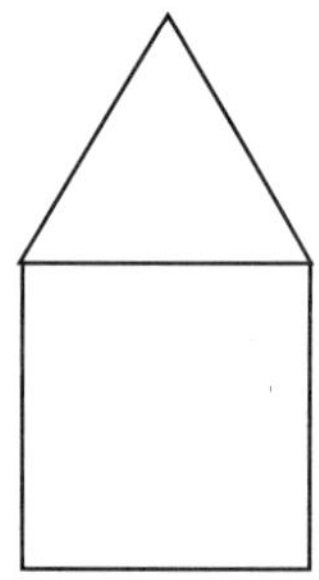

ANSWER For the square to have area 3, it must have side length $\sqrt{3}$. Since an equilateral triangle with side length s has height $\frac{s\sqrt{3}}{2}$, the this equilateral triangle has height

$$\frac{\sqrt{3}\sqrt{3}}{2} = \frac{3}{2}.$$

Thus, the total height of the figure is

$$\sqrt{3} + \frac{3}{2} = \frac{3+2\sqrt{3}}{2} = \frac{3+\sqrt{12}}{2}.$$

Since $m = 3$ and $p = 2$ are relatively prime positive integers, the requested sum is $3 + 12 + 2 = 17$.

What makes the Purple Comet*!* Math Meet successful? Its strengths are easy to list:

- The meet is free and easy to join.
- Homeschooled students can form teams.
- The team format makes the competition fun.
- Some problems are specifically designed for teams to do case analysis.
- The problems span a wide range of mathematics knowledge and difficulty levels.
- Some teachers register their entire class, separating the students into teams of six.
- Team members do not all have to be working at the same location to compete together.
- Girls from some cultures who sometime shy away from competitions are attracted to the team format of the contest.
- Students have a rare opportunity to enter a contest where they can compare their results with similar teams from around the world.

The Purple Comet! Math Meet

I enjoyed mathematics competitions, but up until the spring of 2003 my involvement had been limited to coaching the Putnam team where I worked at the University of Wisconsin - Whitewater and occasionally helping to run a Whitewater Math Meet for area high school students who would bus to Whitewater for a morning of problem solving. But in May of 2003 I received a telephone call from Titu Andreescu. Titu's involvement in mathematics competitions was by that time legendary. He had been an active Mathematical Olympiad participant when he was a student in Romania and later coached the Romanian Mathematical Olympiad team. After immigrating to the United States in 1990, his mathematics team coaching at the Illinois Mathematics and Science Academy soon attracted the attention of the American Mathematics Competitions (AMC) after which he became head coach of the USA Mathematical Olympiad team leading the team to their 1994 victory at the International Mathematical Olympiad where each of the six members of the team received a perfect score. Subsequently, Titu became the director of the AMC itself. To follow was a long list of publications about mathematical problem solving and problem collections bearing his name.

I had met Titu Andreescu a few times and even had the opportunity to bring him to Wisconsin in April of 2002 as a speaker at the Wisconsin Section meeting of the Mathematical Association of America the year I served as program chair for that meeting. But the May 2003 phone call I received from Titu took me quite by surprise. Titu had completed his Ph.D. in Number Theory which he had begun in Romania and was using that opportunity to step down as director of the AMC to seek an academic appointment. He was placing calls to acquaintances hoping to find a position for the next year.

Unfortunately, the only open positions available at the University of Wisconsin - Whitewater were lectureships which would not have come with a suitable salary. Undeterred, I placed a call to my department chairperson, Bennette Harris, and explained how bringing Dr. Andreescu to Whitewater, even for a short-term appointment, could mean a great deal to the department. Bennette discussed the possibility with our campus administrators, and we were encouraged to write a proposal to bring Dr. Andreescu to campus. With Titu's input, we wrote a proposal which was accepted, and, as a result, Titu joined our faculty as a Visiting Scholar for the academic year 2003-2004.

The proposal gave Dr. Andreescu a light teaching load in exchange for our lining up several outreach activities for him that would take advantage of his expertise in mathematical problem solving. Titu and I worked together to set up a workshop for high school teachers interested in learning more about problem solving, to have him contribute a problem of the week column to the student newspaper, and to help coach our Whitewater Putnam team. By November it was time to consider running a mathematics competition, but it was getting too late to plan the logistics needed to bring area high school students to campus for a day. So we approached our department chair, Bennette Harris. Bennette had recently taken a year's leave of absence from teaching to work professionally as a web developer and teach web development courses for IBM. Bennette agreed that if Titu and I could supply the problems, then he could run a UW - Whitewater Math Meet on the web, thus releasing us from having to bring students to our campus for the event.

The first on-line University of Wisconsin - Whitewater Math Meet took place on Thursday December 11, 2003 from 3:30 to 5:00 PM CST. We had sent flyers advertising the event to many area middle and high schools, and we had mentioned the competition to a few associates from around the country. The format of the event was that teams would log onto our web site and register. Then at the time of the competition, 25 problems would be presented to them on the screen. The medium of an on-line contest determined the format of the contest. Because it would be impossible for us to monitor whether students were working together or not, we chose to make the competition a team contest, allowing teams of up to six students each. Since it would be impossible for us to monitor whether students were using calculators or computers to solve problems, we allowed them to use calculating devices. Because it would be easy to have the computer check submitted answers if they were all whole numbers, we designed each problem so it would have a nonnegative integer answer.

Each team was allowed to use whatever strategies they wanted in order to solve as many problems as they could within the allotted 90 minute time frame. About 90 teams participated in the first competition which was only slightly marred by a technical problem that prevented teams from being able to see the problems for several minutes at the beginning of the contest requiring us to add an extra ten minutes to the length of the competition.

Just fourteen weeks later, on March 11, 2004 Titu, Bennette, and I had put together and advertised our second on-line competition. In our first competition, middle and high school teams worked on the same set of problems. In order to make the problems challenging enough for high school teams, we had written problems that were out of reach for most middle school students. To offer middle school students a more appropriate contest, we prepared separate lists of problems for the middle and high school teams.

By the spring of 2005 Titu had found himself a permanent teaching position at the University of Texas at Dallas. We had had so much fun putting on our on-line competition while Titu was visiting us in Whitewater, that we decided to continue the tradition by sponsoring the University of Wisconsin - Whitewater/ University of Texas at Dallas Math Meet. This name was clearly cumbersome at best and only lasted one year. For our third contest, which took place on Thursday April 14, 2005, we reduced both the number of problems and the amount of allotted time for the middle school level of the competition. Now the middle school teams would only need to compete for 60 minutes to try to solve 15 problems while the high school teams would continue to work on 25 problems for 90 minutes. We tried to advertise more widely hoping to attract teams from across North America. When a team from Europe expressed interest in joining the competition in spite of the fact that they would need to participate in the middle of their night, we made arrangements for them to participate.

Many changes were made for our April 2006 meet. First of all, we needed a better name, so we combined the Purple/White color scheme of UW - Whitewater with the comet mascot of UT at Dallas to get the "Purple Comet*!* Math Meet". Bennette added the exclamation point, usually written in a larger font in italics because the symbol looked like a comet. We had always allowed teams to register as a "small school" team or a "large school" team with the idea that larger schools were able to recruit students from a broader spectrum of mathematics talent and could usually compete at a higher level. Since associating each team with a particular school prevented participation by mathematics clubs and homeschoolers, we introduced a new category of "Mixed Team" for such teams.

But perhaps the biggest change we introduced in 2006 was to expand the competition window from a fixed 90 minute period to one where each team could choose to compete during any 90 minute period over several days. This flexibility was welcome by teams that could not easily compete at the time we chose to be good for those in the central time zone of the United States. It allowed us to open up the competition to teams from around the world. We originally chose a 36 hour period on April 26 and 27, but were later talked into adding another half day for a full 48 hours. Because teams would now be able to see our contest problems over a two-day period, we needed a way to ensure that teams would take the competition seriously and not announce our problems before the end of the contest. We accomplished this by requiring every team to have an adult supervisor. The supervisors would register on our web site and then sign up as many teams as they wanted. The supervisors were given the responsibility of enforcing our contest rules and encouraging student participants not to discuss or publicize the problems before the meet had ended. In 2007 we expanded the contest to a four-day format. We had signed up more than 600 teams, but we were frustrated that we had not been able to attract even more students. The software we were using could certainly handle a much larger number of teams, but we were limited in how many participants we could attract because we had no budget to pay for advertising. So in preparation of our 2008 meet, I consulted the Office of Research and Sponsored Programs at the University of Wisconsin - Whitewater. They were amazingly helpful and guided me through the filing of several grant proposals. One of these proposals landed a generous grant from the Lynde and Harry Bradley Foundation of Milwaukee, Wisconsin. This grant allowed us to send out advertising to about 6,000 mathematics teachers. The university's Marketing and Media Department helped us produce a flyer as well as a contest poster which we made available on our web site so that mathematics teachers could print a poster to help attract attention in their schools. We hired student workers to staff our contest helpline so that any participant who had a contest-related question during the contest could have immediate responses to their e-mail or telephone call. As a result, in 2008 we had over 1700 teams registered from 13 countries.

In 2009, we expanded the contest window to a full seven days. In order to attract participants from a wider variety of countries, I found mathematicians who helped us translate our contest problems into 15 different languages. Bennette Harris took a one-semester sabbatical in the fall of 2008 specifically to do a complete redesign of our contest website, making it easier to use by supervisors and teams as well as providing more functionality to those of us administering the contest. One addition was the ability for a registered team to run a practice contest by having their team work problems from one of our earlier contests. This practice showed both teams and supervisors exactly what the actual competition would be like. We had participants from 25 countries.

Another change in 2009 was the providing of certificates of participation and award certificates for all students participating in the contest. When the results of the meet were announced, supervisors were able to log onto our site and download printable copies of these certificates. Within an hour of becoming available, certificates for 160 teams had already been downloaded, so we felt that this was a worthwhile addition.

We had noticed that many teams received scores of zero on the contest, and felt this might be discouraging for them. We did not want the reputation of being an impossibly hard contest. To combat this, for our 2010 contest we added five additional problems to both the middle school and the high school contests. The five problems were specifically designed to be easy for just about any participating team. The difficulty of the most challenging problems was not changed. The change had the desired effect, and we now see only a small number of teams getting a score of zero.

In 2011, the contest was run for a full 9 days, allowing schools with a spring break during the contest week to compete on at least one of the contest days. Breaking all previous records, 2200 teams registered for the contest from 39 countries. In 2012, the contest ran for ten days and saw over 2700 registered teams from 43 countries.

Although the contest has been run for ten years by Drs. Andreescu, Harris, and I, we gladly acknowledge the help we receive. There is a team of volunteer proofreaders who help ensure that our problems are clearly formulated, and their solutions are understandable and correct. There are many volunteer translators who have translated our contest problems into more than 15 languages each year. Every year, about a dozen undergraduate students work our contest help line round the clock to provide assistance to contest participants. The University of Wisconsin - Whitewater has supported us with fund-raising and advertising. And, of course, there have been several contributors supporting these efforts financially. We also send a special thank you to the Lynde and Harry Bradley Foundation which has continued to provide substantial financial support for the project.

In May of 2011 Bennette Harris retired from his teaching position at the University of Wisconsin - Whitewater to take a job in industry. In May of 2012 I also chose to retire. Although we are very grateful to the University of Wisconsin - Whitewater for their many years of support of the Purple Comet! Math Meet, it no longer makes sense for the contest to be run as a program at that institution. Instead, the contest will now be administered through AwesomeMath, LLC and sponsored by their charitable arm, The American Foundation for Excellence in Mathematics. AwesomeMath is best known for the summer mathematics camps it has been running since 2006. Titu Andreescu directs AwesomeMath, and I have been teaching at their summer camp since its inception.

Jonathan Kane, November 2012

Why Purple Comet!

I am an associate professor of mathematics in the Science and Mathematics Education department at the University of Texas at Dallas (UTD). When I came to UTD in January of 2005, I was already well known in the international mathematical community. I served as head coach and leader of the USA International Mathematical Olympiad (IMO) team for eight years, and director of the Mathematical Association of America (MAA) American Mathematics Competitions for five years.

I am an experienced mathematics educator. I have been a high school teacher, a professor, a math coach, and a mentor for more than 30 years. I had the privilege to work with many of the brightest mathematical minds in the USA and around the world. I have given scores of presentations, keynote addresses, seminars, guest lectures, and workshops all over the globe. My service and work have been recognized by a Certificate of appreciation presented by the president of the MAA for "Outstanding service as Coach of the USA Mathematical Olympiad Program in preparing the USA Team for its perfect performance in Hong Kong at the 1994 IMO". I was also the first place winner of the Edith May Sliffe Award for "Distinguished High School Mathematics Teaching", bestowed by the president of the MAA.

I am the author or co-author of 24 books and editor of 18 others. My books employ a problem-solving approach to mathematics. Some of them are structured so that they can be used as textbooks for undergraduate and graduate studies. Others can be utilized as resources for talented mathematical middle and high school students and teachers to supplement their Olympiad training or purely to expand their mathematical horizons.

My original investigation work primarily addresses the preparation of students for mathematics contests and advanced competitions. It is well known in the mathematical community that there are only a few resources for gifted math students seeking to take their ability to the next level. Therefore, in my books I use a problem-based approach to teaching these students. I use it as a tool to help enhance students' interests and skills in mathematics. Certain advanced topics are best introduced to young students by motivating the concepts through problems that encourage investigation.

I first met Jonathan Kane and his wife, Janet Mertz, in June 1999, when they accompanied their young son Dani to the three-week Mathematical Olympiad Summer Program (MOSP) in Lincoln, Nebraska. I was running this prestigious program for exceptional students who had made it to the top in mathematics competitions – these students were winners of the United States of America Mathematical Olympiad (USAMO) or had received honorable mentions. The top six in the remarkable 30-student group went on to represent our country in the International Mathematical Olympiad (IMO), which was held in Romania that year. The 7th grade rookie, Daniel Kane, did not make the team in 1999, but he made it in 2002 and 2003, bringing two gold medals for Team USA.

I saw Jon and Janet again at the 2001, 2002, and 2003 award ceremonies in Washington DC, honoring the top 12 USAMO winners. Daniel Kane was one of the winners, with a perfect score in 2002. We had many good conversations and in 2003, when I was about to end my tenure as Director of the AMC and started to seek a position in academia, I asked Jon if his university had an opening. It was already mid June but two months later I was a visiting professor at University of Wisconsin-Whitewater (UWW). I collaborated with Jon on many interesting math problems and the UWW Math Meet was a natural thing to start. Both Jon and I write challenging questions easily. He is better at coming up with discrete math and combinatorial questions, I am better when it comes to inventing hard algebra or number theory problems. We both like to write geometry and trigonometry questions. Every fall Jon urges me to send him questions for the next year's meet. We usually set the final version of the exam at the MAA-AMS Joint Meetings in January.

The quality of the problems we wrote and the technical expertise of Bennette made this mathematics meet so successful. I don't want to repeat Jon's detailed and accurate historical account of what has become Purple Comet*!* Math Meet. Instead, I would like to point out that the association of Purple Comet*!* with AwesomeMath made both programs even more successful. In 2012, the AwesomeMath Summer Program (AMSP) ran three-week residential summer camps at the University of Texas at Dallas, Cornell University, and University of California, Santa Cruz. AwesomeMath's main goal is to significantly boost mathematics skills and creativity in gifted middle school and high school students. The program debuted in 2006 with just over 100 participants. Since then, the number of participating students increased each year. In 2012, we had more than 450 students. The camps attract students from across the country and around the world and inspire a love for the subject that will last into college years and farther. AwesomeMath takes students beyond the limits of the traditional classroom and gives them the unique opportunity to work with world renowned mathematics coaches, interact with an international student body, and develop advanced problem solving skills. The AwesomeMath Summer Program provides an intellectually stimulating environment for a broad base of gifted students to study meaningful mathematics. My belief is that all students who have the motivation and ability to charge into the frontiers of mathematics should be given the mentoring and guidance they need. This is exactly what AwesomeMath and Purple Comet*!* Math Meet offer.

I worked with Jon, Janet, and Joe Gallian, then president of the Mathematical Association of America, on a report published Oct. 10, 2008, in the Notices of the American Mathematical Society (Titu Andreescu, Joseph A. Gallian, Jonathan M. Kane, and Janet E. Mertz. Cross-Cultural Analysis of Students with Exceptional Talent in Mathematical Problem Solving). The report was a comprehensive analysis of decades of data on students identified as having profound ability in math portraying a culturally constricted pipeline that puts American leadership in the mathematical sciences and related fields at risk. The report was very well received and reviewed in over 100 of the top newspapers and journals from the USA and around the globe.

I look forward to continuing working with Jon and Bennette in order to make Purple Comet*!* even more popular. I greatly enjoy teaching mathematics both inside and outside the classroom. In the latter setting, I seek opportunities to work with students individually and in groups to help them prepare for math competitions. Providing a stimulating mathematical environment for students can be as important to their education as providing them with quality classroom instruction. Indeed, no matter the environment and techniques employed, if we always keep the students in mind and they put in honest effort, then significant learning will take place.

Titu Andreescu, November 2012

Contents

Chapter 1

Problems

1.1 2003 Middle and High School

Problem 1. In eight years Henry will be three times the age that Sally was last year. Twenty five years ago their ages added to 83. How old is Henry now?

Problem 2. What is the least number that could be the date of the first Saturday after the second Monday following the second Thursday of a month?

Problem 3. What is the greatest integer whose prime factors add to 14?

Problem 4. The lengths in inches of the diagonals of a rhombus are two consecutive integers. The area of the rhombus is 210 square inches. Find its perimeter, in inches.

Problem 5. Let a, b, c be nonzero real numbers such that $a + \frac{1}{b} = 5$, $b + \frac{1}{c} = 12$, and $c + \frac{1}{a} = 13$. Find $abc + \frac{1}{abc}$.

Problem 6. Evaluate

$$\frac{1}{\log_2\left(\frac{1}{6}\right)} - \frac{1}{\log_3\left(\frac{1}{6}\right)} - \frac{1}{\log_4\left(\frac{1}{6}\right)}.$$

Problem 7. Find the least n such that every subset of $\{1, 2, 3, \ldots, 2004\}$ with n elements contains at least two elements that are relatively prime.

Problem 8. Let $ABCDEFGHIJKL$ be a regular dodecagon. Find

$$\frac{AB}{AF} + \frac{AF}{AB}.$$

Problem 9. Let f be a real-valued function of real and positive argument such that

$$f(x) + 3xf\left(\frac{1}{x}\right) = 2(x+1)$$

for all real numbers $x > 0$. Find $f(2003)$.

Problem 10. How many gallons of a solution which is 15% alcohol do we have to mix with a solution that is 35% alcohol to make 250 gallons of a solution that is 21% alcohol?

Problem 11. If
$$\frac{1}{1+2}+\frac{1}{1+2+3}+\cdots+\frac{1}{1+2+\cdots+20}=\frac{m}{n},$$
where m and n are positive integers with no common divisor, find $m+n$.

Problem 12. How many triangles appear in the diagram below?

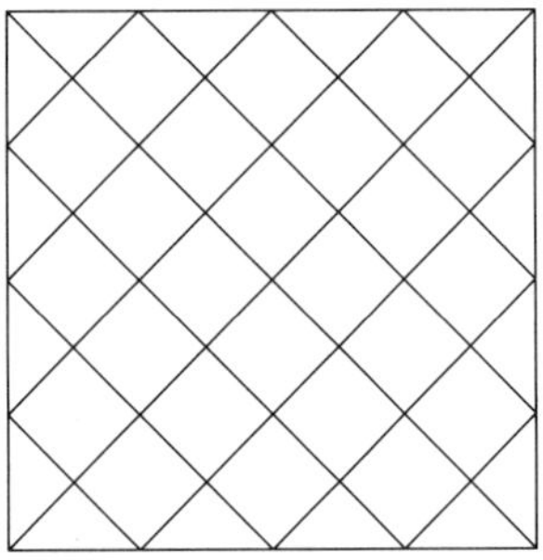

Problem 13. Let $P(x)$ be a polynomial such that, when divided by $x-2$, the remainder is 3 and, when divided by $x-3$, the remainder is 2. If, when divided by $(x-2)(x-3)$, the remainder is $ax+b$, find a^2+b^2.

Problem 14. Let a, b, c be real numbers such that $a^2-2=3b-c$, $b^2+4=3c+a$, and $c^2+4=3a-b$. Find $a^4+b^4+c^4$.

Problem 15. Let r be a real number such that $\sqrt[3]{r}-\dfrac{1}{\sqrt[3]{r}}=2$. Find
$$r^3-\frac{1}{r^3}.$$

Problem 16. Find the greatest real number x such that
$$\left(\frac{x}{x-1}\right)^2+\left(\frac{x}{x+1}\right)^2=\frac{325}{144}.$$

Problem 17. Given that $3\sin x+4\cos x=5$, where x is in $\left(0,\dfrac{\pi}{2}\right)$, find
$$2\sin x+\cos x+4\tan x.$$

Problem 18. A circle of radius 320 is tangent to the inside of a circle of radius 1000. The smaller circle is tangent to a diameter of the larger circle at a point P. How far is P from the outside of the larger circle?

Problem 19. Let x_1 and x_2 be the roots of the equation $x^2 + 3x + 1 = 0$. Evaluate

$$\left(\frac{x_1}{x_2+1}\right)^2 + \left(\frac{x_2}{x_1+1}\right)^2.$$

Problem 20. In how many ways can we form three teams of four players each from a group of 12 participants?

Problem 21. Let $a_n = \sqrt{1+\left(1-\frac{1}{n}\right)^2} + \sqrt{1+\left(1+\frac{1}{n}\right)^2}$, $n \geq 1$. Evaluate

$$\frac{1}{a_1} + \frac{1}{a_2} + \cdots + \frac{1}{a_{20}}.$$

Problem 22. In triangle ABC, $\max(\angle A, \angle B) = \angle C + 30^\circ$ and $\frac{R}{r} = \sqrt{3} + 1$, where R is the radius of the circumcircle and r is the radius of the incircle. Find $\angle C$ in degrees.

Problem 23. For all positive integers m and n define $f(m,n)$ by $f(1,1) = 1$, $f(m+1,n) = f(m,n) + m$ and $f(m,n+1) = f(m,n) - n$. Find the sum of all p such that $f(p,q) = 2004$ for some q.

Problem 24. In triangle ABC, $\angle A = 30^\circ$ and $AB = AC = 16$ inches. Let D lie on segment $\overline{BC}$ such that $\frac{DB}{DC} = \frac{2}{3}$, and let E and F be the orthogonal projections of D onto $\overline{AB}$ and $\overline{AC}$, respectively. Find $DE + DF$ in inches.

Problem 25. Given that $(1+\tan 1^\circ)(1+\tan 2^\circ)\cdots(1+\tan 45^\circ) = 2^n$, find n.

1.2 2004 High School

Problem 1. How many different positive integers divide 10! ?

Problem 2. In triangle ABC, three lines are drawn parallel to side $\overline{BC}$ dividing the altitude of the triangle into four equal parts. If the area of the second largest part is 35, what is the area of the whole triangle ABC?

Problem 3. How many real numbers are zeros of the polynomial

$$x^9 - 37x^8 - 2x^7 + 74x^6 + x^4 - 37x^3 - 2x^2 + 74x?$$

Problem 4. Find x so that $2^{2^{3^{2^2}}} = 4^{4^x}$.

Problem 5. The number $2.5081081081081\ldots$ can be written as $\frac{m}{n}$, where m and n are positive integers with no common factors. Find $m+n$.

Problem 6. Evaluate the product

$$\left(1+\frac{2}{3}\right)\left(1+\frac{2}{4}\right)\left(1+\frac{2}{5}\right)\cdots\left(1+\frac{2}{98}\right).$$

Problem 7. How many positive integers less that 200 are relatively prime to either 15 or 24?

Problem 8. One rainy afternoon you write the number 1 once, the number 2 twice, the number 3 three times, and so forth until you have written the number 99 ninety-nine times. What is the 2005th digit that you write?

Problem 9. Let M and m be the greatest and least values of x, respectively, satisfying

$$4x(x-5) \leq 375.$$

Find $M - m$.

Problem 10. Three lines are drawn parallel to each of the sides of triangle ABC so that they intersect in the interior of ABC. The resulting three smaller triangles have areas 1, 4, and 9. Find the area of triangle ABC.

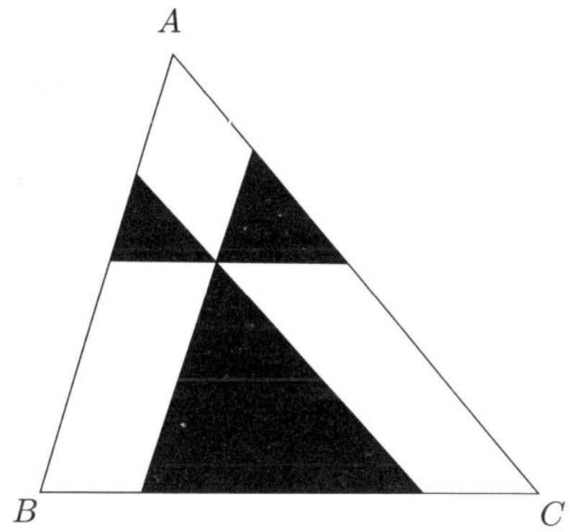

Problem 11. How far is it from the point $(9, 17)$ to its reflection across the line

$$3x + 4y = 15?$$

Problem 12. The diagram shows a circle with radius 24 which contains two circles with radius 12 tangent to each other and the larger circle. The smallest circle is tangent to the three other circles. What is the radius of the smallest circle?

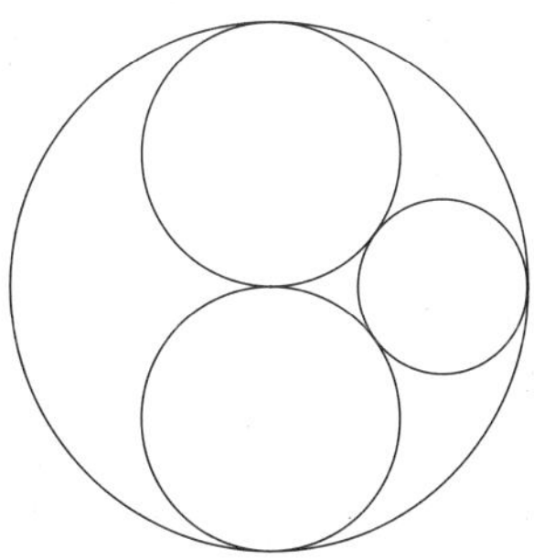

Problem 13. A cubic block with dimensions n by n by n is made up of a collection of 1 by 1 by 1 unit cubes. What is the least n so that

if the outer two layers of unit cubes are removed from the block, more than half the original unit cubes will still remain?

Problem 14. Two circles have radii 15 and 95. If the two external tangents to the circles intersect at 60 degrees, how far apart are the centers of the circles?

Problem 15. Jerry purchased some stock for \$14,400 at the same time that Susan purchased a bond for \$6,250. Jerry's investment went up 20 percent the first year, fell 10 percent the second year, and rose another 20 percent the third year. Susan's investment grew at a constant rate of compound interest for three years. If both investments are worth the same after three years, what was the annual percentage increase of Susan's investment?

Problem 16. Find the total length of the set of real numbers satisfying

$$\frac{x^2 - 80x + 1500}{x^2 - 55x + 700} < 0.$$

Problem 17. We want to paint some identically-sized cubes so that each face of each cube is painted a solid color and each cube is painted with six different colors. If we have seven different colors to choose from, how many distinguishable cubes can we produce?

Problem 18. As x ranges over the interval $(0, \infty)$, the function

$$\sqrt{9x^2 + 173x + 900} - \sqrt{9x^2 + 77x + 900}$$

ranges over the interval $(0, M)$. Find M.

Problem 19. There are three bags. One bag contains three green candies and one red candy. One bag contains two green candies and two red candies. One bag contains one green candy and three red candies. A child randomly selects one of the bags, randomly chooses a first candy from that bag, and eats the candy. If the first candy had been green, the child randomly chooses one of the other two bags and randomly selects a second candy from that bag. If the first candy had been red, the child randomly selects a second candy from the same bag as the first candy. If the probability that the second candy is green is given by the fraction $\frac{m}{n}$ in lowest terms, find $m + n$.

Problem 20. A 70 foot pole stands vertically in a plane supported by three 490 foot wires, all attached to the top of the pole, pulled taut, and anchored to three equally spaced points in the plane. How many feet apart are any two of those anchor points?

Problem 21. Define $a_k = (k^2+1)k!$ and $b_k = a_1 + a_2 + \cdots + a_k$. Let

$$\frac{a_{100}}{b_{100}} = \frac{m}{n},$$

where m and n are relatively prime positive integers. Find $n-m$.

Problem 22. How many non-overlapping 2 by 2 squares fit into a circle with radius 8?

Problem 23. Let a and b be real numbers satisfying

$$a^4 + 8b = 4(a^3 - 1) - 16\sqrt{3}$$

and

$$b^4 + 8a = 4(b^3 - 1) + 16\sqrt{3}.$$

Find $a^4 + b^4$.

Problem 24. The determinant

$$\begin{vmatrix} 3 & -2 & 5 \\ 7 & 1 & -4 \\ 5 & 2 & 3 \end{vmatrix}$$

has the same value as the determinant

$$\begin{vmatrix} x & 1+x & 2+x \\ 3 & 0 & 1 \\ 1 & 1 & 0 \end{vmatrix}$$

Find x.

Problem 25. In the addition

$$\begin{array}{cccccc} & \mathrm{W} & \mathrm{H} & \mathrm{I} & \mathrm{T} & \mathrm{E} \\ + & \mathrm{W} & \mathrm{A} & \mathrm{T} & \mathrm{E} & \mathrm{R} \\ \hline \mathrm{P} & \mathrm{I} & \mathrm{C} & \mathrm{N} & \mathrm{I} & \mathrm{C} \end{array}$$

each distinct letter represents a different digit. Find the number represented by the answer PICNIC.

1.3 2004 Middle School

Problem 1. This year February 29 fell on a Sunday. In what year will February 29 next fall on a Sunday?

Problem 2. If $h(a,b,c) = \dfrac{abc}{a+b+c}$, find $h(3\sqrt{5}, 6\sqrt{5}, 9\sqrt{5})$.

Problem 3. In triangle ABC, three lines are drawn parallel to side $\overline{BC}$ dividing the altitude of the triangle into four equal parts. If the area of the second largest part is 35, what is the area of the whole triangle ABC?

Problem 4. If the numbers $2a+2$ and $2b+2$ add up to 2004, find the sum of the numbers $\frac{a}{2} - 2$ and $\frac{b}{2} - 2$.

Problem 5. Write the number $2004_{(5)}$ [2004 base 5] as a number in base 6.

Problem 6. How many different positive integers divide 10! ?

Problem 7. A rectangle has area 1100. If the length is increased by ten percent and the width is decreased by ten percent, what is the area of the new rectangle?

Problem 8. The number $2.5081081081081\ldots$ can be written as $\frac{m}{n}$, where m and n are positive integers with no common factors. Find $m+n$.

Problem 9. How many positive integers less that 200 are relatively prime to either 15 or 24?

Problem 10. One rainy afternoon you write the number 1 once, the number 2 twice, the number 3 three times, and so forth until you have written the number 99 ninety-nine times. What is the 2005th digit that you write?

Problem 11. Find the sum of all integers x satisfying

$$1 + 8x \leq 358 - 2x \leq 6x + 94.$$

Problem 12. If $f(x, y) = xy + 2x + y + 1$, find $f(f(2, f(3, 4)), 5)$.

Problem 13. How many three-digit numbers are made up of three distinct digits?

Problem 14. A regular polygon has five times as many diagonals as it has sides. How many vertices does the polygon have?

Problem 15. Find the prime number p for which $p + 2500$ is a perfect square.

Problem 16. A week ago, Sandy's seasonal Little League batting average was 360. After five more at bats this week, Sandy's batting average improved to 400. What is the least number of hits that Sandy could have had this season?

Problem 17. Find x so that $2^{2^{3^{2^2}}} = 4^{4^x}$.

Problem 18. Find the number of addition problems in which a two digit number is added to a second two digit number to give a two digit answer, such as in the three examples:

$$\begin{array}{r} 23 \\ 42 \\ \hline 65 \end{array}, \quad \begin{array}{r} 36 \\ 36 \\ \hline 72 \end{array}, \quad \begin{array}{r} 42 \\ 23 \\ \hline 65 \end{array}.$$

Problem 19. Find n such that $n - 76$ and $n + 76$ are both cubes of positive integers.

Problem 20. A circle of area 40 is tangent to a circle of area 10. Let R be the rectangle containing both circles as shown. The area of R is $\frac{n}{\pi}$. Find n.

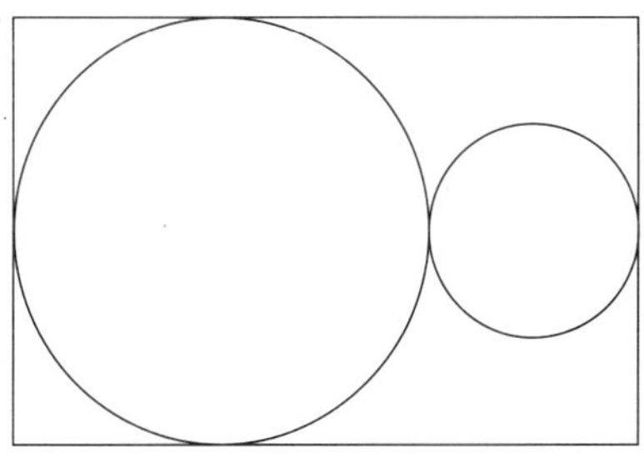

Problem 21. Find the number of different quadruples (a, b, c, d) of positive integers such that

$$ab = cd = a + b + c + d - 3.$$

Problem 22. Two circles have radii 15 and 95. If the two external tangents to the circles intersect at 60 degrees, how far apart are the centers of the circles?

Problem 23. A cubic block with dimensions n by n by n is made up of a collection of 1 by 1 by 1 unit cubes. What is the least n so that if the outer two layers of unit cubes are removed from the block, more than half the original unit cubes will still remain?

Problem 24. Let a be a real number greater than 1 such that

$$\frac{20a}{a^2+1} = \sqrt{2}.$$

Find $\dfrac{14a}{a^2-1}$.

Problem 25. In the addition

$$\begin{array}{cccccc} & W & H & I & T & E \\ + & W & A & T & E & R \\ \hline P & I & C & N & I & C \end{array}$$

each distinct letter represents a different digit. Find the number represented by the answer PICNIC.

1.4 2005 High School

Problem 1. The cost of producing each item is inversely proportional to the square root of the number of items produced. The cost of producing ten items is \$2100. If items sell for \$30 each, how many items need to be sold so that the producers break even?

Problem 2. We glue together 990 one inch cubes into a $9 \times 10 \times 11$ inch rectangular solid. Then we paint the outside of the solid. How many of the original 990 cubes have just one of their sides painted?

Problem 3. Four rectangular strips, each measuring 4 by 16 inches, are laid out with two vertical strips crossing two horizontal strips forming a single polygon which looks like a tic-tack-toe pattern. What is the perimeter of this polygon?

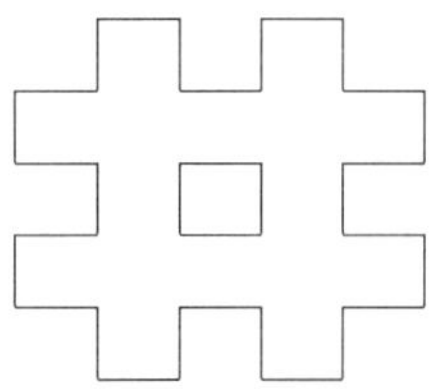

Problem 4. A palindrome is a number that reads the same forwards and backwards such as 3773 or 42924. What is the least 9-digit palindrome which is a multiple of 3 and has at least two digits that are 5's and two digits that are 7's?

Problem 5. In January, Jeff's investment went up by three quarters. In February, it went down by one quarter. In March, it went up by one third. In April, it went down by one fifth. In May, it went up by one seventh. In June, Jeff's investment fell by $\frac{m}{n}$, where m and n are relatively prime positive integers. If Jeff's investment was worth the same amount at the end of June as it had been at the beginning of January, find $m + n$.

Problem 6. Pentagon $ABCDE$ is regular. What is the degree measure of the acute angle at the intersection of line segments $\overline{AC}$ and $\overline{BD}$?

Problem 7. The graph of the equation $y = 5x + 24$ intersects the graph of the equation $y = x^2$ at two points. The two points are a distance $\sqrt{N}$ apart. Find N.

Problem 8. Find x if $\dfrac{1}{\dfrac{1}{\dfrac{1}{\frac{1}{x}+\frac{1}{2}}+\dfrac{1}{\frac{1}{x}+\frac{1}{2}}}+\dfrac{1}{\dfrac{1}{\frac{1}{x}+\frac{1}{2}}+\dfrac{1}{\frac{1}{x}+\frac{1}{2}}}}=\dfrac{x}{36}$.

Problem 9. Let T be a 30-60-90 triangle with hypotenuse of length 20. Three circles, each externally tangent to the other two, have centers at the three vertices of T. The area of the union of the circles intersected with T is $(m+n\sqrt{3})\pi$ for some rational numbers m and n. Find $m+n$.

Problem 10. A jar contains two yellow candies, four red candies, and six blue candies. Candies are randomly drawn out of the jar one-by-one and eaten. The probability that the two yellow candies will be eaten before any of the red candies are eaten is given by the fraction $\frac{m}{n}$, where m and n are relatively prime positive integers. Find $m+n$.

Problem 11. The straight river is one and a half kilometers wide and has a current of 8 kilometers per hour. A boat capable of traveling 10 kilometers per hour in still water, sets out across the water. How many minutes will it take the boat to reach a point directly across from where it started?

Problem 12. For how many nonnegative integers n is $(n^2-3n+1)^2+1$ a prime number?

Problem 13. The work team was working at a rate fast enough to process 1250 items in ten hours. But after working for six hours, the team was given an additional 165 items to process. By what percent does the team need to increase its rate so that it can still complete its work within the ten hours?

Problem 14. Four mathletes and two coaches sit at a circular table. How many distinct arrangements are there of these six people if the two coaches sit opposite each other?

Problem 15. And it came to pass that Jeb owned over a thousand chickens. So Jeb counted his chickens. And Jeb reported the count to Hannah. And Hannah reported the count to Joshua. And Joshua reported the count to Caleb. And Caleb reported the count to Rachel. But as fate would have it, Jeb had over-counted his chickens by nine chickens. Then Hannah interchanged the last two digits of the count before reporting it to Joshua. And

Joshua interchanged the first and the third digits of the number reported to him before reporting it to Caleb. Then Caleb doubled the number reported to him before reporting it to Rachel. Now it so happens that the count reported to Rachel was the correct number of chickens that Jeb owned. How many chickens was that?

Problem 16. A tailor met a tortoise sitting under a tree. When the tortoise was the tailor's age, the tailor was only a quarter of his current age. When the tree was the tortoise's age, the tortoise was only a seventh of its current age. If the sum of their ages is now 264, how old is the tortoise?

Problem 17. Functions f and g are defined so that $f(1) = 4$, $g(1) = 9$, and for each integer $n \geq 1$, $f(n+1) = 2f(n) + 3g(n) + 2n$ and $g(n+1) = 2g(n) + 3f(n) + 5$. Find $f(2005) - g(2005)$.

Problem 18. The side lengths of a trapezoid are $\sqrt[4]{3}$, $\sqrt[4]{3}$, $\sqrt[4]{3}$, and $2\sqrt[4]{3}$. Its area is the ratio of two relatively prime positive integers, m and n. Find $m + n$.

Problem 19. Let x and y be integers satisfying both $x^2 - 16x + 3y = 20$ and $y^2 + 4y - x = -12$. Find $x + y$.

Problem 20. If $\displaystyle\sum_{k=1}^{360} \frac{1}{k\sqrt{k+1} + (k+1)\sqrt{k}}$ is the ratio of two relatively prime positive integers m and n, find $m + n$.

Problem 21. In the diagram below angle CAB, angle CBD, and angle CDE are all right angles with side lengths $AC = 3$, $BC = 5$, $BD = 12$, and $DE = 84$. The distance from point E to the line $\overline{AB}$ can be expressed as the ratio of two relatively prime positive integers, m and n. Find $m + n$.

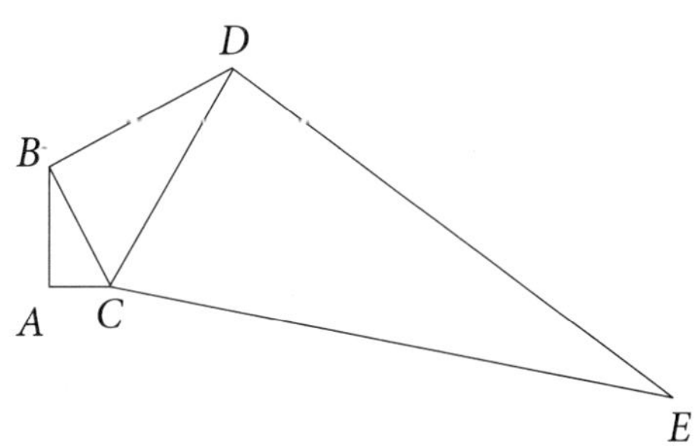

Problem 22. Let d_k be the greatest odd divisor of k for $k = 1, 2, 3, \ldots$. Find $d_1 + d_2 + d_3 + \cdots + d_{1024}$.

Problem 23. Let $a = \sqrt[401]{4} - 1$ and for each $n \geq 2$ let

$$b_n = \binom{n}{1} + \binom{n}{2}a + \cdots + \binom{n}{n}a^{n-1}.$$

Find $b_{2006} - b_{2005}$.

Problem 24. Triangle ABC has area 240. Points X, Y, Z lie on sides $\overline{AB}$, $\overline{BC}$, $\overline{CA}$, respectively. Given that $\dfrac{AX}{BX} = 3$, $\dfrac{BY}{CY} = 4$, $\dfrac{CZ}{AZ} = 5$, find the area of triangle XYZ.

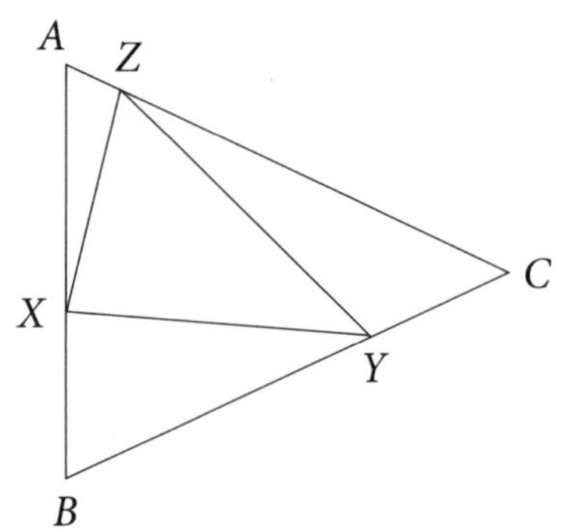

Problem 25. Find the number of quadruples (a, b, c, d) of non-zero integers satisfying both

$$\frac{1}{a} + \frac{1}{b} + \frac{1}{c} + \frac{1}{d} = \frac{1}{2}$$

and

$$2(a + b + c + d) = ab + cd + (a + b)(c + d) + 1.$$

1.5 2005 Middle School

Problem 1. A cubic inch of the newly discovered material madelbromium weighs 5 ounces. How many pounds will a cubic yard of madelbromium weigh?

Problem 2. Jerry is mowing a rectangular lawn which is 77 feet north to south by 83 feet east to west. His lawn mower cuts a path 18 inches wide. Jerry mows the grass by cutting a path from west to east across the north side of the lawn and then making a right turn cutting a path along the east side of the lawn. When he completes mowing each side of the lawn, he continues by making right turns to mow a path along the next side. How many right turns will he make?

Problem 3. Four rectangular strips each measuring 4 by 16 inches are laid out with two vertical strips crossing two horizontal strips forming a single polygon which looks like a tic-tack-toe pattern. What is the perimeter of this polygon?

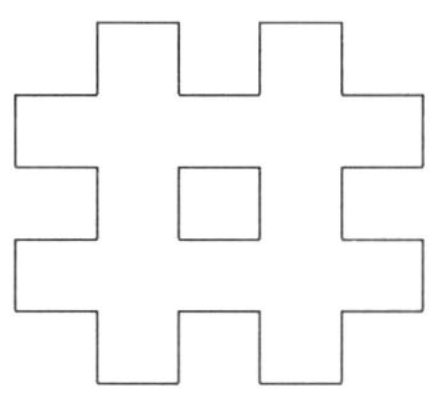

Problem 4. Fill in numbers in the boxes below so that the sum of the entries in each three consecutive boxes is 2005. What is the number that goes into the leftmost box?

		999					888	

Problem 5. A palindrome is a number that reads the same forwards and backwards such as 3773 or 42924. Find the sum of the twelve smallest five digit palindromes.

Problem 6. We glue together 990 one inch cubes into a $9 \times 10 \times 11$ inch rectangular solid. Then we paint the outside of the solid. How many of the original 990 cubes have just one of their sides painted?

Problem 7. Bill's age is one third larger than Tracy's age. In 30 years Bill's age will be one eighth larger than Tracy's age. How many years old is Bill?

Problem 8. The number 1 is special. The number 2 is special because it is relatively prime to 1. The number 3 is not special because it is not relatively prime to the sum of the special numbers less than it, $1+2$. The number 4 is special because it is relatively prime to the sum of the special numbers less than it. So, a number bigger than 1 is special only if it is relatively prime to the sum of the special numbers less than it. Find the twentieth special number.

Problem 9. For how many nonnegative integers n is $(n^2-3n+1)^2+1$ a prime number?

Problem 10. What is the 1000th digit to the right of the decimal point in the decimal representation of $\frac{37}{5500}$?

Problem 11. The work team was working at a rate fast enough to process 1250 items in ten hours. But after working for six hours, the team was given an additional 150 items to process. By what percent does the team need to increase its rate so that it can still complete its work within the ten hours?

Problem 12. Four mathletes and two coaches sit at a circular table. How many distinct arrangements are there of these six people if the two coaches sit opposite each other?

Problem 13. Find x such that $\dfrac{\dfrac{5}{x-50}+\dfrac{7}{x+25}}{\dfrac{2}{x-50}-\dfrac{3}{x+25}}=17.$

Problem 14. Eight identical $1\times1\times1$ cubes each have the numbers 1 through 6 written on their faces with the number 1 written on the face opposite number 2, number 3 written on the face opposite number 5, and number 4 written on the face opposite number 6. The eight cubes are stacked into a single $2\times2\times2$ cube. Add all of the numbers appearing on the outer surface of the new cube. Let M be the maximum possible value for this sum, and N be the minimum possible value for this sum. Find $M-N$.

Problem 15. And it came to pass that Jeb owned over a thousand chickens. So Jeb counted his chickens. And Jeb reported the count to Hannah. And Hannah reported the count to Joshua. And Joshua reported the count to Caleb. And Caleb reported the count to Rachel. But as fate would have it, Jeb had over-counted his chickens by nine chickens. Then Hannah interchanged the last two digits of the count before reporting it to Joshua. And Joshua interchanged the first and the third digits of the number reported to him before reporting it to Caleb. Then Caleb doubled the number reported to him before reporting it to Rachel. Now it so happens that the count reported to Rachel was the correct number of chickens that Jeb owned. How many chickens was that?

1.6 2006 High School

Problem 1. The sizes of the freshmen class and the sophomore class are in the ratio 5:4. The sizes of the sophomore class and the junior class are in the ratio 7:8. The sizes of the junior class and the senior class are in the ratio 9:7. If these four classes together have a total of 2158 students, how many of the students are freshmen?

Problem 2. At a movie theater tickets for adults cost 4 dollars more than tickets for children. One afternoon the theater sold 100 more child tickets than adult tickets for a total sales amount of 1475 dollars. How many dollars would the theater have taken in if the same tickets were sold, but the costs of the child tickets and adult tickets were reversed?

Problem 3. Point P lies outside a circle and two rays drawn from P intersect the circle as shown. One ray intersects the circle at points A and B while the other ray intersects the circle at M and N. Lines $\overline{AN}$ and $\overline{MB}$ intersect at X. Given that $\angle AXB$ measures $127°$ and the minor arc $\overset{\frown}{AM}$ measures $14°$, find the measure of the angle at P.

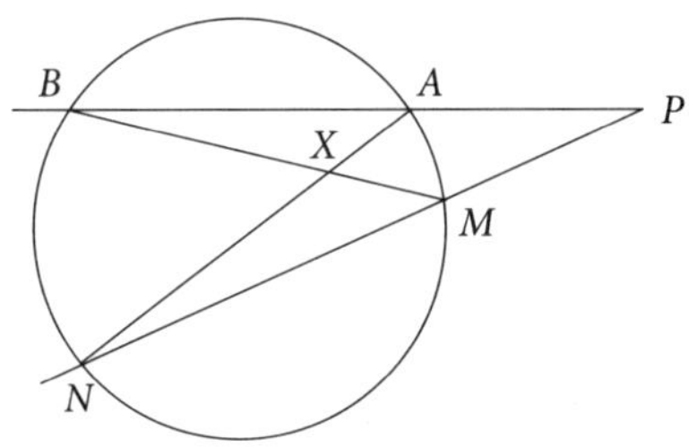

Problem 4. A rogue spaceship escapes, and 54 minutes later the police leave in a spaceship in hot pursuit. If the police spaceship travels 12% faster than the rogue spaceship along the same route, how many minutes will it take for the police to catch up with the rogues?

Problem 5. Find the sum of all positive integers less than 2006 which are both multiples of six and one more than a multiple of seven.

Problem 6. The positive integers v, w, x, y, and z satisfy the equation

$$v + \cfrac{1}{w + \cfrac{1}{x + \cfrac{1}{y + \cfrac{1}{z}}}} = \frac{222}{155}.$$

Compute $10^4 v + 10^3 w + 10^2 x + 10^1 y + 10^0 z$.

Problem 7. Heather and Kyle need to mow a lawn and paint a room. If Heather does both jobs by herself, it will take her a total of nine hours. If Heather mows the lawn and, after she finishes, Kyle paints the room, it will take them a total of eight hours. If Kyle mows the lawn and, after he finishes, Heather paints the room, it will take them a total of seven hours. If Kyle does both jobs by himself, it will take him a total of six hours. It takes Kyle twice as long to paint the room as it does for him to mow the lawn. The number of hours it would take the two of them to complete the two tasks if they worked together to mow the lawn and then worked together to paint the room is $\frac{m}{n}$, where m and n are relatively prime positive integers. Find $m + n$.

Problem 8. Evaluate

$$\frac{2i}{\frac{1}{30-5i} - \frac{1}{30+5i}}$$

where $i^2 = -1$.

Problem 9. How many rectangles are there in the diagram below such that the sum of the numbers within the rectangle is a multiple of 7?

43	44	45	46	47	48	49
36	37	38	39	40	41	42
29	30	31	32	33	34	35
22	23	24	25	26	27	28
15	16	17	18	19	20	21
8	9	10	11	12	13	14
1	2	3	4	5	6	7

Problem 10. An equilateral triangle with side length 6 has a square of side length 6 attached to each of its edges as shown. The distance

between the two farthest vertices of two squares (marked A and B in the figure) can be written as $m + \sqrt{n}$ where m and n are positive integers. Find $m + n$.

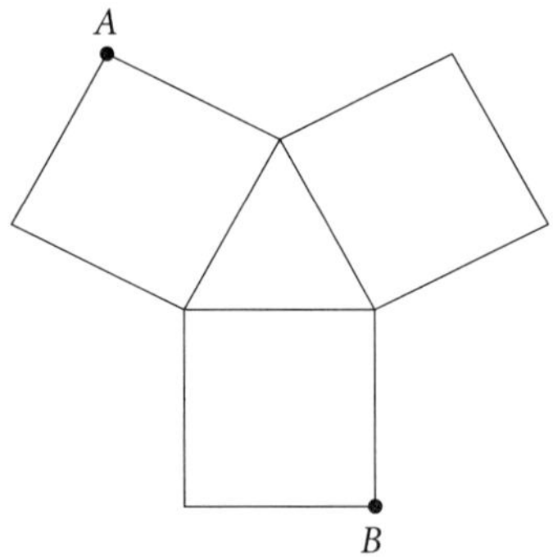

Problem 11. Let k be the product of every third positive integer from 2 to 2006, that is

$$k = 2 \cdot 5 \cdot 8 \cdot 11 \cdot \cdots \cdot 2006.$$

Find the number of trailing zeros that k has.

Problem 12. We draw a triangle inside of a circle with one vertex at the center of the circle and the other two vertices on the circumference of the circle. The angle at the center of the circle measures 75 degrees. We draw a second triangle, congruent to the first, also with one vertex at the center of the circle and the other vertices on the circumference of the circle rotated 75 degrees clockwise from the first triangle so that it shares a side with the first triangle. We draw a third, fourth, and fifth such triangle each rotated 75 degrees clockwise from the previous triangle. The base of the fifth triangle will intersect the base of the first triangle. What is the degree measure of the obtuse angle formed by the intersection?

Problem 13. Twelve students need to form five study groups. They will form three study groups with two students each and two study groups with three students each. In how many ways can these groups be formed?

Problem 14. Consider all ordered pairs (m, n) of positive integers satisfying $59m - 68n = mn$. Find the sum of all the possible values of n in these ordered pairs.

Problem 15. A snowman is built on a level plane by placing a ball radius 6 on top of a ball radius 8 on top of a ball radius 10 as shown. If the average height above the plane of a point in the snowman is $\frac{m}{n}$, where m and n are relatively prime positive integers, find $m+n$.

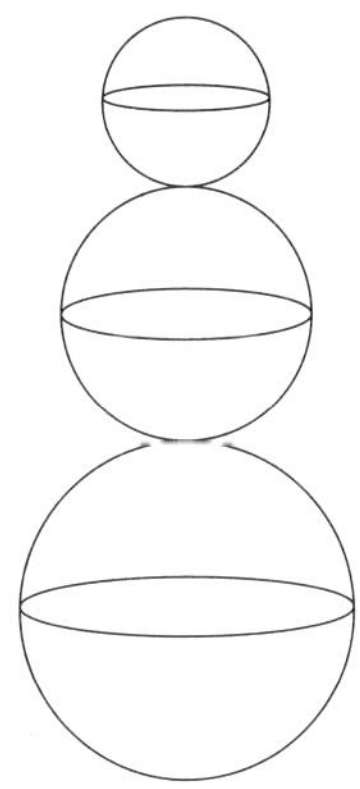

Problem 16. Let $f(x)$ and $g(x)$ be linear functions such that for all x

$$f(g(x)) = g(f(x)) = x.$$

If $f(0) = 4$ and $g(5) = 17$, compute $f(2006)$.

Problem 17. A concrete sewer pipe fitting is shaped like a cylinder with diameter 48 with a cone on top. A cylindrical hole of diameter 30 is bored all the way through the center of the fitting as shown. The cylindrical portion has height 60 while the conical top portion has height 20. Find n such that the volume of the concrete is $n\pi$.

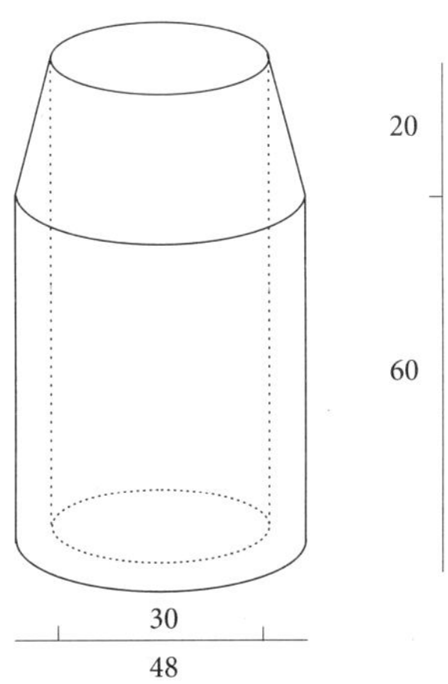

Problem 18. In how many ways can 100 be written as the sum of three positive integers x, y, z satisfying $x < y < z$?

Problem 19. There is a popular race course where runners frequently go for a daily run. Assume that all runners randomly select a start time, a starting position on the course, and a direction to run. Also assume that all runners make exactly one complete circuit of the race course, all runners run at the same speed, and all runners complete the circuit in one hour. Suppose that one afternoon you go for a run on this race course, and you count 300 runners who pass you going in the opposite direction, although some of those runners you count twice since they pass you twice. What is the expected value of the number of different runners that you pass not counting duplicates?

Problem 20. Find the sum of all positive integers having at most three, not necessarily distinct, prime factors where the primes are from the set $\{2, 3, 5, 7\}$.

Problem 21. In triangle ABC, $AB = 52$, $BC = 56$, $CA = 60$. Let D be the foot of the altitude from A, and let E be the intersection of the internal angle bisector of $\angle BAC$ with $\overline{BC}$. Find DE.

Problem 22. Let $F_0 = 0$, $F_1 = 1$, and for $n \geq 1$, $F_{n+1} = F_n + F_{n-1}$. Define

$$a_n = \left(\frac{1+\sqrt{5}}{2}\right)^n F_n.$$

Then there are rational numbers A and B such that

$$\frac{a_{30} + a_{29}}{a_{26} + a_{25}} = A + B\sqrt{5}.$$

Find $A + B$.

Problem 23. We have two positive integers both less than 1000. The arithmetic mean and the geometric mean of these numbers are consecutive odd integers. Find the maximum possible value of the difference of the two integers we have.

Problem 24. A semicircle with diameter of length 16 contains a circle of radius 3 tangent both to the inside of the semicircle and its diameter as shown. A second larger circle is tangent to the inside of the semicircle, the outside of the the circle, and the diameter of the semicircle. The diameter of the second circle can be written as

$\dfrac{n+k\sqrt{2}}{m}$, where m, n, k are positive integers and m and n have no factors in common. Find $m+n+k$.

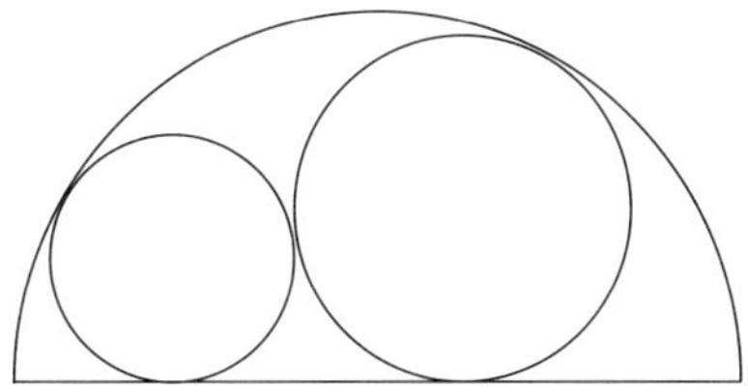

Problem 25. Let x and y be real numbers such that

$$2\sin x\sin y + 3\cos y + 6\cos x\sin y = 7.$$

Find $\tan^2 x + 2\tan^2 y$.

1.7 2006 Middle School

Problem 1. Michael is celebrating his fifteenth birthday today. How many Sundays have there been in his lifetime?

[This problem was originally stated for the contest held on Wednesday and Thursday, April 26 & 27, 2006.]

Problem 2. Evaluate $\dfrac{\dfrac{1}{\frac{1}{10}-\frac{1}{12}}}{\dfrac{1}{\frac{1}{8}-\frac{1}{6}}+\dfrac{1}{\frac{1}{5}-\frac{1}{6}}}$.

Problem 3. Find the sum of all prime numbers less than 100 that are one more than a multiple of six.

Problem 4. At the beginning of each hour from 1 o'clock AM to 12 NOON and from 1 o'clock PM to 12 MIDNIGHT a coo-coo clock's coo-coo bird coo-coos the number of times equal to the number of the hour. In addition, the coo-coo clock's coo-coo bird coo-coos a single time at 30 minutes past each hour. How many times does the coo-coo bird coo-coo from 12:42 PM on Monday until 3:42 AM on Wednesday?

Problem 5. The sizes of the freshmen class and the sophomore class are in the ratio 5:4. The sizes of the sophomore class and the junior class are in the ratio 7:8. The sizes of the junior class and the senior class are in the ratio 9:7. If these four classes together have a total of 2158 students, how many of the students are freshmen?

Problem 6. We draw a radius of a circle. We draw a second radius 23 degrees clockwise from the first radius. We draw a third radius 23 degrees clockwise from the second. This continues until we have drawn 40 radii each 23 degrees clockwise from the one before it. What is the measure in degrees of the smallest angle between any two of these 40 radii?

Problem 7. At a movie theater tickets for adults cost 4 dollars more than tickets for children. One afternoon the theater sold 100 more child tickets than adult tickets for a total sales amount of 1475 dollars. How much money would the theater have taken in if the same tickets were sold, but the costs of the child tickets and adult tickets were reversed?

Problem 8. A rogue spaceship escapes, and 54 minutes later the police leave in a spaceship in hot pursuit. If the police spaceship travels 12% faster than the rogue spaceship along the same route, how many minutes will it take for the police to catch up with the rogues?

Problem 9. Moving horizontally and vertically from point to point along the lines in the diagram below, how many routes are there from point A to point B consisting of six horizontal moves and six vertical moves?

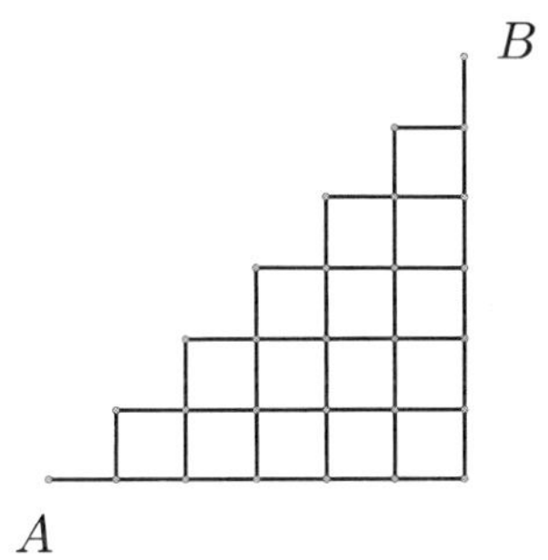

Problem 10. How many rectangles are there in the diagram below such that the sum of the numbers within the rectangle is a multiple of 7?

43	44	45	46	47	48	49
36	37	38	39	40	41	42
29	30	31	32	33	34	35
22	23	24	25	26	27	28
15	16	17	18	19	20	21
8	9	10	11	12	13	14
1	2	3	4	5	6	7

Problem 11. Consider the polynomials

$$P(x) = (x + \sqrt{2})(x^2 - 2x + 2)$$
$$Q(x) = (x - \sqrt{2})(x^2 + 2x + 2)$$
$$R(x) = (x^2 + 2)(x^8 + 16).$$

Find the coefficient of x^4 in $P(x) \cdot Q(x) \cdot R(x)$.

Problem 12. How many positive integers divide the number

$$10! = 1 \times 2 \times 3 \times 4 \times 5 \times 6 \times 7 \times 8 \times 9 \times 10?$$

Problem 13. An equilateral triangle with side length 6 has a square of side length 6 attached to each of its edges as shown. The distance between the two farthest vertices of this figure (marked A and B in the figure) can be written as $m + \sqrt{n}$ where m and n are positive integers. Find $m + n$.

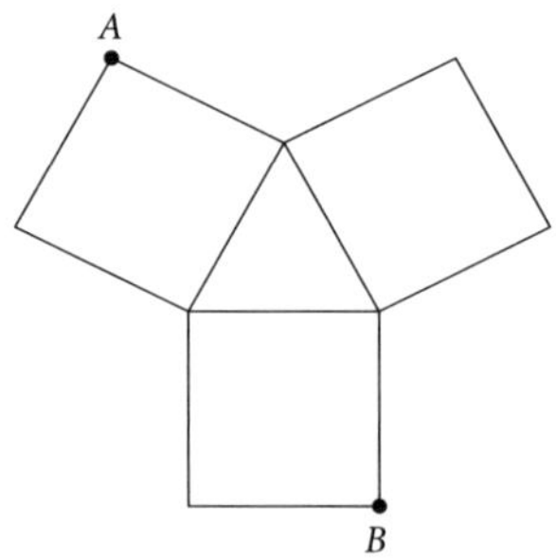

Problem 14. The rodent control task force went into the woods one day and caught 200 rabbits and 18 squirrels. The next day they went into the woods and caught three fewer rabbits and two more squirrels than the day before. Each day they went into the woods and caught three fewer rabbits and two more squirrels than the day before. This continued through the day when they caught more squirrels than rabbits. Up through that day how many rabbits did they catch in all?

Problem 15. A concrete sewer pipe fitting is shaped like a cylinder with diameter 48 with a cone on top. A cylindrical hole of diameter 30 is bored all the way through the center of the fitting as shown. The cylindrical portion has height 60, while the conical top portion has height 20. Find n such that the volume of the concrete is $n\pi$.

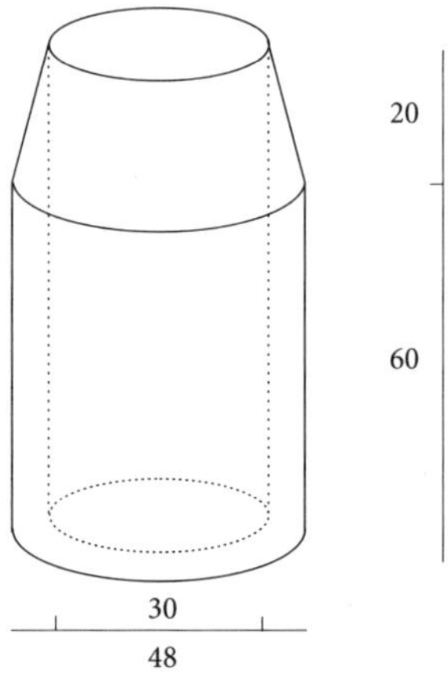

1.8 2007 High School

Problem 1. The sum of nine consecutive odd integers is 2007. Find the greatest of these nine integers.

Problem 2. A positive number $\frac{m}{n}$ has the property that it is equal to the ratio of 7 plus the number's reciprocal to 65 minus the number's reciprocal. Given that m and n are relatively prime positive integers, find $2m + n$.

Problem 3. A bowl contained 10% blue candies and 25% red candies. A bag containing three quarters red candies and one quarter blue candies was added to the bowl. Now the bowl has 16% blue candies. What percentage of the candies in the bowl are now red?

Problem 4. To the nearest degree, find the measure of the largest angle in a triangle with side lengths 3, 5, and 7.

Problem 5. Let $F(0) = 3$ and $F(n) = F(n-1) + 4$ when n is positive. Find $F(F(F(5)))$.

Problem 6. Find the sum of all positive integers that are divisors of either 96 or 180.

Problem 7. If x is a real number, what is the greatest value that can be attained by the function $25 \sin 4x - 60 \cos 4x$?

Problem 8. You know that the Jones family has five children, and the Smith family has three children. Of the eight children you know that there are five girls and three boys. Let $\frac{m}{n}$ be the probability that at least one of the families has only girls for children. Given that m and n are relatively prime positive integers, find $m + n$.

Problem 9. Sets A, B, C, D each have 400 elements. The intersection of any two of the sets has 115 elements. The intersection of any three of the sets has 53 elements. The intersection of all four sets has 28 elements. How many elements are there in the union of the four sets?

Problem 10. For each value of the angle θ we take the product of the two complex numbers

$$(8+i)\sin\theta + (7+4i)\cos\theta \quad \text{and} \quad (1+8i)\sin\theta + (4+7i)\cos\theta$$

to get a complex number in the form $a+bi$, where a and b are real numbers. Find the greatest possible value for $a+b$.

Problem 11. A dart board consists of three concentric circles with radii of 4, 6, and 8. Three darts are thrown at the board so that they stick at three random locations on the board. The probability that one dart sticks in each of the three regions of the dart board is $\frac{m}{n}$, where m and n are relatively prime positive integers. Find $m+n$.

Problem 12. Find the maximum possible value of

$$8 \cdot 27^{\log_6 x} + 27 \cdot 8^{\log_6 x} - x^3$$

as x varies over the positive real numbers.

Problem 13. Find the circumradius of the triangle with side lengths 104, 112, and 120.

Problem 14. Square $ABCD$ has side length 1000. Squares with side lengths 400 are cut from each of the corners C and D as shown. If m and n are relatively prime positive integers such that the centroid (the center of mass or balance point) of the resulting figure is at P at a distance $\frac{m}{n}$ from side $\overline{AB}$, find $m+n$.

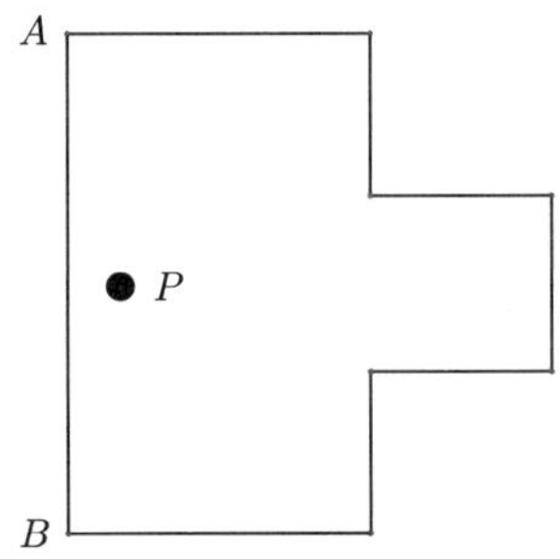

Problem 15. The alphabet in its natural order

$$ABCDEFGHIJKLMNOPQRSTUVWXYZ$$

is T_0. We apply a permutation to T_0 to get T_1 which is

$$JQOWIPANTZRCVMYEGSHUFDKBLX.$$

If we apply the same permutation to T_1, we get T_2 which is

$$ZGYKTEJMUXSODVLIAHNFPWRQCB.$$

We continually apply this permutation to each T_m to get T_{m+1}. Find the least positive integer n so that $T_n = T_0$.

Problem 16. We have some identical paper squares which are black on one side of the sheet and white on the other side. We can join nine squares together to make a 3 by 3 sheet of squares by placing each of the nine squares either white side up or black side up. Two of these 3 by 3 sheets are distinguishable if neither can be made to look like the other by rotating the sheet or by turning it over. How many distinguishable 3 by 3 squares can we form?

Problem 17. A circle with diameter 20 has points A, B, C, D, E, and F equally spaced along its circumference. A second circle is tangent to the lines $\overline{AB}$ and $\overline{AF}$ and internally tangent to the first circle. If the second circle has diameter $\sqrt{m}+n$ for some integers m and n, find $m+n$.

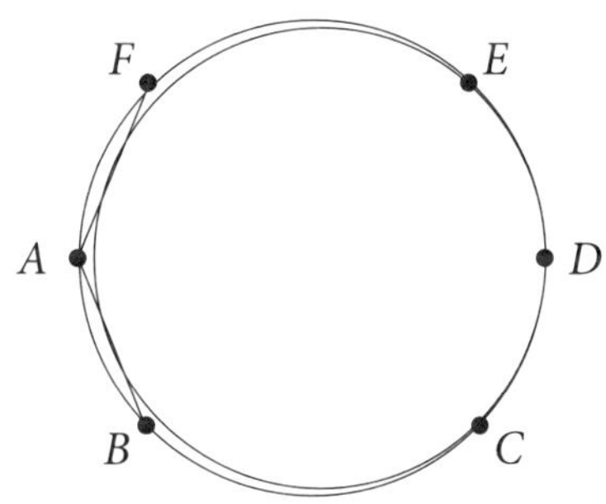

Problem 18. Let S be the graph of $y = x^3$, and T be the graph of $y = \sqrt[3]{x}$. Let S^* be S rotated around the origin 15 degrees clockwise, and T^* be T rotated around the origin 45 degrees counterclockwise. Curves S^* and T^* intersect at a point in the first quadrant a distance $m + \sqrt{n}$ from the origin, where m and n are positive integers. Find $m + n$.

Problem 19. Six chairs sit in a row. Six people randomly seat themselves in the chairs. Each person randomly chooses either to set their feet on the floor, to cross their legs to the right, or to cross their legs to the left. There is a problem only if two people sitting next to each other have the person on the right crossing their legs to the left and the person on the left crossing their legs to the right. The probability that this will **not** happen is given by $\frac{m}{n}$, where m and n are relatively prime positive integers. Find $m + n$.

Problem 20. Three congruent ellipses are mutually tangent. Their major axes are parallel. Two of the ellipses are tangent at the end points of their minor axes as shown. The distance between the centers of these two ellipses is 4. The distances from those two centers to the center of the third ellipse are both 14. There are positive integers m and n so that the area between the three ellipses is $\sqrt{n} - m\pi$. Find $m + n$.

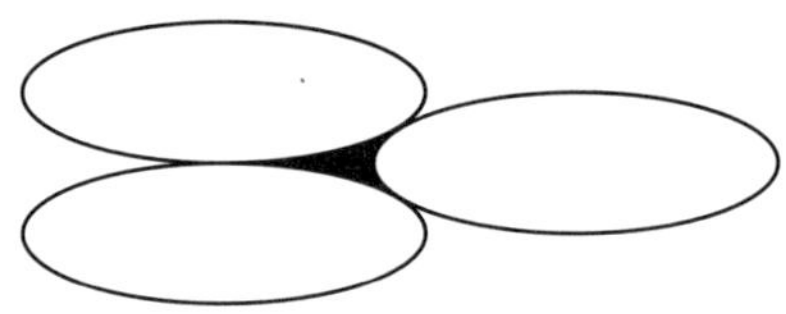

Problem 21. What is the greatest positive integer m such that

$$n^2(1 + n^2 - n^4) \equiv 1 \pmod{2^m}$$

for all odd integers n?

Problem 22. Let $a = 3^{1/223}$ and for all $n \geq 3$ let

$$f(n) = \binom{n}{0}a^{n-1} - \binom{n}{1}a^{n-2} + \binom{n}{2}a^{n-3} - \cdots + (-1)^{n-1}\binom{n}{n-1}a^0.$$

Find $f(2007) + f(2008)$.

Problem 23. Two circles with radius 2 and radius 4 have a common center at P. Points A, B, and C on the larger circle are the vertices of an equilateral triangle. Point D is the intersection of the smaller circle and the line segment $\overline{PB}$. Find the square of the area of triangle ADC.

Problem 24. Starting with a sequence of n 1's, you can insert plus signs to get various sums. For example, when $n = 10$, you can get the sum $1+1+1+11+11+111 = 136$ and the sum $1+1+11+111+111 = 235$. Find the number of values of n so that the sum of 1111 is possible.

Problem 25. Let x be a positive integer less than 200 and let y be obtained by writing the base 10 digits of x in reverse order. Given that x and y satisfy $11x^2 + 363y = 7xy + 6571$, find x.

1.9 2007 Middle School

Problem 1. Last Sunday at noon the date on the calendar was 15 (April 15, 2007). What will be the date on the calendar one million minutes after that time?

Problem 2. How many numbers n have the property that both $\frac{n}{2}$ and $2n$ are four-digit whole numbers?

Problem 3. Square $ABCD$ has side length 36. Point E is on side $\overline{AB}$ a distance 12 from B, point F is the midpoint of side $\overline{BC}$, and point G is on side $\overline{CD}$ a distance 12 from C. Find the area of the region that lies inside triangle EFG and outside triangle AFD.

Problem 4. Terry drove along a scenic road using 9 gallons of gasoline. Then Terry went onto the freeway and used 17 gallons of gasoline. Assuming that Terry gets 6.5 miles per gallon better gas mileage on the freeway than on the scenic road, and Terry's average gas mileage for the entire trip was 30 miles per gallon, find the number of miles Terry drove.

Problem 5. The repeating decimal $0.328181818181\ldots$ can equivalently be expressed as $\frac{m}{n}$, where m and n are relatively prime positive integers. Find $m+n$.

Problem 6. The product of two positive real numbers is equal to 50 times their sum and 75 times their difference. Find their sum.

Problem 7. There is an interval $[a, b]$ that is the solution to the inequality

$$|3x - 80| \leq |2x - 105|.$$

Find $a + b$.

Problem 8. Penelope plays a game where she adds 25 points to her score each time she wins a game and deducts 13 points from her score each time she loses a game. Starting with a score of zero, Penelope plays m games and has a total score of 2007 points. What is the least possible value for m?

Problem 9. Purple College keeps a careful count of its students as they progress each year from the freshman class to the sophomore class to the junior class and, finally, to the senior class. Each year at the college one third of the freshman class drops out of school, 40 students in the sophomore class drop out of school, and one tenth of the junior class drops out of school. Given that the college only admits new freshman students, and that it wants to begin each school year with 3400 students enrolled, how many students does it need to admit into the freshman class each year?

Problem 10. Tom can run to Beth's house in 63 minutes. Beth can run to Tom's house in 84 minutes. At noon, Tom starts running from his house toward Beth's house while at the same time Beth starts running from her house toward Tom's house. When they meet, they both run at Beth's speed back to Beth's house. At how many minutes after noon will they arrive at Beth's house?

Problem 11. The alphabet in its natural order

$$ABCDEFGHIJKLMNOPQRSTUVWXYZ$$

is T_0. We apply a permutation to T_0 to get T_1 which is

$$JQOWIPANTZRCVMYEGSHUFDKBLX.$$

If we apply the same permutation to T_1, we get T_2 which is

$$ZGYKTEJMUXSODVLIAHNFPWRQCB.$$

We continually apply this permutation to each T_m to get T_{m+1}. Find the least positive integer n so that $T_n = T_0$.

Problem 12. If you alphabetize all of the distinguishable rearrangements of the letters in the word *PURPLE*, find the number n such that the word *PURPLE* is the n-th item in the list.

Problem 13. Evaluate the sum

$$1^2 + 2^2 - 3^2 - 4^2 + 5^2 + 6^2 - 7^2 - 8^2 + \cdots - 1000^2 + 1001^2.$$

Problem 14. A rectangular storage bin measures 10 feet by 12 feet, is 3 feet tall, and sits on a flat plane. A pile of dirt is pushed up against the outside of the storage bin so that it slants down from the top of the storage bin to points on the ground 4 feet away from the base of the storage bin, as shown.

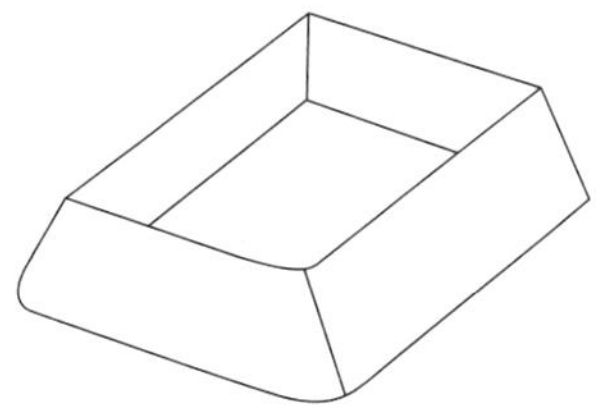

The number of cubic feet of dirt needed to form the pile can be written as $m + n\pi$ where m and n are positive integers. Find $m + n$.

Problem 15. We have some identical paper squares which are black on one side of the sheet and white on the other side. We can join nine squares together to make a 3 by 3 sheet of squares by placing each of the nine squares either white side up or black side up. Two of these 3 by 3 sheets are distinguishable if neither can be made to look like the other by rotating the sheet or by turning it over. How many distinguishable 3 by 3 squares can we form?

1.10 2008 High School

Problem 1. What is the least positive integer with the property that the product of its digits is 9! ?

Problem 2. A canister contains two and a half cups of flour. Greg and Sally have a brownie recipe which calls for one and one third cups of flour. Greg and Sally want to make one and a half recipes of brownies. To the nearest whole percent, what percent of the flour in the canister would they use?

Problem 3. Find the least integer n greater than 345 such that

$$\frac{3n+4}{5}, \frac{4n+5}{3} \text{ and } \frac{5n+3}{4}$$

and are all integers.

Problem 4. The student population at one high school consists of freshmen, sophomores, juniors, and seniors. There are 25 percent more freshmen than juniors, 10 percent fewer sophomores than freshmen, and 20 percent of the students are seniors. If there are 144 sophomores, how many students attend the school?

Problem 5. Find the sum of all digits in the decimal representations of all positive integers less than 1000.

Problem 6. Three friends who live in the same building leave the building at the same time to go rock climbing on a cliff nearby. Susan walks to the cliff and climbs to the top of the cliff. Fred runs to the cliff twice as fast as Susan walks and climbs the top of the cliff at a rate that is only two-thirds as fast as Susan climbs. Ralph bikes to the cliff at a speed twice as fast as Fred runs and takes two hours longer to climb to the top of the cliff than Susan does. If all three friends reach the top of the cliff at the same time, how many minutes after they left home is that?

Problem 7. A line through the origin passes through the curve whose equation is $y = 2x^2 - 9x + 10$ at two points whose x-coordinates add up to 77. Find the slope of the line.

Problem 8. The shape of a container is a square-based pyramid where the base has side length 23 centimeters and the height is 120 centimeters. The container is open at the base of the pyramid and stands in an open field with its vertex pointing down. One afternoon 5 centimeters of rain falls in the open field partially filling the previously empty container. Find the depth in centimeters of the rainwater in the bottom of the container after the rain.

Problem 9. Find the sum of all the integers $N > 1$ with the properties that each prime factor of N is either 2, 3, or 5, and N is not divisible by any perfect cube greater than 1.

Problem 10. There are three men and eleven women taking a dance class. In how many different ways can each man be paired with a woman partner and then have the eight remaining women be paired into four pairs of two?

Problem 11. Six regular octagons each with sides of length 2 are placed in a two-by-three array and inscribed in a square as shown. The area of the square can be written in the form $m + n\sqrt{2}$, where m and n are positive integers. Find $m + n$.

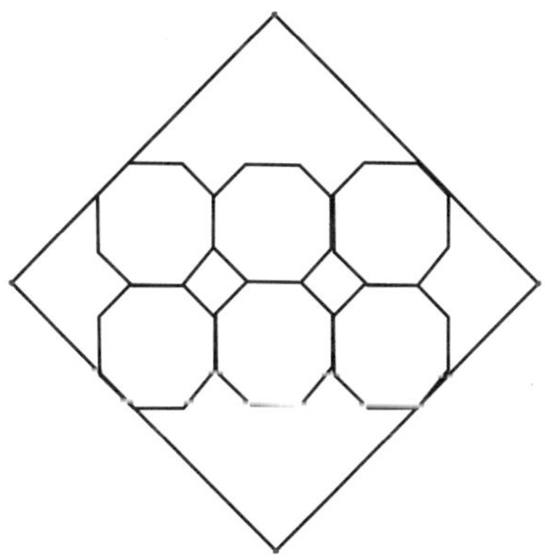

Problem 12. We record one number every workday: Monday, Tuesday, Wednesday, Thursday, and Friday. On the first Monday, the number we record is ten and a half. On every Tuesday and every Thursday, the number we record is one third of what it was on the previous workday. On every Monday, Wednesday, and Friday, the number we record is double what it was on the previous workday. This goes on forever. What is the sum of all of the numbers we will record?

Problem 13. If you roll six fair dice, let p be the probability that exactly five different numbers appear on the upper faces of the six dice. If

$p = \dfrac{m}{n}$, where m and n are relatively prime positive integers, find $m+n$.

Problem 14. A circular track with diameter 500 is externally tangent at a point A to a second circular track with diameter 1700. Two runners start at point A at the same time and run at the same speed. The first runner runs clockwise along the smaller track, while the second runner runs clockwise along the larger track. There is a first time after they begin running when the positions of the two runners together with the point A are three distinct collinear points. At that time each runner will have run a distance of $\dfrac{m\pi}{n}$, where m and n are relatively prime positive integers. Find $m+n$.

Problem 15. For each positive integer n let $f(n)$ be the sum of the digits of n. For example, $f(16)=7$, $f(f(78))=6$, and $f(f(f(5978)))=2$. Find the least positive integer n such that $f(f(f(n)))$ is not a one-digit number.

Problem 16. Square $ABCD$ has side length 7. Let A_1, B_1, C_1, D_1 be points on rays $\overrightarrow{AB}$, $\overrightarrow{BC}$, $\overrightarrow{CD}$, $\overrightarrow{DA}$, respectively, where each point is 3 units from the end of the ray so that $A_1B_1C_1D_1$ forms a second square, as shown. Similarly, let A_2, B_2, C_2, D_2 be points on segments $\overline{A_1B_1}$, $\overline{B_1C_1}$, $\overline{C_1D_1}$, $\overline{D_1A_1}$, respectively, forming another square, where A_2 divides segment $\overline{A_1B_1}$ into two pieces whose lengths are in the same ratio as AA_1 is to A_1B. Continue this process to construct square $A_nB_nC_nD_n$ for each positive integer n. Find the total of all the perimeters of all of the squares.

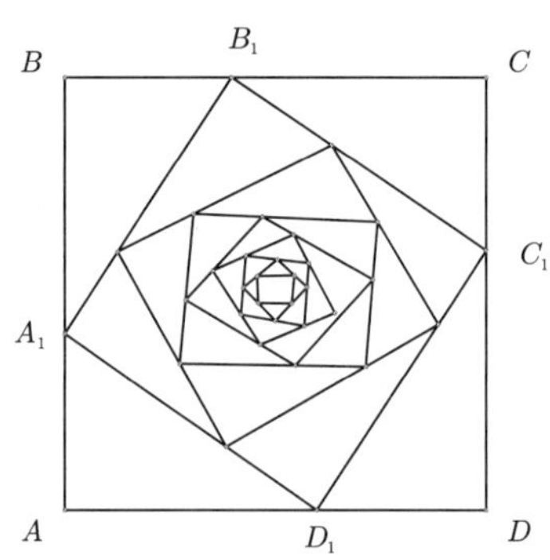

Problem 17. The number $24! = 620{,}448{,}401{,}733{,}239{,}439{,}360{,}000$ ends in four zeros, and $25! = 15{,}511{,}210{,}043{,}330{,}985{,}984{,}000{,}000$ ends in six zeros. Thus there is no integer n such that $n!$ ends

in exactly five zeros. Let S be the set of all k such that for no integer n does $n!$ end in exactly k zeros. If the numbers in S are listed in increasing order, 5 will be the first number. Find the 100th number in that list.

Problem 18. The diagram below contains eight line segments, all the same length. Each of the angles formed by the intersections of two segments is either a right angle or a 45 degree angle. If the outside square has area 1000, find the largest integer less than or equal to the area of the inside square.

Problem 19. One side of a triangle has length 75. Of the other two sides, the length of one is double the length of the other. What is the maximum possible area for this triangle?

Problem 20. Find the least positive integer n such that the decimal representation of the binomial coefficient $\binom{2n}{n}$ ends in four zeros.

Problem 21. The area of the quadrilateral with vertices at the four points in three dimensional space (0,0,0), (2,6,1), $(-3,0,3)$, $(-4,2,5)$ is the number $\frac{m}{n}$, where m and n are relatively prime positive integers. Find $m+n$.

Problem 22. Let $*$ be a binary operation on the positive integers satisfying

$$(a+b)*c = (a*c)+(b*c) \quad \text{and} \quad a*(b+c) = (a*b)*c$$

for all a, b, c. Given that $5*5 = 160$, find the value of $7*7$.

Problem 23. Circle B has radius 2008 and is tangent to horizontal line A at point P. Circle C_1 has radius 1 and is tangent both to circle B and to line A at a point to the right of point P. Circle C_2 has radius larger than 1 and is tangent to line A and both circles

B and C_1. For $n > 1$, circle C_n is tangent to line A and both circles B and C_{n-1}. Find the greatest n such that this sequence of circles can be constructed through circle C_n where the n circles are all tangent to line A at points to the right of P.

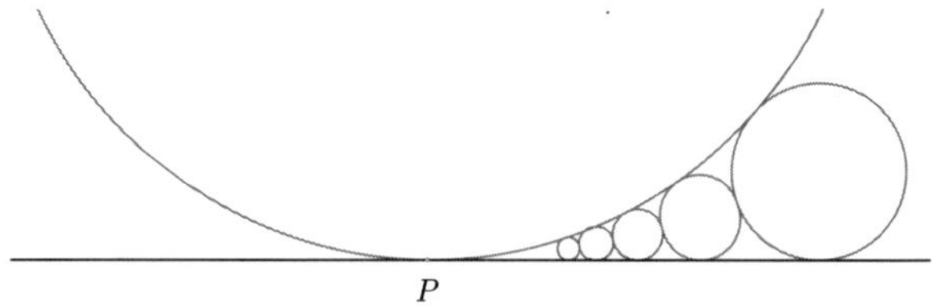

Problem 24. Each of the distinct letters in the following addition problem represents a different digit. Find the number represented by the word $MEET$.

$$\begin{array}{cccccc} P & U & R & P & L & E \\ & & C & O & M & E & T \\ & & + & M & E & E & T \\ \hline Z & Z & Z & Z & Z & Z \end{array}$$

Problem 25. Let S be a point chosen at random in the interior of the square $ABCD$. Let P be the probability that segments $\overline{AS}$, $\overline{SB}$, $\overline{AC}$ are congruent to the sides of a triangle. Then P can be written as $\dfrac{a - \pi\sqrt{b} - \sqrt{c}}{d}$, where a, b, c, d are positive integers and d is as small as possible. Find $ab + cd$.

1.11 2008 Middle School

Problem 1. Find the greatest prime factor in the sum of the two largest two-digit prime numbers.

Problem 2. Starting on April 15, 2008, you can go one day backward and one day forwards to get the calendar dates 14 and 16. If you go 15 days backward and 15 days forward, you get the calendar dates 31 (from March) and 30 (from April). Find the least positive integer k so that if you go k days backward and k days forward you get two calendar dates that are the same.

[This problem was originally given in 2008, a leap year.]

Problem 3. There were 891 people voting at precinct 91. There were 20 percent more female voters than male voters. How many female voters were there?

Problem 4. While driving his car, Ken pulled off the road to get gasoline when he was $\frac{7}{12}$ of the way through his trip. After driving another eleven miles, he noticed that he was $\frac{13}{20}$ of the way through his trip. How many miles long was his entire trip?

Problem 5. What is the measurement in degrees of the angle formed by the minute and hour hands when a clock reads 12:12?

Problem 6. The fraction

$$\frac{2}{\dfrac{2}{\dfrac{2}{\dfrac{2}{2+1}+1}+1}+1+\dfrac{2}{1+\dfrac{2}{1+\dfrac{2}{2+1}}}}$$

can be written in the form $\frac{m}{n}$, where m and n are relatively prime positive integers.

Find $2m + n$.

Problem 7. The diagram below shows an isosceles triangle with base 21 and height 28. Inscribed in the triangle is a square.

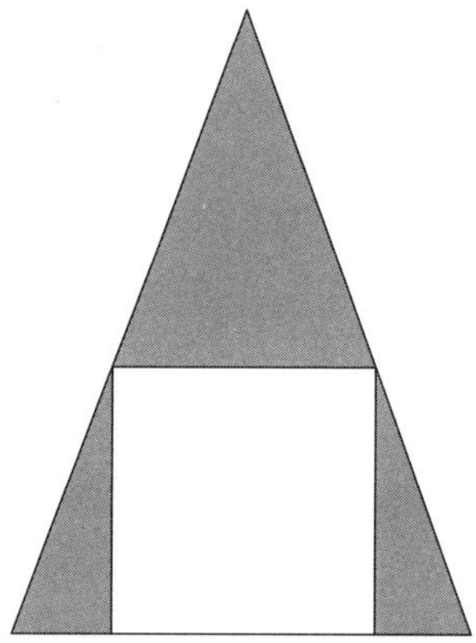

Find the area of the shaded region inside the triangle and outside of the square.

Problem 8. At Mallard High School there are three intermural sports leagues: football, basketball, and baseball. There are 427 students participating in these sports: 128 play on football teams, 291 play on basketball teams, and 318 play on baseball teams. If exactly 36 students participate in all three of the sports, how many students participate in exactly two of the sports?

Problem 9. One container of paint is exactly enough to cover the inside of an old rectangle which is three times as long as it is wide. If we make a new rectangle by shortening the old rectangle by 18 feet and widening it by 8 feet as shown below, one container of paint is also exactly enough to cover the inside of the new rectangle. Find the length in feet of the perimeter of the new rectangle.

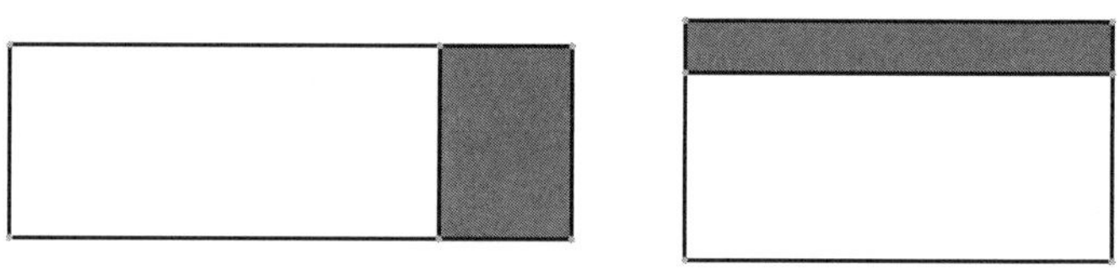

Problem 10. A 16×16 square sheet of paper is folded once in half horizontally and once in half vertically to make an 8×8 square. This square is again folded in half twice to make a 4×4 square. This square is folded in half twice to make a 2×2 square. This square is folded in half twice to make a 1×1 square. Finally, a scissor is used to make cuts through both diagonals of all the layers of the 1×1 square. How many pieces of paper result?

Problem 11. When Tim was Jim's age, Kim was twice as old as Jim. When Kim was Tim's age, Jim was 30. When Jim becomes Kim's age, Tim will be 88. When Jim becomes Tim's age, what will be the sum of the ages of Tim, Jim, and Kim?

Problem 12. A city is laid out with a rectangular grid of roads with 10 streets numbered from 1 to 10 running east-west and 16 avenues numbered from 1 to 16 running north-south. All streets end at First and Sixteenth Avenues, and all avenues end at First and Tenth Streets. A rectangular city park is bounded on the north and south by Sixth Street and Eighth Street, and bounded on the east and west by Fourth Avenue and Twelfth Avenue. Although there are no breaks in the roads that bound the park, no road goes through the park. The city paints a crosswalk from every street corner across any adjacent road. Thus, where two roads cross such as at Second Street and Second Avenue, there are four crosswalks painted, while at corners such as First Street and First Avenue, there are only two crosswalks painted. How many crosswalks are there painted on the roads of this city?

Problem 13. Let $A_1A_2A_3 \dots A_{12}$ be a regular dodecagon. Find the number of right triangles whose vertices are in the set

$$\{A_1, A_2, A_3, \dots, A_{12}\}.$$

Problem 14. Ralph is standing along a road which heads straight east. If you go nine miles east, make a left turn, and travel seven miles north, you will find Pamela with her mountain bike. At exactly the same time that Ralph begins running eastward along the road at 6 miles per hour, Pamela begins biking in a straight line at 10 miles per hour. Pamela's direction is chosen so that she will reach a point on the road where Ralph is running at exactly the same time Ralph reaches that same point. Let m and n be relatively prime positive integers such that $\frac{m}{n}$ is the number of hours that it takes Pamela and Ralph to meet. Find $m + n$.

Problem 15. Each of the distinct letters in the following subtraction represents a different digit. Find the number represented by the word *TEAM*.

$$\begin{array}{cccccc} P & U & R & P & L & E \\ - & C & O & M & E & T \\ \hline & & T & E & A & M \end{array}$$

1.12 2009 High School

Problem 1. In five years, Tom will be twice as old as Cindy. Thirteen years ago, Tom was three times as old as Cindy. How many years ago was Tom four times as old as Cindy?

Problem 2. Find the least positive integer n such that for every prime number p, $p^2 + n$ is never a prime.

Problem 3. In the diagram $ABCDEFG$, is a regular heptagon. Shown is the star $AEBFCGD$. The degree measure of the obtuse angle formed by $\overline{AE}$ and $\overline{CG}$ is $\frac{m}{n}$, where m and n are relatively prime positive integers. Find $m + n$.

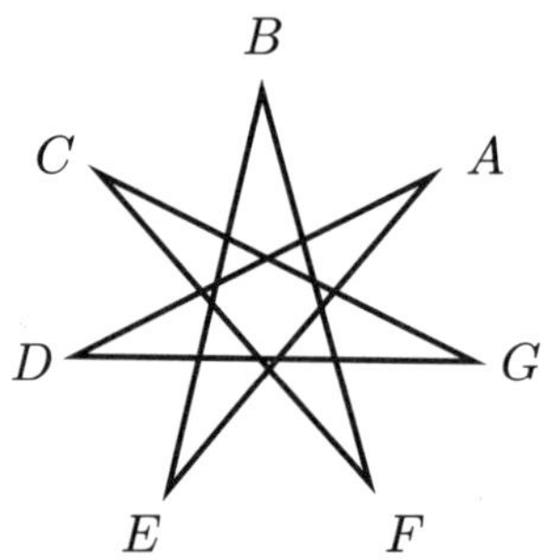

Problem 4. There are three bags of marbles. Bag two has twice as many marbles as bag one. Bag three has three times as many marbles as bag one. Half the marbles in bag one, one third the marbles in bag two, and one fourth the marbles in bag three are green. If all three bags of marbles are dumped into a single pile, $\frac{m}{n}$ of the marbles in the pile would be green, where m and n are relatively prime positive integers. Find $m + n$.

Problem 5. Find n so that $\left(4^{n+7}\right)^3 = \left(2^{n+23}\right)^4$.

Problem 6. Wiles county contains eight townships as shown on the map.

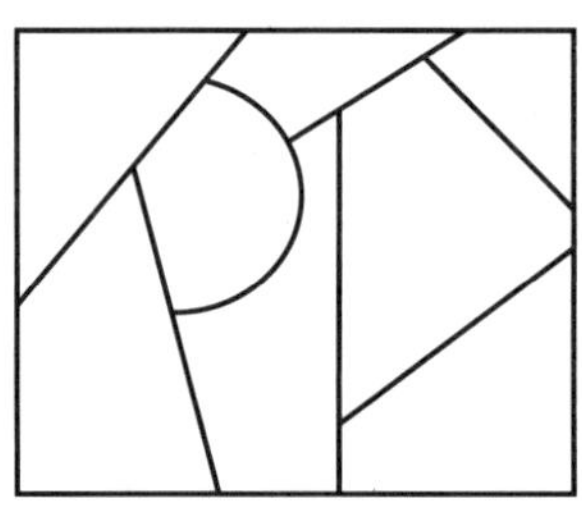

If there are four colors available, in how many ways can the the map be colored so that each township is colored with one color and no two townships that share a border are colored with the same color?

Problem 7. Figure $ABCD$ is bounded by a semicircle CDA and a quarter circle ABC. Given that the distance from A to C is 18, find the area of the figure.

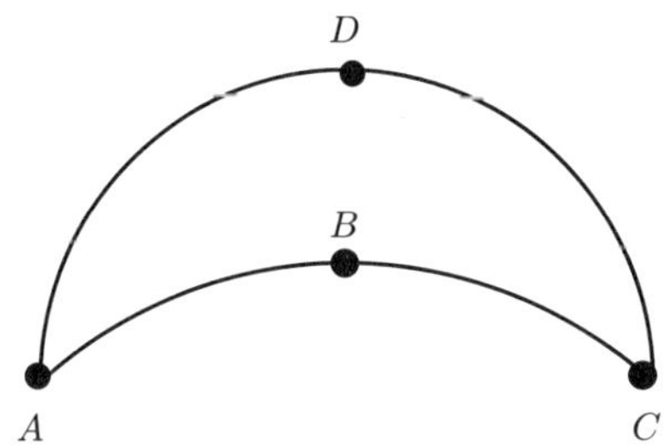

Problem 8. Find the least positive integer that has exactly 20 positive integer divisors.

Problem 9. Bill bought 13 notebooks, 26 pens, and 19 markers for 25 dollars. Paula bought 27 notebooks, 18 pens, and 31 markers for 31 dollars. How many dollars would it cost Greg to buy 24 notebooks, 120 pens, and 52 markers?

Problem 10. Towers grow at points along a line. All towers start with height 0 and grow at the rate of one meter per second. As soon as any two adjacent towers are each at least 1 meter tall, a new tower begins to grow at a point along the line exactly half way between those two adjacent towers. Before time 0 there are no towers, but at time 0 the first two towers begin to grow at two points along the line. Find the total of all the heights of all towers at time 10 seconds.

Problem 11. The four points $A(-1,2)$, $B(3,-4)$, $C(5,-6)$, $D(-2,8)$ lie in the coordinate plane. Compute the minimum possible value of $PA + PB + PC + PD$ over all points P.

Problem 12. What is the least possible sum of two positive integers a and b where $a \cdot b = 10!$?

Problem 13. Greta is completing an art project. She has twelve sheets of paper: four red, four white, and four blue. She also has twelve paper stars: four red, four white, and four blue. She randomly places one star on each sheet of paper. The probability that no star will be placed on a sheet of paper that is the same color as the star is $\frac{m}{n}$, where m and n are relatively prime positive integers. Find $n - 100m$.

Problem 14. Let $ABCD$ be a trapezoid with $\overline{AB}$ parallel to $\overline{CD}$, base $\overline{AB}$ has length 1, and base $\overline{CD}$ has length 41. Let points X and Y lie on sides $\overline{AD}$ and $\overline{BC}$, respectively, such that $\overline{XY}$ is parallel to $\overline{AB}$ and $\overline{CD}$, and $\overline{XY}$ has length 31. Let m and n be relatively prime positive integers such that the ratio of the area of $ABYX$ to the area of $CDXY$ is $\frac{m}{n}$. Find $m + 2n$.

Problem 15. What is the remainder when 7^{8^9} is divided by 1000?

Problem 16. Let

$$z = \cos\frac{1}{1000} + i\sin\frac{1}{1000}.$$

Find the least positive integer n so that z^n has an imaginary part that exceeds $\frac{1}{2}$.

Problem 17. How many ordered triples (a, b, c) of odd positive integers satisfy

$$a + b + c = 25\ ?$$

Problem 18. In triangle ABC let D be the point on $\overline{AB}$ so that $\overline{CD}$ is an altitude of the triangle, and let E be the point on $\overline{BC}$ so that $\overline{AE}$ bisects angle BAC. Let G be the intersection of $\overline{AE}$ and $\overline{CD}$, and let point F be the intersection of side $\overline{AC}$ and the ray $\overrightarrow{BG}$. If $\overline{AB}$ has length 28, $\overline{AC}$ has length 14, and $\overline{CD}$ has length 10, then the length of $\overline{CF}$ can be written as $\frac{k-m\sqrt{p}}{n}$, where k, m, n, p are positive integers, k and n are relatively prime, and p is not divisible by the square of any prime. Find $k - m + n + p$.

Problem 19. If a and b are complex numbers such that

$$a^2 + b^2 = 5 \text{ and } a^3 + b^3 = 7,$$

then their sum, $a + b$, is real. The greatest possible value for the sum $a + b$ is $\frac{m+\sqrt{n}}{2}$ where m and n are integers. Find n.

Problem 20. Five men and seven women stand in a line in random order. Let m and n be relatively prime positive integers so that $\frac{m}{n}$ is the probability that each man stands next to at least one woman. Find $m + n$.

Problem 21. A cylinder radius 12 and a cylinder radius 36 are held tangent to each other with a tight band. The length of the band is $m\sqrt{k} + n\pi$ where m, k, and n are positive integers and k is not divisible by the square of any prime. Find $m + k + n$.

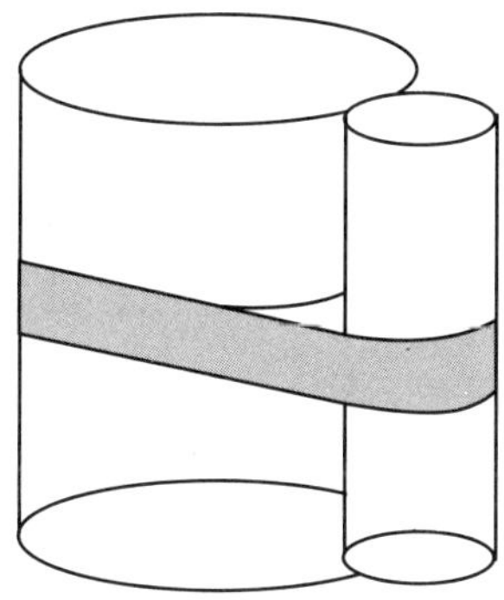

Problem 22. A parabola, a line perpendicular to the parabola's axis of symmetry, and three similar isosceles triangles each with a base on the line and vertex on the parabola are shown in the following diagram. The two smaller triangles are congruent and each have one base vertex on the parabola and one base vertex shared with the larger triangle. The ratio of the height of the larger triangle to the height of the smaller triangles is $\frac{a+\sqrt{b}}{c}$ where a, b, c are positive integers, and a and c are relatively prime. Find $a+b+c$.

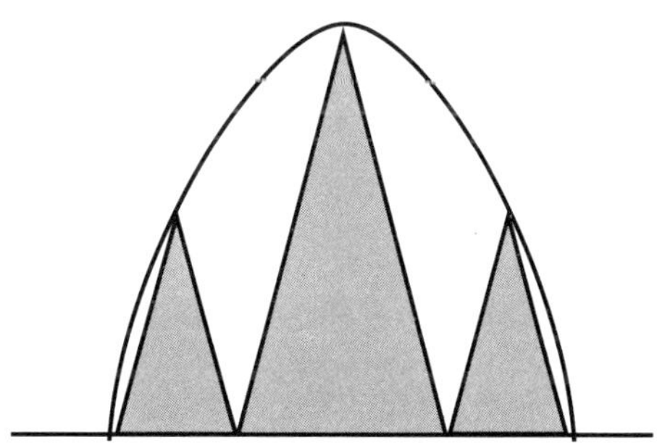

Problem 23. Square $ABCD$ has side length 4. Points E and F are the midpoints of sides $\overline{AB}$ and $\overline{CD}$, respectively. Eight 1 by 2 rectangles are placed inside the square so that no two of the eight rectangles overlap (see diagram). If the arrangement of eight rectangles is chosen randomly, then there are relatively prime positive integers m and n so that $\frac{m}{n}$ is the probability that none of the

rectangles crosses the line segment $\overline{EF}$ (as in the arrangement on the right). Find $m+n$.

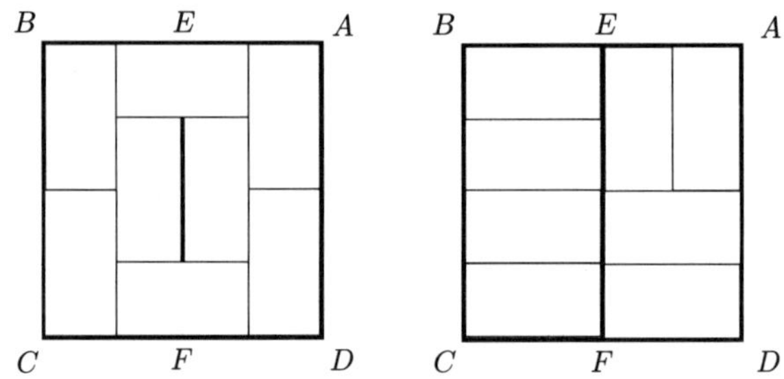

Problem 24. A right circular cone pointing downward forms an angle of 60° at its vertex. Sphere S with radius 1 is set into the cone so that it is tangent to the side of the cone. Three congruent spheres are placed in the cone on top of S so that they are all tangent to each other, to sphere S, and to the side of the cone. The radius of these congruent spheres can be written as $\frac{a+\sqrt{b}}{c}$ where a, b, c are positive integers such that a and c are relatively prime. Find $a+b+c$.

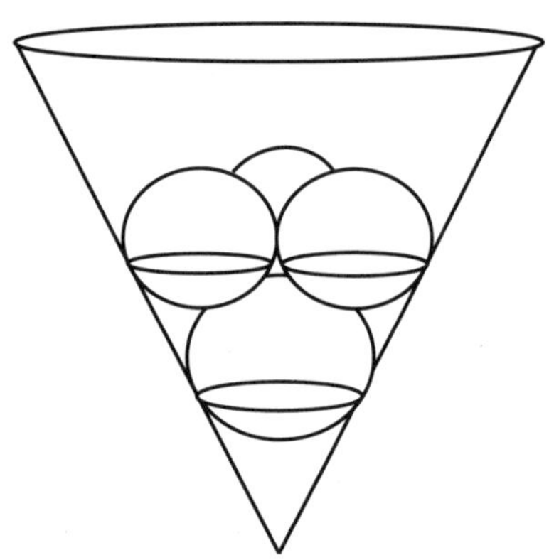

Problem 25. The polynomial $P(x) = a_0 + a_1x + a_2x^2 + \cdots + a_8x^8 + 2009x^9$ has the property that $P\left(\frac{1}{k}\right) = \frac{1}{k}$ for $k = 1,2,3,4,5,6,7,8,9$. There are relatively prime positive integers m and n such that $P\left(\frac{1}{10}\right) = \frac{m}{n}$. Find $n - 10m$.

1.13 2009 Middle School

Problem 1. The pentagon below has three right angles. Find its area.

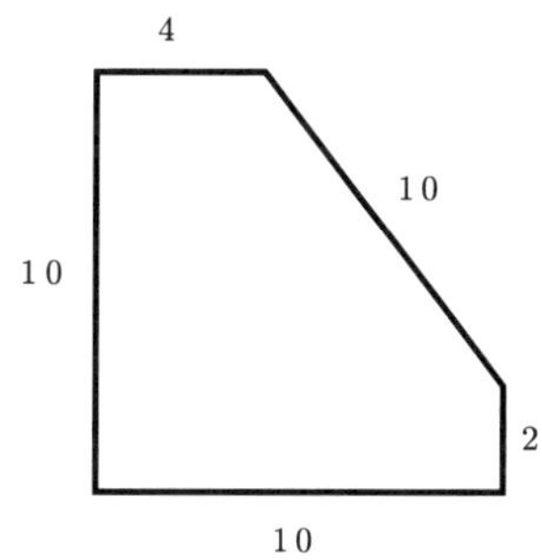

Problem 2. Let $p_1 = 2, p_2 = 3, p_3 = 5, \ldots$ be the sequence of prime numbers. Find the least positive even integer n so that $p_1+p_2+p_3+\cdots+p_n$ is not a prime.

Problem 3. The Purple Comet*!* Math Meet runs from April 27 through May 3, so the sum of the calendar dates for these seven days is $27 + 28 + 29 + 30 + 1 + 2 + 3 = 120$. What is the largest sum of the calendar dates for seven consecutive Fridays occurring at any time in any year?

Problem 4. John, Paul, George, and Ringo baked a circular pie. Each cut a piece that was a sector of the circle. John took one-third of the whole pie. Paul took one-fourth of the whole pie. George took one-fifth of the whole pie. Ringo took one-sixth of the whole pie. At the end the pie had one sector remaining. Find the measure in degrees of the angle formed by this remaining sector.

Problem 5. A train car held 6000 pounds of mud which was 88 percent water. Then the train car sat in the sun, and some of the water evaporated so that now the mud is only 82 percent water. How many pounds does the mud weigh now?

Problem 6. Find n so that $20^{2009} = 10^{2000} \cdot 40^9 \cdot 2^n$.

Problem 7. How many distinct four letter arrangements can be formed by rearranging the letters found in the word **FLUFFY**? For example, FLYF and ULFY are two possible arrangements.

Problem 8. Find the number of non-congruent scalene triangles whose sides all have integral length, and the longest side has length 11.

Problem 9. One plant is now 44 centimeters tall and will grow at a rate of 3 centimeters every two years. A second plant is now 80 centimeters tall and will grow at a rate of 5 centimeters every six years. In how many years will the plants be the same height?

Problem 10. The diagram shows a 20 by 20 square $ABCD$. Points E, F, G are equally spaced on side $\overline{BC}$, while points H, I, J, K on side $\overline{DA}$ are placed so that the triangles BKE, EJF, FIG, and GHC are isosceles. Points L and M are midpoints of the sides $\overline{AB}$ and $\overline{CD}$, respectively. Find the total area of the shaded regions.

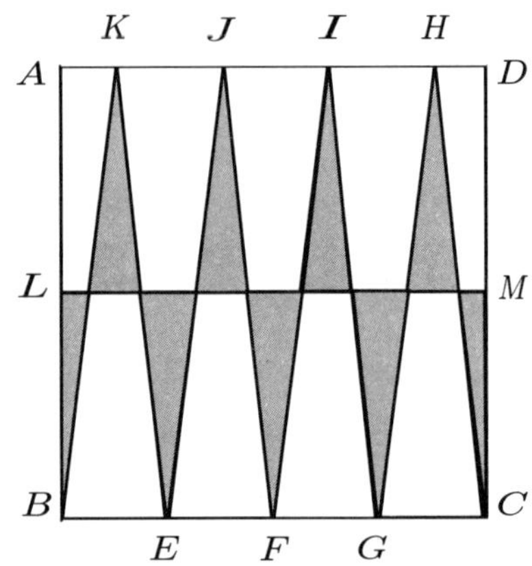

Problem 11. Aisha went shopping. At the first store she spent 40 percent of her money plus four dollars. At the second store she spent 50 percent of her remaining money plus 5 dollars. At the third store she spent 60 percent of her remaining money plus six dollars. When Aisha was done shopping at the three stores, she had two dollars left. How many dollars did she have with her when she started shopping?

Problem 12. In isosceles triangle ABC sides $\overline{AB}$ and $\overline{BC}$ have length 125, while side $\overline{AC}$ has length 150. Point D is the midpoint of side $\overline{AC}$. Point E is on side $\overline{BC}$ so that $\overline{BC}$ and $\overline{DE}$ are perpendicular. Similarly, F is on side $\overline{AB}$ so that $\overline{AB}$ and $\overline{DF}$ are perpendicular. Find the area of triangle DEF.

Problem 13. How many subsets of the set $\{1, 2, 3, \ldots, 12\}$ contain exactly one or two prime numbers?

Problem 14. Rectangle $ABCD$ measures 70 by 40. Eighteen points (including A and C) are marked on the diagonal $\overline{AC}$, dividing the diagonal into 17 congruent pieces. Twenty-two points (including A and B) are marked on the side $\overline{AB}$, dividing the side into 21 congruent pieces. Seventeen non-overlapping triangles are constructed as shown. Each triangle has two vertices that are two of these adjacent marked points on the side of the rectangle, and one vertex that is one of the marked points along the diagonal of the rectangle. Only the left 17 of the 21 congruent pieces along the side of the rectangle are used as bases of these triangles. Find the sum of the areas of these 17 triangles.

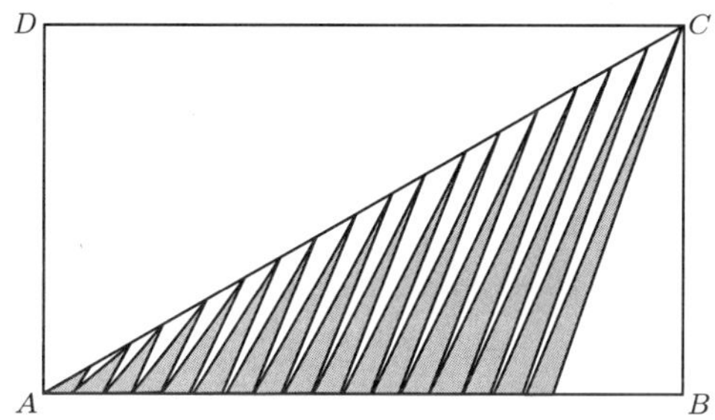

Problem 15. We have twenty-seven 1 by 1 cubes. Each face of every cube is marked with a positive integer so that two opposite faces (top and bottom, front and back, left and right) are always marked with an even number and an odd number where the even number is twice that of the odd number. The twenty-seven cubes are put together to form one 3 by 3 cube as shown. When two cubes are placed face-to-face, adjoining faces are always marked with an odd number and an even number where the even number is one greater than the odd number. Find the sum of all of the numbers on all of the faces of all the 1 by 1 cubes.

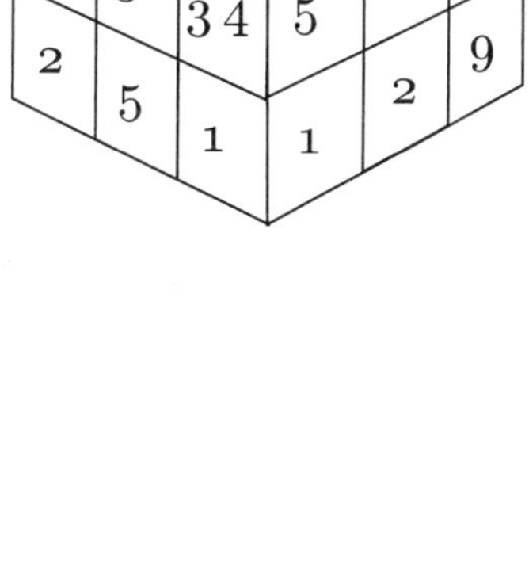

1.14 2010 High School

Problem 1. Let x satisfy $(6x+7)+(8x+9) = (10+11x)+(12+13x)$. There are relatively prime positive integers m and n so that $x = -\frac{m}{n}$. Find $m+n$.

Problem 2. The prime factorization of $12 = 2 \cdot 2 \cdot 3$ has three prime factors. Find the number of prime factors in the factorization of

$$12! = 12 \cdot 11 \cdot 10 \cdot 9 \cdot 8 \cdot 7 \cdot 6 \cdot 5 \cdot 4 \cdot 3 \cdot 2 \cdot 1.$$

Problem 3. The grid below contains six rows with six points in each row. Points that are adjacent either horizontally or vertically are a distance two apart. Find the area of the irregularly shaped ten sided figure shown.

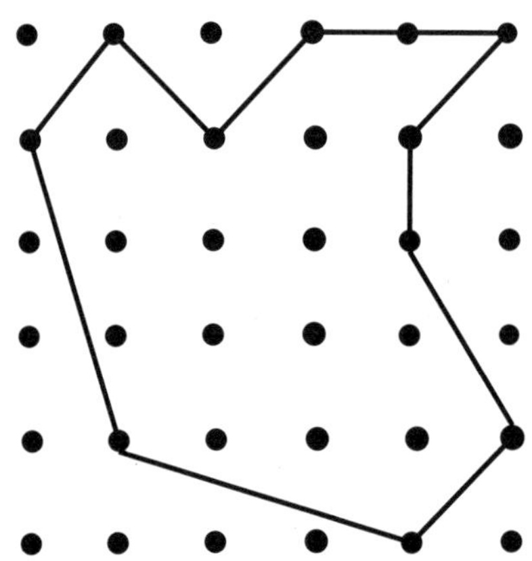

Problem 4. Sally's salary in 2006 was \$37,500. For 2007 she got a salary increase of x percent. For 2008 she got another salary increase of x percent. For 2009 she got a salary decrease of $2x$ percent. Her 2009 salary is \$34,825. Suppose instead, Sally had gotten a $2x$ percent salary decrease for 2007, an x percent salary increase for 2008, and an x percent salary increase for 2009. What would her 2009 salary be then?

Problem 5. If a and b are positive integers such that $a \cdot b = 2400$, find the least possible value of $a+b$.

Problem 6. Evaluate the sum $1+2-3+4+5-6+7+8-9+\cdots+208+209-210$.

Problem 7. Find the sum of the digits in the decimal representation of the number $5^{2010} \cdot 16^{502}$.

Problem 8. The diagram below shows some small squares each with area 3 enclosed inside a larger square. Squares that touch each other do so with the corner of one square coinciding with the midpoint of a side of the other square. Find the integer n such that the area of the shaded region inside the larger square but outside the smaller squares is $\sqrt{n}$.

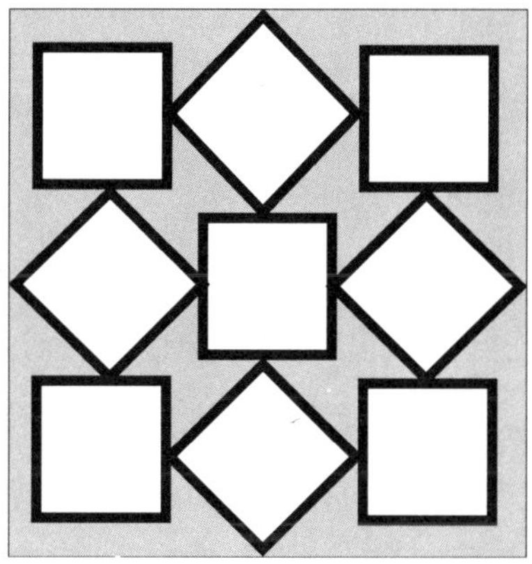

Problem 9. Find the positive integer n such that $\frac{80-6\sqrt{n}}{n}$ is the reciprocal of $\frac{80+6\sqrt{n}}{n}$.

Problem 10. The set S contains nine numbers. The mean of the numbers in S is 202. The mean of the five smallest of the numbers in S is 100. The mean of the five largest numbers in S is 300. What is the median of the numbers in S?

Problem 11. A jar contains one white marble, two blue marbles, three red marbles, and four green marbles. If you select two of these marbles without replacement, the probability that both marbles will be the same color is $\frac{m}{n}$, where m and n are relatively prime positive integers. Find $m+n$.

Problem 12. A good approximation of π is 3.14. Find the least positive integer d such that if the area of a circle with diameter d is calculated using the approximation 3.14, the error will exceed 1.

Problem 13. Let S be the set of all 10-term arithmetic progressions that include the numbers 4 and 10. For example,

$$(-2, 1, 4, 7, 10, 13, 16, 19, 22, 25)$$

and

$$\left(10, 8\frac{1}{2}, 7, 5\frac{1}{2}, 4, 2\frac{1}{2}, 1, -\frac{1}{2}, -2, -3\frac{1}{2}\right)$$

are both members of S. Find the sum of all values of a_{10} for each

$$(a_1, a_2, a_3, a_4, a_5, a_6, a_7, a_8, a_9, a_{10}) \in S,$$

that is,

$$\sum_{(a_1,a_2,a_3,\ldots,a_{10})\in S} a_{10}.$$

Problem 14. There are positive integers b and c such that the polynomial $2x^2 + bx + c$ has two real roots which differ by 30. Find the least possible value of $b + c$.

Problem 15. Find the least possible sum $a + b + c + d + e$, where a, b, c, d, e are positive integers satisfying the conditions

- each of the pairs of integers (a, b), (b, c), (c, d), (d, e) are **not** relatively prime
- all other pairs of the five integers **are** relatively prime.

Problem 16. Triangle ABC has sides lengths $AB = 39$, $BC = 57$, $CA = 70$, as shown. Median $\overline{AD}$ is divided into three congruent segments by points E and F. Lines $\overline{BE}$ and $\overline{BF}$ intersect side $\overline{AC}$ at G and H, respectively. Find the distance from G to H.

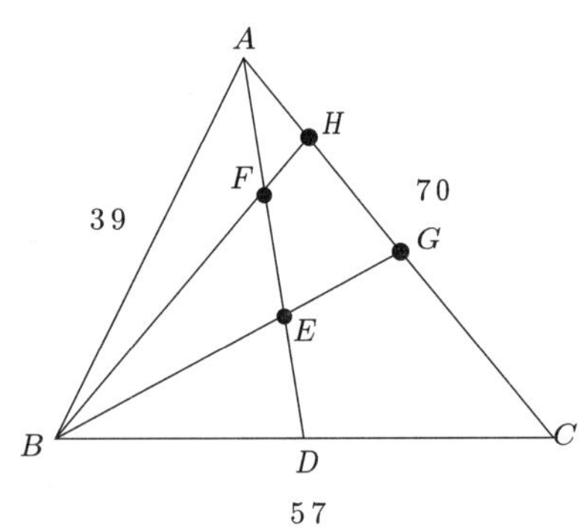

Problem 17. Alan, Barb, Cory, and Doug are on the golf team; Doug, Emma, Fran, and Greg are on the swim team; and Greg, Hope, Inga, and Alan are on the tennis team. These nine people sit in a circle in random order. The probability that no two people from the same team sit next to each other is $\frac{m}{n}$, where m and n are relatively prime positive integers. Find $m + n$.

Problem 18. When $4\cos\theta - 3\sin\theta = \frac{13}{3}$, it follows that

$$7\cos 2\theta - 24\sin 2\theta = \frac{m}{n},$$

where m and n are relatively prime positive integers. Find $m+n$.

Problem 19. The centers of the three circles A, B, and C are collinear with the center of circle B lying between the centers of circles A and C. Circles A and C are both externally tangent to circle B, and the three circles share a common tangent line. Given that circle A has radius 12 and circle B has radius 42, find the radius of circle C.

Problem 20. How many of the rearrangements of the digits 123456 have the property that, for each digit, no more than two digits smaller than that digit appear to the right of that digit? For example, the rearrangement 315426 has this property because digits 1 and 2 are the only digits smaller than 3 which follow 3, digits 2 and 4 are the only digits smaller than 5 which follow 5, and digit 2 is the only digit smaller than 4 which follows 4.

Problem 21. Let a be the sum of the numbers:

$$\begin{aligned} &99 \times 0.9 \\ &999 \times 0.9 \\ &9999 \times 0.9 \\ &\vdots \\ &999\cdots 9 \times 0.9, \end{aligned}$$

where the final number in the list is 0.9 times a number written as a string of 101 digits all equal to 9. Find the sum of the digits in the number a.

Problem 22. Ten distinct points are placed on a circle. All ten of the points are paired so that the line segments connecting the pairs do not intersect. In how many different ways can this pairing be done?

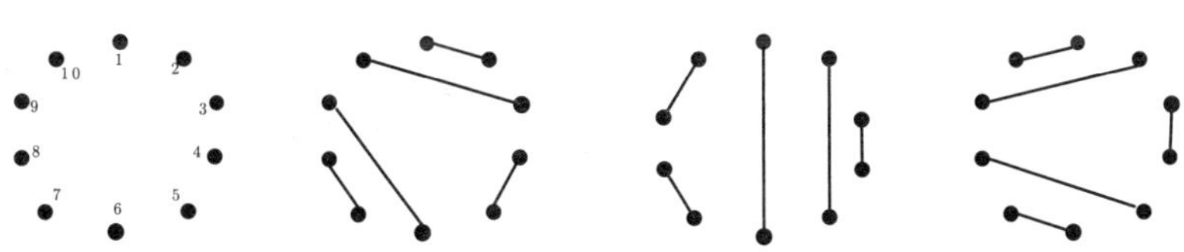

Problem 23. A disk with radius 10 and a disk with radius 8 are drawn so that the distance between their centers is 3. Two congruent small circles lie in the intersection of the two disks so that they are tangent to each other and to each of the larger circles, as shown. The radii of the smaller circles are both $\frac{m}{n}$ where m and n are relatively prime positive integers. Find $m + n$.

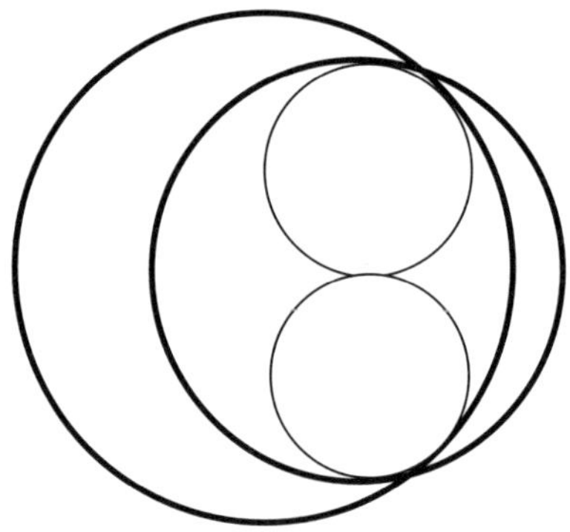

Problem 24. Find the number of ordered pairs of integers (m, n) that satisfy

$$20m - 10n = mn.$$

Problem 25. Let x_1, x_2, x_3 be the three roots of the polynomial $x^3 + 3x + 1$. There are relatively prime positive integers m and n so that

$$\frac{m}{n} = \frac{x_1^2}{(5x_2+1)(5x_3+1)} + \frac{x_2^2}{(5x_3+1)(5x_1+1)} + \frac{x_3^2}{(5x_1+1)(5x_2+1)}.$$

Find $m + n$.

Problem 26. In the coordinate plane a parabola passes through the points (7,6), (7,12), (18,19), (18,48). The axis of symmetry of the parabola is a line with slope $\frac{r}{s}$ where r and s are relatively prime positive integers. Find $r + s$.

Problem 27. Let a and b be real numbers satisfying

$$2(\sin a + \cos a)\sin b = 3 - \cos b.$$

Find $3\tan^2 a + 4\tan^2 b$.

Problem 28. There are relatively prime positive integers m and n such that

$$\frac{m}{n} = \sum_{k=3}^{\infty} \frac{1}{k^5 - 5k^3 + 4k}.$$

Find $m + n$.

Problem 29. Square $ABCD$ is shown in the diagram below. Points E, F, G are on sides $\overline{AB}$, $\overline{BC}$, $\overline{DA}$, respectively, so that lengths of $\overline{BE}$, $\overline{BF}$, $\overline{DG}$ are equal. Points H and I are the midpoints of segments $\overline{EF}$ and $\overline{CG}$, respectively. Segment $\overline{GJ}$ is the perpendicular bisector of segment $\overline{HI}$. The ratio of the areas of pentagon $AEHJG$ and quadrilateral $CIHF$ can be written as $\frac{m}{n}$, where m and n are relatively prime positive integers. Find $m+n$.

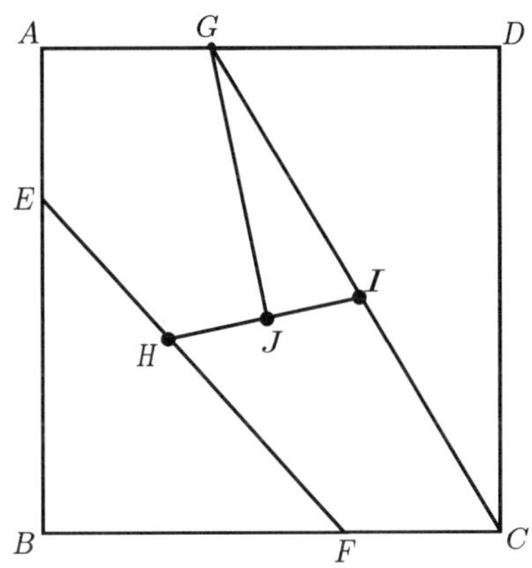

Problem 30. Let x and y be real numbers satisfying

$$(x^2+x-1)(x^2-x+1)=2(y^3-2\sqrt{5}-1)$$

and

$$(y^2+y-1)(y^2-y+1)=2(x^3+2\sqrt{5}-1)$$

Find $8x^2+4y^3$.

1.15 2010 Middle School

Problem 1. If $125 + n + 135 + 2n + 145 = 900$, find n.

Problem 2. Three boxes each contain four bags. Each bag contains five marbles. How many marbles are there altogether in the three boxes?

Problem 3. The sum

$$\frac{1}{1} + \frac{1}{2} + \frac{1}{3} + \frac{1}{4} + \frac{1}{5} + \frac{1}{6} = \frac{m}{n},$$

where m and n are relatively prime positive integers. Find $m+n$.

Problem 4. The grid below contains five rows with six points in each row. Points that are adjacent either horizontally or vertically are a distance one apart. Find the area of the pentagon shown.

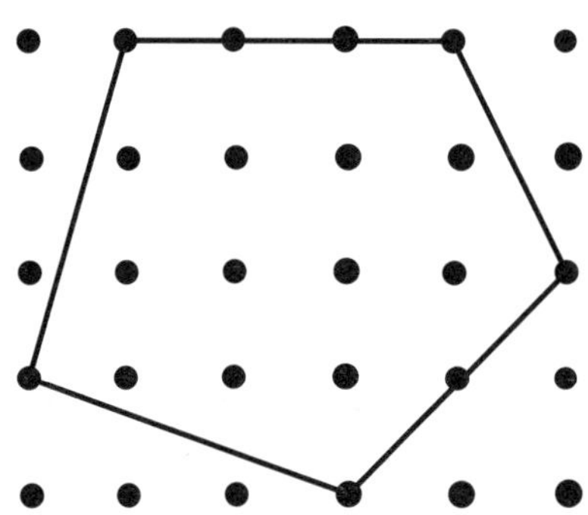

Problem 5. Find the least positive integer k so that $k+25973$ is a palindrome (a number which reads the same forward and backwards).

Problem 6. Find the sum of the prime factors of 777.

Problem 7. If x and y are positive real numbers where x is p percent of y and y is $4p$ percent of x, find p.

Problem 8. There are exactly two four-digit numbers that are multiples of 3 and their first digit is double their second digit, their third digit is three more than their fourth digit, and their second digit is 2 less than their fourth digit. Find the difference of these two numbers.

Problem 9. What percent of the numbers 1, 2, 3,. . . , 1000 are divisible by exactly one of the numbers 4 and 5?

Problem 10. A baker uses $6\frac{2}{3}$ cups of flour when she prepares $\frac{5}{3}$ recipes of rolls. She will use $9\frac{3}{4}$ cups of flour when she prepares $\frac{m}{n}$ recipes of rolls, where m and n are relatively prime positive integers. Find $m + n$.

Problem 11. There are two rows of seats with three side-by-side seats in each row. Two little boys, two little girls, and two adults sit in the six seats so that neither little boy sits to the side of either little girl. In how many different ways can these six people be seated?

Problem 12. The diagram below shows twelve 30-60-90 triangles placed in a circle so that the hypotenuse of each triangle coincides with the longer leg of the next triangle. The fourth and last triangle in this diagram are shaded. The ratio of the perimeters of these two triangles can be written as $\frac{m}{n}$, where m and n are relatively prime positive integers. Find $m + n$.

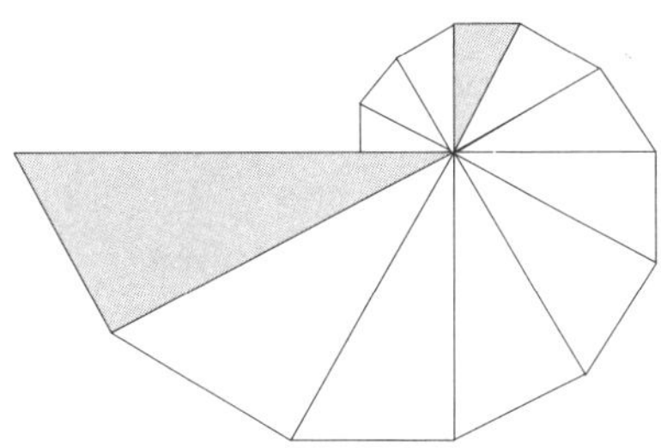

Problem 13. Find the number of sets A that satisfy the three conditions:

- A is a set of two positive integers
- each of the numbers in A is at least 22 percent the size of the other number
- A contains the number 30.

Problem 14. Let $ABCD$ be a trapezoid where $\overline{AB}$ is parallel to $\overline{CD}$. Let P be the intersection of diagonal $\overline{AC}$ and diagonal $\overline{BD}$. If the area of triangle PAB is 16, and the area of triangle PCD is 25, find the area of the trapezoid.

Problem 15. In the arrangement

1				
2	3			
4	5	6		
7	8	9	10	
11	12	13	14	15
⋮				

what is the number that will appear directly below the number 2010?

Problem 16. Half the volume of a 12 foot high cone-shaped pile is grade A ore while the other half is grade B ore. The pile is worth \$62. One-third of the volume of a similarly shaped 18 foot pile is grade A ore while the other two-thirds is grade B ore. The second pile is worth \$162. Two-thirds of the volume of a similarly shaped 24 foot pile is grade A ore while the other one-third is grade B ore. What is the value in dollars of the 24 foot pile?

Problem 17. The diagram below shows a triangle divided into sections by three horizontal lines which divide the altitude of the triangle into four equal parts, and three lines connecting the top vertex with points that divide the opposite side into four equal parts. If the shaded region has area 100, find the area of the entire triangle.

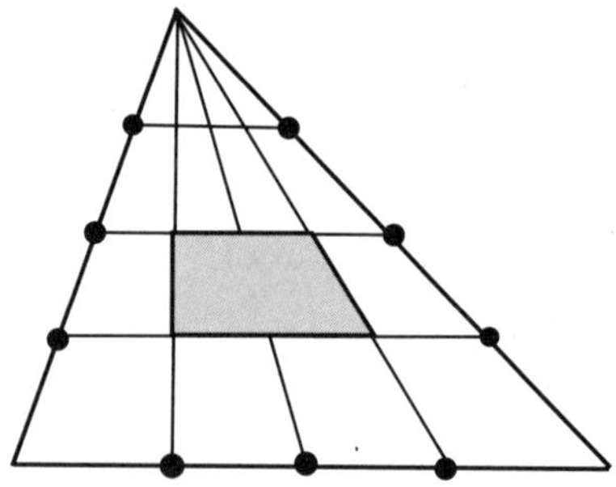

Problem 18. How many three-digit positive integers contain both even and odd digits?

Problem 19. Square A is adjacent to square B, which is adjacent to square C. The three squares all have their bottom sides along a common horizontal line, as shown. The upper left vertices of the three squares are collinear. If square A has area 24 and square B has area 36, find the area of square C.

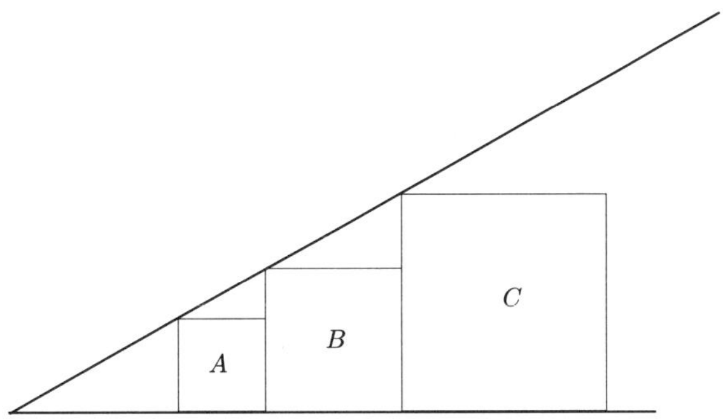

Problem 20. Suppose that f is a function such that

$$3f(x) - 5xf\left(\frac{1}{x}\right) = x - 7$$

for all non-zero real numbers x. Find $f(2010)$.

1.16 2011 High School

Problem 1. The ratio of 3 to the positive number n is the same as the ratio of n to 192. Find n.

Problem 2. The target below is made up of concentric circles with diameters 4, 8, 12, 16, and 20. The area of the dark region is $n\pi$. Find n.

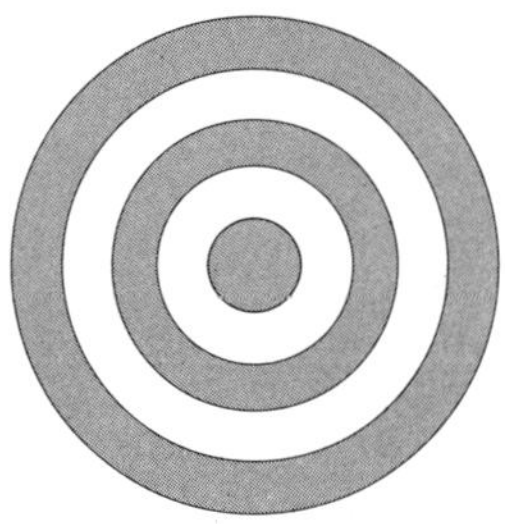

Problem 3. Shirley went to the store planning to buy 120 balloons for 10 dollars. When she arrived, she was surprised to find that the balloons were on sale for 20 percent less than expected. How many balloons could Shirley buy for her 10 dollars?

Problem 4. Five non-overlapping equilateral triangles meet at a common vertex so that the angles between adjacent triangles are all congruent. What is the degree measure of the angle between two adjacent triangles?

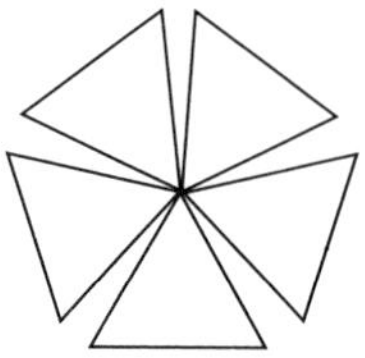

Problem 5. Let $a_1 = 2$, and for $n \geq 1$, let $a_{n+1} = 2a_n + 1$. Find the smallest value of an a_n that is not a prime number.

Problem 6. Working alone, the expert can paint a car in one day, the amateur can paint a car in two days, and the beginner can paint a car in three days. If the three painters work together at these speeds to paint three cars, it will take them $\frac{m}{n}$ days where m and n are relatively prime positive integers. Find $m + n$.

Problem 7. Find the prime number p such that $71p + 1$ is a perfect square.

Problem 8. When 126 is added to its reversal, 621, the sum is $126 + 621 = 747$. Find the greatest integer which when added to its reversal yields 1211.

Problem 9. There are integers m and n so that $9 + \sqrt{11}$ is a root of the polynomial $x^2 + mx + n$. Find $m + n$.

Problem 10. The diagram shows a large circular dart board with four smaller shaded circles each internally tangent to the larger circle. Two of the internal circles have half the radius of the large circle, and are, therefore, tangent to each other. The other two smaller circles are tangent to these circles. If a dart is thrown so that it sticks to a point randomly chosen on the dart board, then the probability that the dart sticks to a point in the shaded area is $\frac{m}{n}$ where m and n are relatively prime positive integers. Find $m + n$.

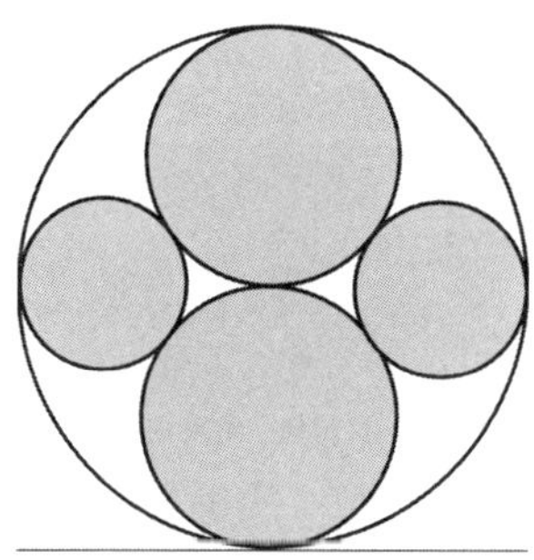

Problem 11. Six distinct positive integers are randomly chosen between 1 and 2011, inclusive. The probability that some pair of the six chosen integers has a difference that is a multiple of 5 is n percent. Find n.

Problem 12. Find the area of the region in the coordinate plane satisfying the three conditions

- $x \leq 2y$
- $y \leq 2x$
- $x + y \leq 60$.

Problem 13. A 3 by 3 determinant has three entries equal to 2, three entries equal to 5, and three entries equal to 8. Find the maximum possible value of the determinant.

Problem 14. The lengths of the three sides of a right triangle form a geometric sequence. The sine of the smallest of the angles in the triangle is $\frac{m+\sqrt{n}}{k}$ where m, n, and k are integers, and n is not divisible by the square of any prime. Find $m+n+k$.

Problem 15. A pyramid has a base which is an equilateral triangle with side length 300 centimeters. The vertex of the pyramid is 100 centimeters above the center of the triangular base. A mouse starts at a corner of the base of the pyramid and walks up the edge of the pyramid toward the vertex at the top. When the mouse has walked a distance of 134 centimeters, how many centimeters above the base of the pyramid is the mouse?

Problem 16. Evaluate $1^3 - 2^3 + 3^3 - 4^3 + 5^3 - \cdots + 101^3$.

Problem 17. In how many distinguishable rearrangements of the letters $ABCCDEEF$ does the A precede both C's, the F appears between the 2 C's, and the D appears after the F?

Problem 18. Let a be a positive real number such that $\frac{a^2}{a^4-a^2+1} = \frac{4}{37}$. Then $\frac{a^3}{a^6-a^3+1} = \frac{m}{n}$, where m and n are relatively prime positive integers. Find $m+n$.

Problem 19. The diagrams below shows a 2 by 2 grid made up of four 1 by 1 squares. Shown are two paths along the grid from the lower left corner to the upper right corner of the grid, one with length 4 and one with length 6. A path may not intersect itself by moving to a point where the path has already been. Find the sum of the lengths of all the paths from the lower left corner to the upper right corner of the grid.

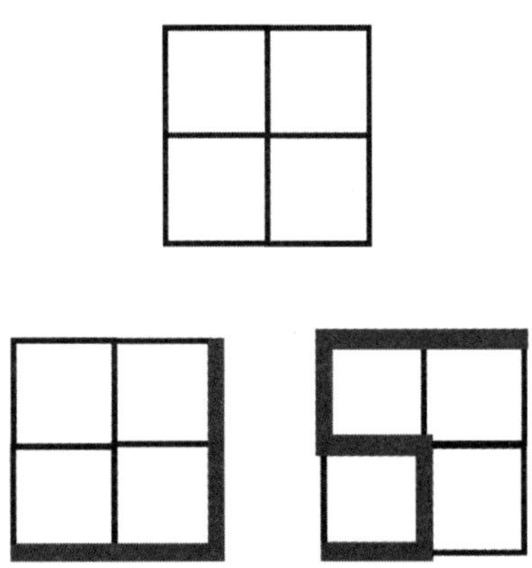

Problem 20. Points A and B are the endpoints of a diameter of a circle with center C. Points D and E lie on the same diameter so that C bisects segment $\overline{DE}$. Let F be a randomly chosen point within the circle. The probability that $\triangle DEF$ has a perimeter less than the length of the diameter of the circle is $\frac{17}{128}$. There are relatively prime positive integers m and n so that the ratio of DE to AB is $\frac{m}{n}$. Find $m+n$.

Problem 21. If a, b, and c are non-negative real numbers satisfying

$$a+b+c=400,$$

find the maximum possible value of $\sqrt{2a+b}+\sqrt{2b+c}+\sqrt{2c+a}$.

Problem 22. Five congruent circles have centers at the vertices of a regular pentagon so that each of the circles is tangent to its two neighbors. A sixth circle (shaded in the diagram below) congruent to the other five is placed tangent to two of the five. If this sixth circle is allowed to roll without slipping around the exterior of the figure formed by the other five circles, then it will turn through an angle of k degrees before it returns to its starting position. Find k.

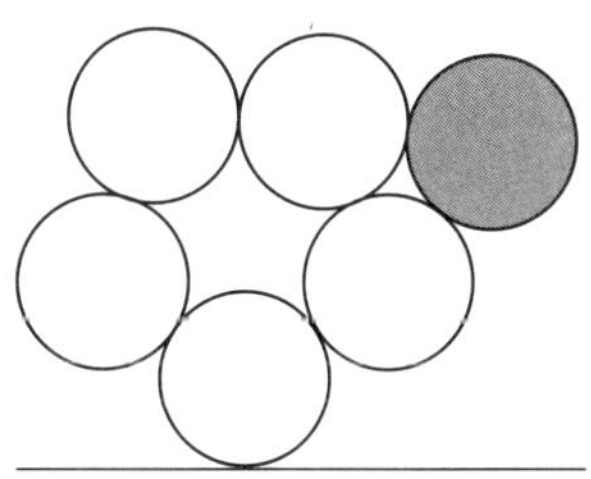

Problem 23. Let x be a real number in the interval $\left(0, \frac{\pi}{2}\right)$ such that

$$\frac{1}{\sin x \cos x} + 2\cot 2x = \frac{1}{2}.$$

Evaluate

$$\frac{1}{\sin x \cos x} - 2\cot 2x.$$

Problem 24. The diagram below shows a regular hexagon with an inscribed square where two sides of the square are parallel to two sides of the hexagon. There are positive integers m, n, and p such that the ratio of the area of the hexagon to the area of the square can be written as $\frac{m+\sqrt{n}}{p}$ where m and p are relatively prime.

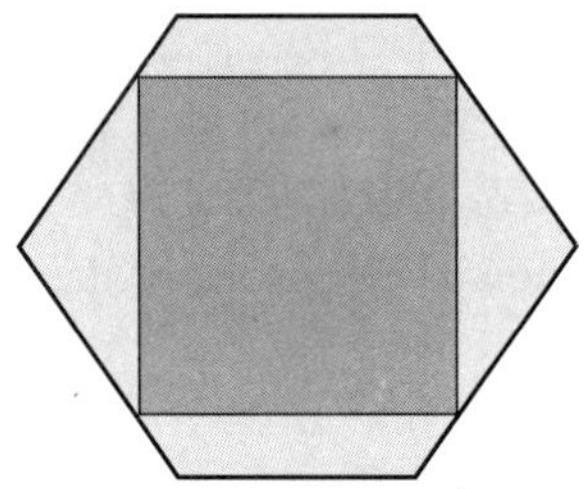

Find $m + n + p$.

Problem 25. Find the remainder when $A = 3^3 \cdot 33^{33} \cdot 333^{333} \cdot 3333^{3333}$ is divided by 100.

Problem 26. The diagram below shows two parallel rows with seven points in the upper row and nine points in the lower row. The points in each row are spaced one unit apart, and the two rows are two units apart. How many trapezoids which are not parallelograms have vertices in this set of 16 points and have area of at least six square units?

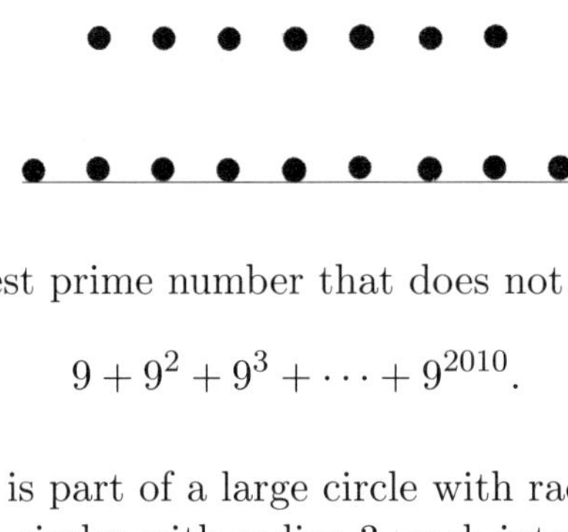

Problem 27. Find the smallest prime number that does not divide

$$9 + 9^2 + 9^3 + \cdots + 9^{2010}.$$

Problem 28. Pictured below is part of a large circle with radius 30. There is a chain of three circles with radius 3, each internally tangent to the large circle and each tangent to its neighbors in the chain. There are two circles with radius 2 each tangent to two of the radius 3 circles. The distance between the centers of the two circles with radius 2 can be written as $\frac{a\sqrt{b}-c}{d}$ where a, b, c, and d are positive integers, c and d are relatively prime, and b is not divisible by the square of any prime. Find $a + b + c + d$.

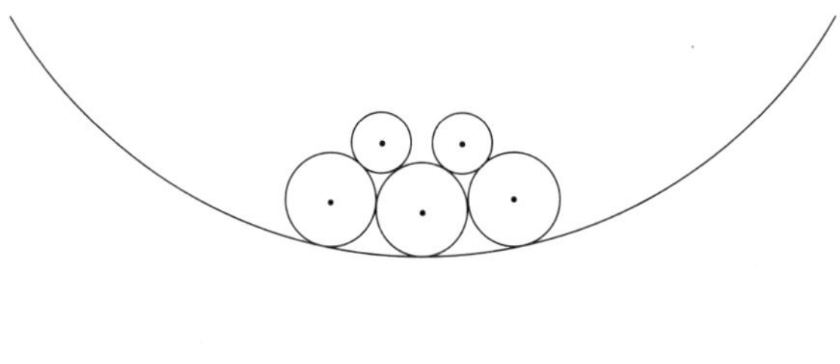

Problem 29. Let S be a randomly selected four-element subset of $\{1,2,3,4,5,6,7,8\}$. Let m and n be relatively prime positive integers so that the expected value of the maximum element in S is $\frac{m}{n}$. Find $m+n$.

Problem 30. Four congruent spheres are stacked so that each is tangent to the other three. A larger sphere, R, contains the four congruent spheres so that all four are internally tangent to R. A smaller sphere, S, sits in the space between the four congruent spheres so that all four are externally tangent to S. The ratio of the surface area of R to the surface area of S can be written $m+\sqrt{n}$ where m and n are positive integers. Find $m+n$.

1.17 2011 Middle School

Problem 1. There are relatively prime positive integers m and n so that

$$\frac{\frac{1}{2}}{\frac{\frac{1}{3}}{\frac{1}{4}+\frac{1}{4}+\frac{1}{4}}+\frac{\frac{1}{3}}{\frac{1}{4}+\frac{1}{4}+\frac{1}{4}}}=\frac{m}{n}.$$

Find $m+2n$.

Problem 2. The diagram below shows a 12-sided figure made up of three congruent squares. The figure has total perimeter 60. Find its area.

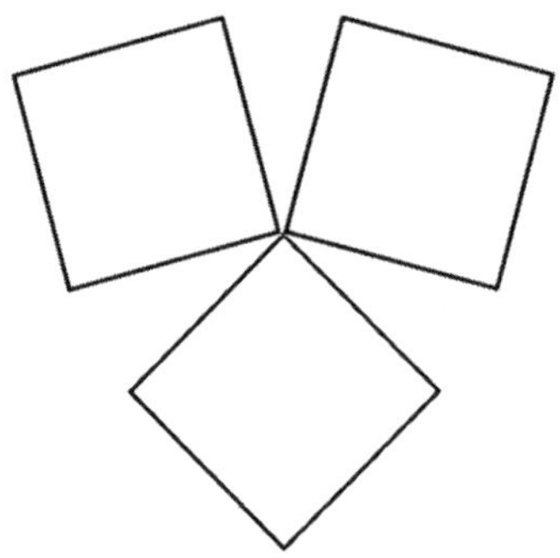

Problem 3. Find the sum of all two-digit integers which are both prime and are 1 more than a multiple of 10.

Problem 4. Jerry buys a bottle of 150 pills. Using a standard 12 hour clock, he sees that the clock reads exactly 12 when he takes the first pill. If he takes one pill every five hours, what hour will the clock read when he takes the last pill in the bottle?

Problem 5. Given that $\frac{6}{11}-\frac{10}{19}=\frac{9}{19}-\frac{n}{11}$, find n.

Problem 6. The following addition problem is not correct if the numbers are interpreted as base 10 numbers. In what number base is the problem correct?

$$\begin{array}{r} 66 \\ 87 \\ 85 \\ +\ 48 \\ \hline 132 \end{array}$$

Problem 7. When 12^{18} is divided by 18^{12}, the result is $\left(\frac{m}{n}\right)^3$, where m and n are relatively prime positive integers. Find $m - n$.

Problem 8. A square measuring 15 by 15 is partitioned into five rows of five congruent squares as shown below. The small squares are alternately colored black and white as shown. Find the total area of the part colored black.

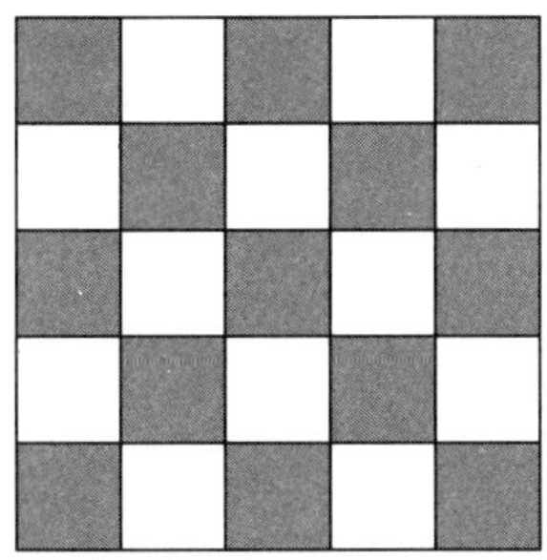

Problem 9. A jar contains one quarter red jelly beans and three quarters blue jelly beans. If Chris removes three quarters of the red jelly beans and one quarter of the blue jelly beans, what percent of the jelly beans remaining in the jar will be red?

Problem 10. Five rays $\overrightarrow{OA}$, $\overrightarrow{OB}$, $\overrightarrow{OC}$, $\overrightarrow{OD}$, and $\overrightarrow{OE}$ radiate in a clockwise order from O forming four non-overlapping angles such that $\angle EOD = 2\angle COB$, $\angle COB = 2\angle BOA$, while $\angle DOC = 3\angle BOA$. If E, O, A are collinear with O between A and E, what is the degree measure of $\angle DOB$?

Problem 11. How many numbers are there that appear both in the arithmetic sequence 10, 16, 22, 28, ... 1000 and the arithmetic sequence 10, 21, 32, 43, ..., 1000?

Problem 12. When Troy writes his digits, his 0, 1, and 8 look the same right-side-up and upside-down as seen in the figure below. His 6 and 9 look like upside-down images of each other. None of his other digits look like digits when they are inverted. How many different five-digit numbers (which do not begin with the digit zero) can Troy write which read the same right-side-up and upside-down?

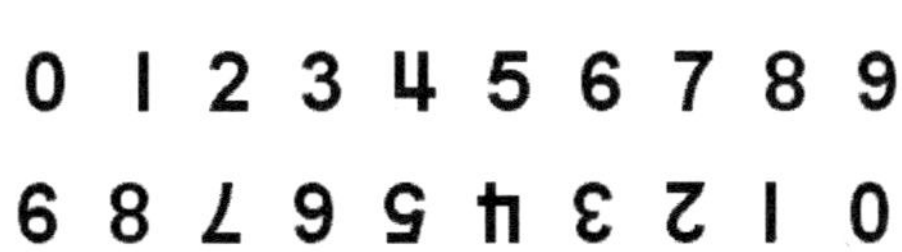

Problem 13. The diagram shows two equilateral triangles with side length 4 mounted on two adjacent sides of a square also with side length 4. The distance between the two vertices marked A and B can be written as $\sqrt{m}+\sqrt{n}$ for two positive integers m and n. Find $m+n$.

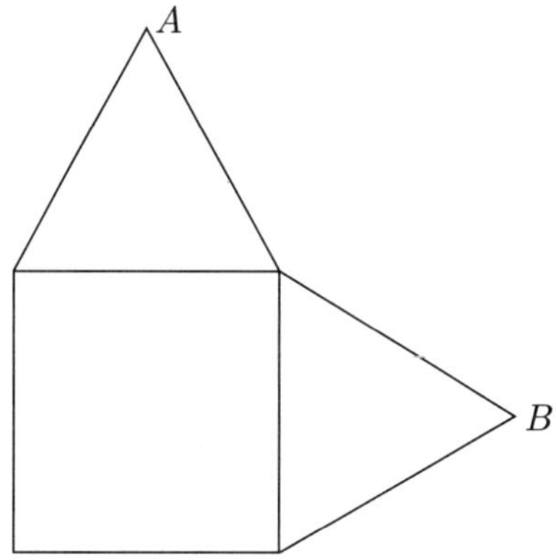

Problem 14. The five-digit number 12110 is divisible by the sum of its digits $1+2+1+1+0=5$. Find the greatest five-digit number which is divisible by the sum of its digits.

Problem 15. In the diagram below, $\overline{AB}$ and $\overline{CD}$ are parallel, $\angle BXY = 45°$, $\angle DZY = 25°$, and $XY = YZ$. What is the degree measure of $\angle YXZ$?

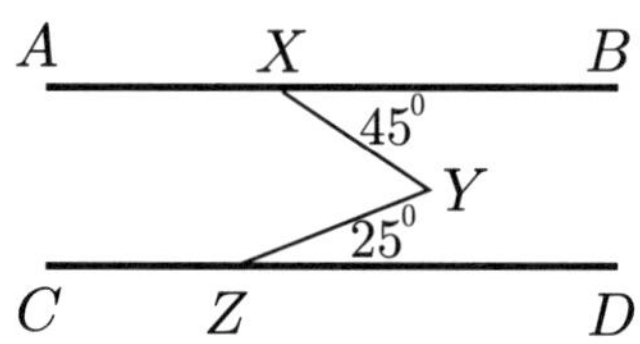

Problem 16. Let a and b be nonzero real numbers such that $\frac{1}{3a}+\frac{1}{b}=2011$ and $\frac{1}{a}+\frac{1}{3b}=1$. What is the quotient when $a+b$ is divided by ab?

Problem 17. Find the number of ordered quadruples (a,b,c,d) where each of a, b, c, and d are (not necessarily distinct) elements of $\{1,2,3,4,5,6,7\}$ and $3abc+4abd+5bcd$ is even. For example, $(2,2,5,1)$ and $(3,1,4,6)$ satisfy the conditions.

Problem 18. Find the positive integer n so that n^2 is the perfect square closest to $8+16+24+\cdots+8040$.

Problem 19. How many ordered pairs of sets (A, B) have the properties:

(a) $A \subseteq \{1, 2, 3, 4, 5, 6\}$

(b) $B \subseteq \{2, 3, 4, 5, 6, 7, 8\}$

(c) $A \cap B$ has exactly 3 elements.

Problem 20. Let V be the set of vertices of a regular 25 sided polygon with center at point C. How many triangles have vertices in V and contain the point C in the interior of the triangle?

1.18 2012 High School

Problem 1. Last month a pet store sold three times as many cats as dogs. If the store had sold the same number of cats but eight more dogs, it would have sold twice as many cats as dogs. How many cats did the pet store sell last month?

Problem 2. What is the greatest three-digit divisor of 111777?

Problem 3. The diagram below shows a large square divided into nine congruent smaller squares. There are circles inscribed in five of the smaller squares. The total area covered by all the five circles is 20π. Find the area of the large square.

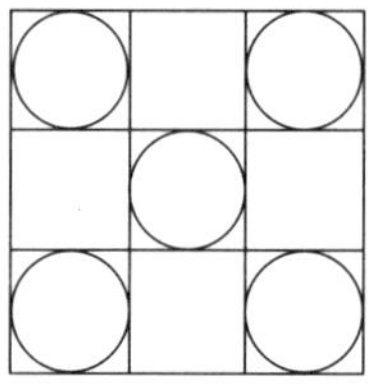

Problem 4. The following diagram shows an equilateral triangle and two squares that share common edges. The area of each square is 75. Find the distance from point A to point B.

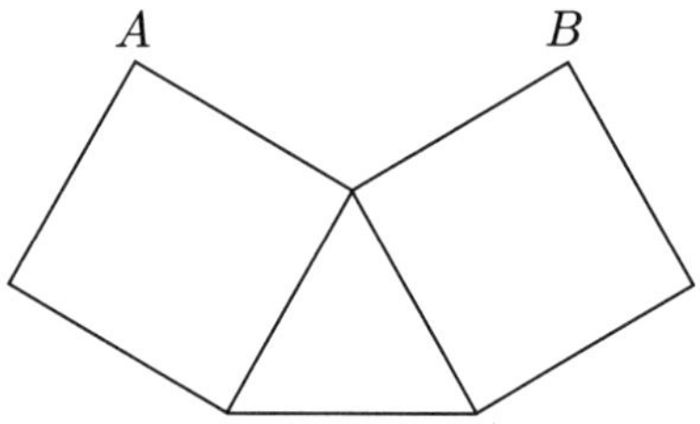

Problem 5. Find the sum of the squares of the values x that satisfy

$$\frac{1}{x} + \frac{2}{x+3} + \frac{3}{x+6} = 1.$$

Problem 6. Find the least positive integer n so that both n and $n+1$ have prime factorizations with exactly four (not necessarily distinct) prime factors.

Problem 7. Two convex polygons have a total of 33 sides and 243 diagonals. Find the number of diagonals in the polygon with the greater number of sides.

Problem 8. In the tribe of Zimmer, being able to hike long distances and knowing the roads through the forest are both extremely important, so a boy who reaches the age of manhood is not designated as a man by the tribe until he completes an interesting rite of passage. The man must go on a sequence of hikes. The first hike is a 5 kilometer hike down the main road. The second hike is a $5\frac{1}{4}$ kilometer hike down a secondary road. Each hike goes down a different road and is a quarter kilometer longer than the previous hike. The rite of passage is completed at the end of the hike where the cumulative distance walked by the man on all his hikes exceeds 1000 kilometers. So in the tribe of Zimmer, how many roads must a man walk down, before you call him a man?

Problem 9. Find the value of x that satisfies $\log_3(\log_9 x) = \log_9(\log_3 x)$.

Problem 10. Consider a sequence of eleven squares that have side lengths $3, 6, 9, 12, \ldots, 33$. Eleven copies of a single square each with area A have the same total area as the total area of the eleven squares of the sequence. Find A.

Problem 11. Define $f(x) = 2x + 3$ and suppose that

$$g(x+2) = f\left(f(x-1) \cdot f(x+1) + f(x)\right).$$

Find $g(6)$.

Problem 12. Ted flips seven fair coins. There are relatively prime positive integers m and n so that $\frac{m}{n}$ is the probability that Ted flips at least two heads given that he flips at least three tails. Find $m + n$.

Problem 13. Find the least n for which $n!(n+1)!(2n+1)! - 1$ ends in thirty digits that are all 9's.

Problem 14. A circle in the first quadrant with center on the curve

$$y = 2x^2 - 27$$

is tangent to the y-axis and the line $4x = 3y$. The radius of the circle is $\frac{m}{n}$ where m and n are relatively prime positive integers. Find $m + n$.

Problem 15. Let N be a positive integer whose digits add up to 23. What is the greatest possible product the digits of N can have?

Problem 16. Let a, b, and c be non-zero real numbers such that $\frac{ab}{a+b} = 3$, $\frac{bc}{b+c} = 4$, and $\frac{ca}{c+a} = 5$. There are relatively prime positive integers m and n so that $\frac{abc}{ab+bc+ca} = \frac{m}{n}$. Find $m+n$.

Problem 17. How many positive integer solutions are there to

$$w + x + y + z = 20$$

where $w + x \geq 5$ and $y + z \geq 5$?

Problem 18. Find the number of three-digit numbers such that its first two digits are each divisible by its third digit.

Problem 19. Find the remainder when $2^{5^9} + 5^{9^2} + 9^{2^5}$ is divided by 11.

Problem 20. Square $ABCD$ has side length 68. Let E be the midpoint of segment $\overline{CD}$, and let F be the point on segment $\overline{AB}$ a distance 17 from point A. Point G is on segment $\overline{EF}$ so that $\overline{EF}$ is perpendicular to segment $\overline{GD}$. The length of segment $\overline{BG}$ can be written as $m\sqrt{n}$ where m and n are positive integers, and n is not divisible by the square of any prime. Find $m+n$.

Problem 21. Each time you click a toggle switch, the switch either turns from *off* to *on* or from *on* to *off*. Suppose that you start with three toggle switches with one of them *on* and two of them *off*. On each move you randomly select one of the three switches and click it. Let m and n be relatively prime positive integers so that $\frac{m}{n}$ is the probability that after four such clicks, one switch will be *on* and two of them will be *off*. Find $m+n$.

Problem 22. The diagram below shows circles radius 1 and 2 externally tangent to each other and internally tangent to a circle radius 3. There are relatively prime positive integers m and n so that a circle radius $\frac{m}{n}$ is internally tangent to the circle radius 3 and externally tangent to the other two circles as shown. Find $m+n$.

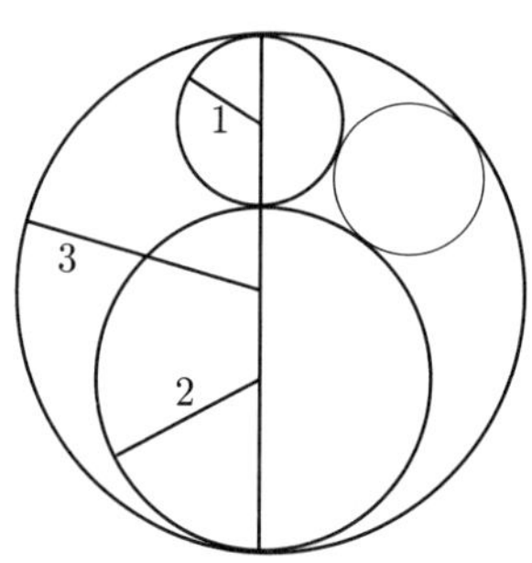

Problem 23. Find the greatest seven-digit integer divisible by 132 whose digits, in order, are 2, 0, x, y, 1, 2, z where x, y, and z are single digits.

Problem 24. There are positive integers m and n so that $x = m + \sqrt{n}$ is a solution to the equation $x^2 - 10x + 1 = \sqrt{x}(x+1)$. Find $m + n$.

Problem 25. Find the largest prime that divides

$$1 \cdot 2 \cdot 3 \ + \ 2 \cdot 3 \cdot 4 \ + \ ... \ + \ 44 \cdot 45 \cdot 46.$$

Problem 26. A paper cup has a base that is a circle with radius r, a top that is a circle with radius $2r$, and sides that connect the two circles with straight line segments as shown below. This cup has height h and volume V. A second cup that is exactly the same shape as the first is held upright inside the first cup so that its base is a distance of $\frac{h}{2}$ from the base of the first cup. The volume of liquid that will fit inside the first cup and outside the second cup can be written $\frac{m}{n} \cdot V$ where m and n are relatively prime positive integers. Find $m + n$.

Problem 27. You have some white one-by-one tiles and some black and white two-by-one tiles as shown below. There are four different color patterns that can be generated when using these tiles to cover a three-by-one rectangle by laying these tiles side by side (WWW, BWW, WBW, WWB). How many different color patterns can be generated when using these tiles to cover a ten-by-one rectangle?

Problem 28. A bag contains 8 green candies and 4 red candies. You randomly select one candy at a time to eat. If you eat five candies, there

are relatively prime positive integers m and n so that $\frac{m}{n}$ is the probability that you do not eat a green candy after you eat a red candy. Find $m+n$.

Problem 29. Let $A = \{1,3,5,7,9\}$ and $B = \{2,4,6,8,10\}$. Let f be a randomly chosen function from the set $A \cup B$ into itself. There are relatively prime positive integers m and n such that $\frac{m}{n}$ is the probability that f is a one-to-one function on $A \cup B$ given that it maps A one-to-one into $A \cup B$ and it maps B one-to-one into $A \cup B$. Find $m+n$.

Problem 30. The diagram below shows four regular hexagons each with side length 1 meter attached to the sides of a square. This figure is drawn onto a thin sheet of metal and cut out. The hexagons are then bent upward along the sides of the square so that A_1 meets A_2, B_1 meets B_2, C_1 meets C_2, and D_1 meets D_2. The resulting dish is set on a table with the square lying flat on the table. If this dish is filled with water, the water will rise to the height of the corner where the A_1 and A_2 meet. There are relatively prime positive integers m and n so that the number of cubic meters of water the dish will hold is $\sqrt{\frac{m}{n}}$. Find $m+n$.

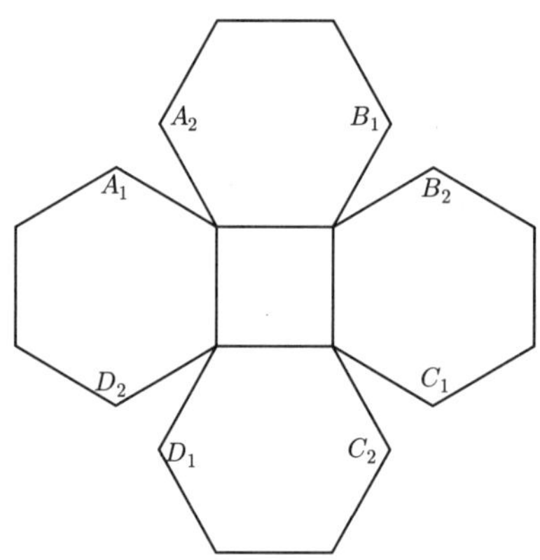

1.19 2012 Middle School

Problem 1. Evaluate $5^4 - 4^3 - 3^2 - 2^1 - 1^0$.

Problem 2. The diagram below shows rectangle $ABDE$ where C is the midpoint of side $\overline{BD}$, and F is the midpoint of side $\overline{AE}$. If $AB = 10$ and $BD = 24$, find the area of the shaded region.

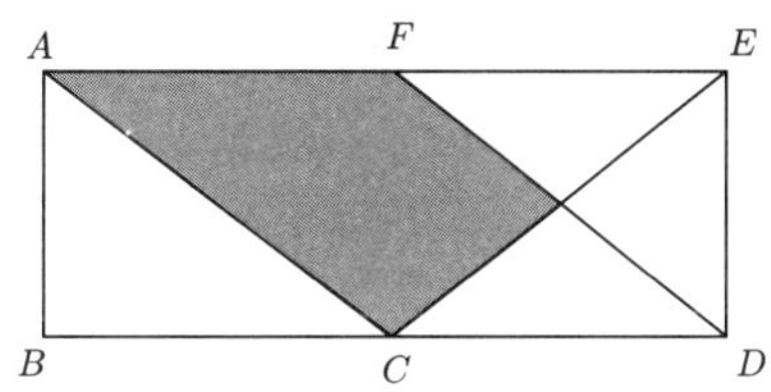

Problem 3. While Peter was driving from home to work, he noticed that after driving 21 miles, the distance he had left to drive was 30 percent of the total distance from home to work. How many miles was his complete trip from home to work?

Problem 4. How many two-digit positive integers contain at least one digit equal to 5?

Problem 5. Meredith drives 5 miles to the northeast, then 15 miles to the southeast, then 25 miles to the southwest, then 35 miles to the northwest, and finally 20 miles to the northeast. How many miles is Meredith from where she started?

Problem 6. Volume A equals one fourth of the sum of the volumes B and C, while volume B equals one sixth of the sum of the volumes A and C. There are relatively prime positive integers m and n so that the ratio of volume C to the sum of the other two volumes is $\frac{m}{n}$. Find $m + n$.

Problem 7. A snail crawls $2\frac{1}{2}$ centimeters in $4\frac{1}{4}$ minutes. At that rate, how many centimeters can the snail crawl in 85 minutes?

Problem 8. Seven boys and three girls are playing basketball. In how many different ways can they make two teams of five players so that both teams have at least one girl?

Problem 9. Points E and F lie inside rectangle $ABCD$ with $AE = DE = BF = CF = EF$. If $AB = 11$ and $BC = 8$, find the area of the quadrilateral $AEFB$.

Problem 10. Find the least positive multiple of 999 that does not have 9 as a digit.

Problem 11. For some integers a and b the function $f(x) = ax + b$ has the properties that $f(f(0)) = 0$ and $f(f(f(4))) = 9$. Find $f(f(f(f(10))))$.

Problem 12. Pentagon $ABCDE$ consists of a square $ACDE$ and an equilateral triangle ABC that share the side $\overline{AC}$. A circle centered at C has area 24. The intersection of the circle and the pentagon has half the area of the pentagon. Find the area of the pentagon.

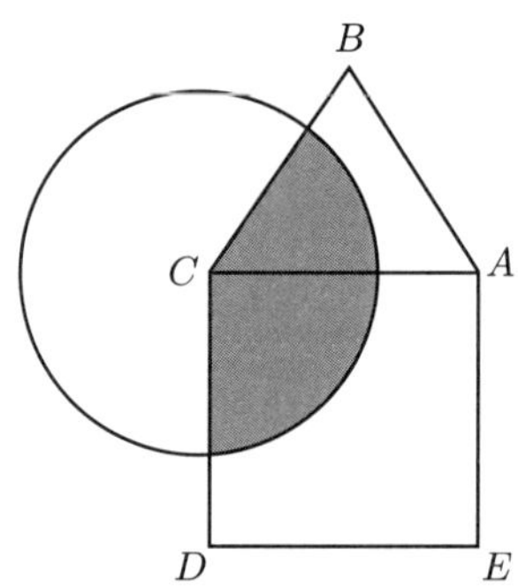

Problem 13. Find the least positive integer N which is both a multiple of 19 and whose digits add to 23.

Problem 14. At the 4 PM show, all the seats in the theater were taken, and 65 percent of the audience was children. At the 6 PM show, again, all the seats were taken, but this time only 50 percent of the audience was children. Of all the people who attended either of these shows, 57 percent were children although there were 12 adults and 28 children who attended both shows. How many people does the theater seat?

Problem 15. The top and bottom of a rectangular tank each have area 48 square inches. The four vertical sides of the tank have a total area of 96 square inches. The longest diagonal of the tank is 13 inches. Find the sum of the height, the width, and the length of this tank in inches.

Problem 16. The following sequence lists all the positive rational numbers that do not exceed $\frac{1}{2}$ by first listing the fraction with denominator 2, followed by the one with denominator 3, followed by the two fractions with denominator 4 in increasing order, and so forth so that the sequence is $\frac{1}{2}, \frac{1}{3}, \frac{1}{4}, \frac{2}{4}, \frac{1}{5}, \frac{2}{5}, \frac{1}{6}, \frac{2}{6}, \frac{3}{6}, \frac{1}{7}, \frac{2}{7}, \frac{3}{7}, \ldots$. Let m and n be relatively prime positive integers so that the 2012th fraction in the list is equal to $\frac{m}{n}$. Find $m+n$.

Problem 17. The diagram below shows nine points on a circle where $AB = BC = CD = DE = EF = FG = GH$. Given that $\angle GHJ = 117°$ and $\overline{BH}$ is perpendicular to $\overline{EJ}$, there are relatively prime positive integers m and n so that the degree measure of $\angle AJB$ is $\frac{m}{n}$. Find $m+n$.

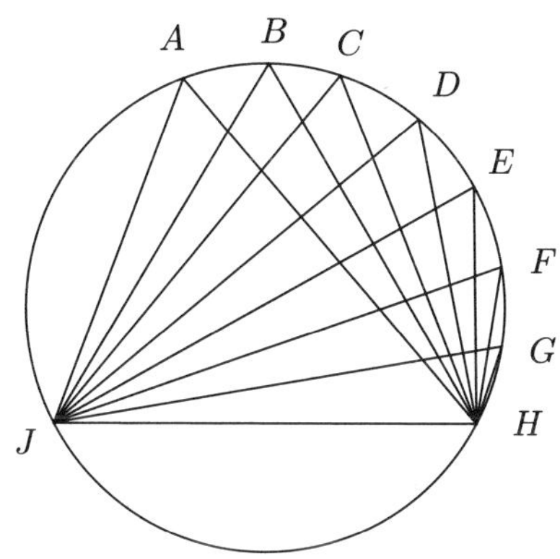

Problem 18. Find the smallest positive integer whose remainder when divided by 5 is 3, when divided by 7 is 5, and when divided by 9 is 1.

Problem 19. A teacher suggests four possible books for students to read. Each of six students selects one of the four books. How many ways can these selections be made if each of the books is read by at least one student?

Problem 20. In the following addition, different letters represent different non-zero digits. What is the 5-digit number $ABCDE$?

$$
\begin{array}{rccccccc}
 & A & B & C & D & E & D & B \\
 & & B & C & D & E & D & B \\
 & & & C & D & E & D & B \\
 & & & & D & E & D & B \\
 & & & & & E & D & B \\
 & & & & & & D & B \\
+ & & & & & & & B \\
\hline
 & A & A & A & A & A & A & A
\end{array}
$$

Chapter 2

Solutions

2.1 2003 Middle and High School

Problem 1. In eight years Henry will be three times the age that Sally was last year. Twenty five years ago their ages added to 83. How old is Henry now?

Answer: 97

Let h represent Henry's current age, and s represent Sally's current age. In eight years Henry will have age $h + 8$. Last year Sally had age $s - 1$. So

$$h + 8 = 3(s - 1).$$

Also, 25 years ago their ages were $h - 25$ and $s - 25$, so the sum of their ages 25 years ago was

$$(h - 25) + (s - 25) = h + s - 50 = 83.$$

These equations simplify to

$$h - 3s = -11 \quad \text{and} \quad h + s = 133.$$

Adding three time the second equation to the first equation yields

$$h - 3s + 3(h + s) = -11 + 3 \cdot 133 \quad \text{or} \quad 4h = 388 \quad \text{or} \quad h = 97.$$

So Henry is currently 97 years old.

Problem 2. What is the least number that could be the date of the first Saturday after the second Monday following the second Thursday of a month?

Answer: 1

If the month in question is a 28-day February starting on a Saturday, the the first Thursday is the 6th. The second Thursday is the 13th. The first Monday after that is the 17th. The second Monday after the second Thursday is then the 24th. The following Saturday is then March 1. Since the date must be a positive integer, the smallest possible value for the date is 1.

Note: This is the only problem in the history of the Purple Comet*!* Math Meet for which no team submitted a correct answer.

Problem 3. What is the greatest integer whose prime factors add to 14?

Answer: 162

Note that $2\times2\times2 = 8$ and $3\times3 = 9$, so you get a larger product by multiplying two threes than by multiplying three twos even though $2+2+2 = 3+3 = 6$. Taking products of larger numbers is less economical. The product of two fives is $5\times5 = 25$ whereas the product of five twos $2\times2\times2\times2\times2 = 32$ is larger. So, to get the largest possible product, you would want to take the product of as many threes as possible to get $3 \times 3 \times 3 \times 3 \times 2 = 162$.

Problem 4. The lengths in inches of the diagonals of a rhombus are two consecutive integers. The area of the rhombus is 210 square inches. Find its perimeter, in inches.

Answer: 58

Let x and $x + 1$ be the lengths of the two diagonals. Then

$$\frac{x(x+1)}{2} = 210,$$

which reduces to

$$x^2 + x - 420 = 0 \quad \text{or} \quad (x - 20)(x + 21) = 0.$$

The only acceptable solution is $x = 20$. The side length s of the rhombus is the side length of the hypotenuse in a right-angled triangle whose legs are 10 and 10.5. The Pythagorean Theorem yields $s = 14.5$, so the perimeter of the rhombus is $4\times14.5 = 58$.

Problem 5. Let a, b, c be nonzero real numbers such that $a + \frac{1}{b} = 5$, $b + \frac{1}{c} = 12$, and $c + \frac{1}{a} = 13$. Find $abc + \frac{1}{abc}$.

Answer: 750

We have $\left(a+\frac{1}{b}\right)\left(b+\frac{1}{c}\right)\left(c+\frac{1}{a}\right) = 5\cdot12\cdot13$, yielding

$$abc + \left(a+\frac{1}{b}\right) + \left(b+\frac{1}{c}\right) + \left(c+\frac{1}{a}\right) + \frac{1}{abc} = 780.$$

Then

$$abc + \frac{1}{abc} = 780 - 5 - 12 - 13 = 750.$$

Problem 6. Evaluate

$$\frac{1}{\log_2\left(\frac{1}{6}\right)} - \frac{1}{\log_3\left(\frac{1}{6}\right)} - \frac{1}{\log_4\left(\frac{1}{6}\right)}.$$

Answer: 1

The given expression is equal to

$$\log_{\frac{1}{6}} 2 - \log_{\frac{1}{6}} 3 - \log_{\frac{1}{6}} 4 = \log_{\frac{1}{6}} \frac{2}{3 \cdot 4} = \log_{\frac{1}{6}} \frac{1}{6} = 1.$$

Problem 7. Find the least n such that every subset of $\{1, 2, 3, \ldots, 2004\}$ with n elements contains at least two elements that are relatively prime.

Answer: 1003

The answer must be larger than 1002 since the set

$$\{2, 4, 6, 8, \ldots, 2004\}$$

has 1002 elements yet all of the elements have a factor of 2, so no two elements are relatively prime. Consider the collection A of 1002 sets

$$\{1, 2\},\ \{3, 4\},\ \{5, 6\}, \ldots, \{2003, 2004\}.$$

If you have 1003 numbers from the set $1, 2, 3, 4, \ldots, 2004$, then by the pigeon hole principal, at least two of the elements must fall into one of the sets in the collection A. Thus, there is a k so that the 1003 numbers include both k and $k+1$. Since these two numbers are relatively prime, all sets of size 1003 must contain two numbers which are relatively prime. Thus, the answer is 1003.

Problem 8. Let $ABCDEFGHIJKL$ be a regular dodecagon. Find

$$\frac{AB}{AF} + \frac{AF}{AB}.$$

Answer: 4

Let R be the radius of the circle circumscribed about the regular dodecagon. From the Law of Sines derive that

$$AB = 2R\sin\frac{\pi}{12} \quad \text{and} \quad AF = 2R\sin\frac{5\pi}{12}.$$

Then

$$\begin{aligned}\frac{AB}{AF}+\frac{AF}{AB} &= \frac{\sin\frac{\pi}{12}}{\sin\frac{5\pi}{12}}+\frac{\sin\frac{5\pi}{12}}{\sin\frac{\pi}{12}} = \frac{\sin\frac{\pi}{12}}{\cos\frac{\pi}{12}}+\frac{\cos\frac{\pi}{12}}{\sin\frac{\pi}{12}}\\ &= \frac{\sin^2\frac{\pi}{12}+\cos^2\frac{\pi}{12}}{\sin\frac{\pi}{12}\cos\frac{\pi}{12}} = \frac{1}{\sin\frac{\pi}{12}\cos\frac{\pi}{12}}\\ &= \frac{2}{2\sin\frac{\pi}{12}\cos\frac{\pi}{12}} = \frac{2}{\sin\frac{\pi}{6}} = \frac{2}{\frac{1}{2}} = 4.\end{aligned}$$

Problem 9. Let f be a real-valued function of real and positive argument such that

$$f(x)+3xf\left(\frac{1}{x}\right)=2(x+1)$$

for all real numbers $x>0$. Find $f(2003)$.

Answer: 1002

We have

$$f(x)+3xf\left(\frac{1}{x}\right)=2(x+1) \text{ for all } x>0,$$

so

$$f\left(\frac{1}{x}\right)+3\cdot\frac{1}{x}\cdot f(x)=2\left(\frac{1}{x}+1\right) \text{ for all } x>0$$

as well. Multiplying the second equality by $3x$ yields

$$3xf\left(\frac{1}{x}\right)+9f(x)=6+6x,$$

and subtracting the first equality now gives

$$8f(x)=4x+4, \text{ and } f(x)=\frac{x+1}{2}.$$

It follows that

$$f(2003)=\frac{2003+1}{2}=1002.$$

Problem 10. How many gallons of a solution which is 15% alcohol do we have to mix with a solution that is 35% alcohol to make 250 gallons of a solution that is 21% alcohol?

Answer: 175

Let x represent the number of gallons of 15% alcohol solution that we need to mix. To make 250 gallons, we need to mix the x gallons of 15% alcohol solution with $(250-x)$ gallons of the 35% alcohol solution. The mix will then contain $.15x + .35(250 - x)$ gallons of alcohol which should be 21% of the total 250 gallons or .21 (250). This gives

$$\begin{aligned} .15x + .35(250 - x) &= .21(250) \\ 15x + 35(250) - 35x &= 21(250) \\ 14(250) &= 20x \\ 7(25) &= 175 = x. \end{aligned}$$

So we need to use 175 gallons of 15% alcohol solution.

Problem 11. If

$$\frac{1}{1+2} + \frac{1}{1+2+3} + \cdots + \frac{1}{1+2+\cdots+20} = \frac{m}{n},$$

where m and n are positive integers with no common divisor, find $m + n$.

Answer: 40

Because

$$1 + 2 + \cdots + k = \frac{k(k+1)}{2},$$

we have

$$\frac{1}{1+2+\cdots+k} = \frac{2}{k(k+1)} = \frac{2}{k} - \frac{2}{k+1}.$$

$$\begin{aligned} \frac{1}{1+2} + \cdots + \frac{1}{1+2+\ldots+20} &= \frac{2}{2} - \frac{2}{3} + \cdots + \frac{2}{20} - \frac{2}{21} \\ &= 1 - \frac{2}{21} = \frac{19}{21} \end{aligned}$$

and so $m + n = 40$.

Problem 12. How many triangles appear in the diagram below?

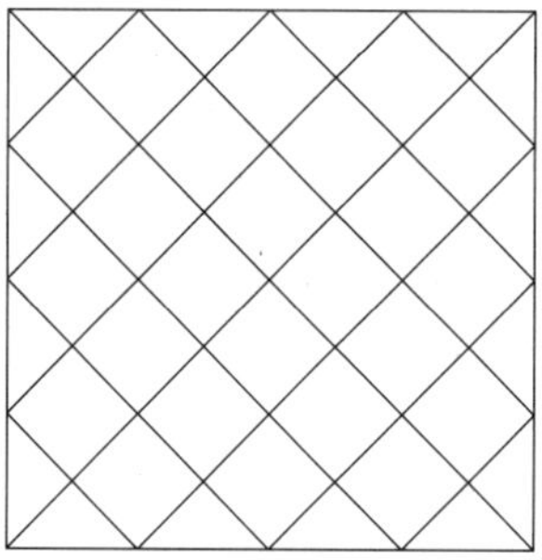

Answer: 56

To count the triangles notice that all triangles are right triangles. There are 10 triangles with a right angle at the intersection of two diagonal lines and a hypotenuse along the left vertical side of the rectangle. By symmetry, there are just as many triangles with a hypotenuse along the top, right, or bottom side of the rectangle. This accounts for 40 triangles. There are 4 triangles with right angle at the lower right corner of the rectangle. Again by symmetry, there are four times this many triangles, or 16. Thus, the diagram contains $40 + 16 = 56$ triangles.

Problem 13. Let $P(x)$ be a polynomial such that, when divided by $x - 2$, the remainder is 3 and, when divided by $x - 3$, the remainder is 2. If, when divided by $(x-2)(x-3)$, the remainder is $ax + b$, find $a^2 + b^2$.

Answer: 26

By the remainder theorem, $P(2) = 3$ and $P(3) = 2$. The division algorithm would give us a polynomial $K(x)$ so that

$$P(x) = (x-2)(x-3)K(x) + ax + b.$$

Letting $x = 2$ gives $3 = 2a + b$. Letting $x = 3$ gives $2 = 3a + b$. Now solve for a and b.

Solving

$$3a + b = 2 \quad \text{and} \quad 2a + b = 3$$

gives $a = -1$ and $b = 5$. So $a^2 + b^2 = (-1)^2 + (5)^2 = 26$.

Problem 14. Let a, b, c be real numbers such that $a^2 - 2 = 3b - c$, $b^2 + 4 = 3c + a$, and $c^2 + 4 = 3a - b$. Find $a^4 + b^4 + c^4$.

Answer: 18

Adding up these three equalities gives $a^2+b^2+c^2+6 = 4a+2b+2c$ or $(a-2)^2 + (b-1)^2 + (c-1)^2 = 0$, yielding $a = 2$, $b = 1$, and $c = 1$. Thus $a^4 + b^4 + c^4 = 16 + 1 + 1 = 18$.

Problem 15. Let r be a real number such that $\sqrt[3]{r} - \frac{1}{\sqrt[3]{r}} = 2$. Find

$$r^3 - \frac{1}{r^3}.$$

Answer: 2786

The identity $x^3 - y^3 = (x-y)^3 + 3xy(x-y)$ yields

$$r - \frac{1}{r} = 8 + 6 = 14.$$

Applying the above again gives

$$r^3 - \frac{1}{r^3} = 14^3 + 42 = 2786.$$

Problem 16. Find the greatest real number x such that

$$\left(\frac{x}{x-1}\right)^2 + \left(\frac{x}{x+1}\right)^2 = \frac{325}{144}.$$

Answer: 5

Adding $\frac{2x^2}{x^2-1}$ to both sides gives

$$\left(\frac{x}{x-1} + \frac{x}{x+1}\right)^2 = \frac{325}{144} + \frac{2x^2}{x^2-1}$$

or

$$\left(\frac{2x^2}{x^2-1}\right)^2 - \frac{2x^2}{x^2-1} = \frac{325}{144}.$$

Setting $\frac{2x^2}{x^2-1} = t$ yields $t^2 - t - \frac{325}{144} = 0$, whose solutions are

$$\frac{1}{2}\left(1 \pm \sqrt{1 + \frac{325}{36}}\right) = \frac{1}{2}\left(1 \pm \frac{19}{6}\right).$$

Since $\frac{2x^2}{x^2-1} > 2$, only $\frac{1}{2}\left(1 + \frac{19}{6}\right) = \frac{25}{12}$ qualifies, hence $\frac{2x^2}{x^2-1} = \frac{25}{12}$ yielding $x^2 = 25$. The largest x with this property is 5.

Problem 17. Given that $3\sin x + 4\cos x = 5$, where x is in $\left(0, \frac{\pi}{2}\right)$, find

$$2\sin x + \cos x + 4\tan x.$$

Answer: 5

Setting $\tan\frac{x}{2} = t$, we have

$$\sin x = \frac{2t}{1+t^2}, \qquad \cos x = \frac{1-t^2}{1+t^2}.$$

Then

$$\frac{6t}{1+t^2} + \frac{4(1-t^2)}{1+t^2} = 5$$

which reduces to $9t^2 - 6t + 1 = 0$ or, equivalently, $(3t-1)^2 = 0$. It follows that $t = \frac{1}{3}$ and

$$\tan x = \frac{2t}{1-t^2} = \frac{\frac{2}{3}}{\frac{8}{9}} = \frac{3}{4}.$$

Then

$$2\sin x + \cos x + 4\tan x = \frac{\frac{4}{3}}{\frac{10}{9}} + \frac{\frac{8}{9}}{\frac{10}{9}} + 3 = 2 + 3 = 5.$$

Alternatively, divide the given equation by 5 to get

$$\frac{3}{5}\sin x + \frac{4}{5}\cos x = 1.$$

There is an angle θ so that $\sin(x+\theta) = 1$. Thus, x and θ are complements, implying

$$\sin\theta = \frac{3}{5}, \quad \cos\theta = \frac{4}{5}, \quad \text{and} \quad \tan\theta = \frac{3}{4}$$

as above.

Problem 18. A circle of radius 320 is tangent to the inside of a circle of radius 1000. The smaller circle is tangent to a diameter of the larger circle at a point P. How far is P from the outside of the larger circle?

Answer: 400

Let the center of the larger circle be A, and the center of the smaller circle be B. Then triangle APB has a right angle at P. By the Pythagorean Theorem

$$(AP)^2 + (PB)^2 = (AB)^2$$

where AB is $1000 - 320 = 680$ and $PB = 320$. Solving for AP gives

$$AP = \sqrt{680^2 - 320^2} = 40\sqrt{17^2 - 8^2} = 40\sqrt{289 - 64}$$
$$= 40\sqrt{225} = 40 \cdot 15 = 600.$$

Therefore, the point P is a distance $1000 - 600 = 400$ from the edge of the circle.

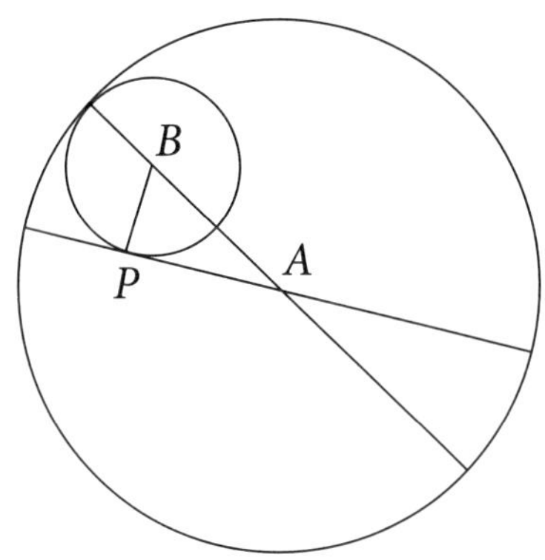

Problem 19. Let x_1 and x_2 be the roots of the equation $x^2 + 3x + 1 = 0$. Evaluate

$$\left(\frac{x_1}{x_2+1}\right)^2 + \left(\frac{x_2}{x_1+1}\right)^2.$$

Answer: 18

We have $x_1^2 + 3x_1 + 1 = 0$ and $x_2^2 + 3x_2 + 1 = 0$, hence

$$\left(\frac{x_1}{x_2+1}\right)^2 + \left(\frac{x_2}{x_1+1}\right)^2 = \frac{x_1^2}{x_2^2+2x_2+1} + \frac{x_2^2}{x_1+2x_1+1}$$
$$= \frac{-3x_1-1}{-x_2} + \frac{-3x_2-1}{-x_1} = \frac{3(x_1^2+x_2^2)+x_1+x_2}{x_1x_2}$$
$$= \frac{3(x_1+x_2)^2 - 6x_1x_2 + (x_1+x_2)}{x_1x_2}.$$

From the coefficients of the polynomial x^2+3x+1, it follows that $x_1+x_2=-3$ and $x_1x_2=1$. Thus, the desired value is

$$\frac{3(-3)^2-6-3}{1}=27-9=18.$$

Problem 20. In how many ways can we form three teams of four players each from a group of 12 participants?

Answer: 5775

If we designate a team 1, a team 2, and a team 3, there would be $\binom{12}{4}=495$ ways of selecting 4 members for team 1 out of the original twelve players. Having selected team 1, there would be $\binom{8}{4}=70$ ways of selecting 4 members for team 2 out of the eight remaining players. So, there are (495)(70) ways of selecting a team 1, team 2, and team 3. But we do not care which team is team 1, which team is team 2, and which team is team 3, so we must have counted each collection of three teams in six different ways because there are six ways to permute the order of the three teams. Thus, there are

$$\frac{(495)(70)}{6}=(165)(35)=5775$$

ways to select the three teams.

Problem 21. Let $a_n=\sqrt{1+\left(1-\frac{1}{n}\right)^2}+\sqrt{1+\left(1+\frac{1}{n}\right)^2}$, $n\geq 1$. Evaluate

$$\frac{1}{a_1}+\frac{1}{a_2}+\cdots+\frac{1}{a_{20}}.$$

Answer: 7

$$\begin{aligned}\frac{1}{a_n}&=\frac{n}{\sqrt{n^2+(n-1)^2}+\sqrt{n^2+(n+1)^2}}\\&=\frac{n\left(\sqrt{n^2+(n+1)^2}-\sqrt{n^2+(n-1)^2}\right)}{(n+1)^2-(n-1)^2}\\&=\frac{\sqrt{n^2+(n+1)^2}-\sqrt{n^2+(n-1)^2}}{4}\\&=\frac{1}{4}(b_{n+1}-b_n),\end{aligned}$$

where $b_n = \sqrt{(n-1)^2 + n^2}$. Then $\dfrac{1}{a_1} + \dfrac{1}{a_2} + \cdots + \dfrac{1}{a_{20}}$ telescopes to

$$\frac{1}{4}(b_{21} - b_1) = \frac{1}{4}\left(\sqrt{20^2 + 21^2} - 1\right) = \frac{1}{4}(29 - 1) = 7.$$

Problem 22. In triangle ABC, $\max(\angle A, \angle B) = \angle C + 30^\circ$ and $\dfrac{R}{r} = \sqrt{3} + 1$, where R is the radius of the circumcircle and r is the radius of the incircle. Find $\angle C$ in degrees.

Answer: 60

From the identity

$$r = 4R \sin\frac{A}{2} \sin\frac{B}{2} \sin\frac{C}{2}$$

it follows that

$$r = 4(\sqrt{3} + 1)r \sin\frac{A}{2} \sin\frac{B}{2} \sin\frac{C}{2}$$

or

$$\frac{1}{2(\sqrt{3} + 1)} = 2\left(\sin\frac{A}{2} \sin\frac{C}{2}\right) \sin\frac{B}{2}.$$

Assume $\angle A \geq \angle B$. Then

$$\frac{\sqrt{3} - 1}{4} = \left(\cos\frac{A - C}{2} - \cos\frac{A + C}{2}\right) \sin\frac{B}{2}$$

and, since $A - C = 30^\circ$, we obtain

$$\frac{\sqrt{3} - 1}{4} = \left(\frac{\sqrt{6} + \sqrt{2}}{4} - \sin\frac{B}{2}\right) \sin\frac{B}{2}.$$

Letting $\sin\dfrac{B}{2} = x$ yields

$$x^2 - \frac{\sqrt{6} + \sqrt{2}}{4}x + \frac{\sqrt{3} - 1}{4} = 0,$$

whose solutions are $x = \dfrac{\sqrt{6} - \sqrt{2}}{4}$ and $x = \dfrac{\sqrt{2}}{2}$. It follows that $\dfrac{B}{2} = 15^\circ$ or $\dfrac{B}{2} = 45^\circ$. The second solution is not acceptable, because $A \geq B$. Hence $B = 30^\circ$, $A = 90^\circ$ and $C = 60^\circ$.

Problem 23. For all positive integers m and n define $f(m,n)$ by $f(1,1)=1$, $f(m+1,n)=f(m,n)+m$ and $f(m,n+1)=f(m,n)-n$. Find the sum of all p such that $f(p,q)=2004$ for some q.

Answer: 3007

We have

$$\begin{aligned} f(1,1) &= 1, \\ f(2,1) &= f(1,1)+1, \\ f(3,1) &= f(2,1)+2 = f(1,1)+1+2, \end{aligned}$$

and so forth. Hence

$$\begin{aligned} f(m,1) &= 1+1+2+3+\cdots+(m-1), \\ f(m,2) &= f(m,1)-1, \\ f(m,3) &= f(m,2)-2 = f(m,1)-1-2, \end{aligned}$$

and so forth. Thus

$$\begin{aligned} f(m,n) &= f(m,1)-1-2-3-\cdots-(n-1) \\ &= 1+(1+2+\cdots+(m-1))-(1+2+\cdots+(n-1)). \end{aligned}$$

When $m \le n$, $f(m,n) \le 1$ and not equal to 2004. Otherwise,

$$f(m,n) = 1+\sum_{j=1}^{m-1} j - \sum_{j=1}^{n-1} j = 1+\sum_{j=n}^{m-1} j.$$

So, summing the arithmetic series yields

$$f(m,n) = 1+\frac{(m-n)(m+n-1)}{2}.$$

If this is 2004, then

$$(m-n)(m+n-1) = 2\cdot 2003.$$

Since both 2 and 2003 are prime, either $m+n-1=2003$ when $m-n=2$ or $m+n-1=2\cdot 2003$ when $m-n=1$. In the first case $m=1003$. In the second case $m=2004$. The solution to the problem is the sum of these two numbers, $1003+2004=3007$.

Problem 24. In triangle ABC, $\angle A = 30°$ and $AB = AC = 16$ inches. Let D lie on segment $\overline{BC}$ such that $\frac{DB}{DC} = \frac{2}{3}$, and let E and F be the orthogonal projections of D onto $\overline{AB}$ and $\overline{AC}$, respectively. Find $DE + DF$ in inches.

Answer: 8

The condition $\dfrac{DB}{DC} = \dfrac{2}{3}$ is not necessary. It was introduced as a distractor. The relation

$$\text{area}(ADB) + \text{area}(ADC) = \text{area}(ABC)$$

yields

$$\frac{DE \cdot AB}{2} + \frac{DF \cdot AC}{2} = \frac{AB \cdot AC \cdot \sin 30^\circ}{2}$$

so

$$DE + DF = \frac{1}{2}AB = \frac{1}{2}AC = 8.$$

Problem 25. Given that $(1 + \tan 1^\circ)(1 + \tan 2^\circ) \cdots (1 + \tan 45^\circ) = 2^n$, find n.

Answer: 23

$$2^n = \left(1 + \frac{\sin 1^\circ}{\cos 1^\circ}\right)\left(1 + \frac{\sin 2^\circ}{\cos 2^\circ}\right) \cdots \left(1 + \frac{\sin 44^\circ}{\cos 44^\circ}\right) \cdot 2$$

so

$$\begin{aligned} 2^{n-1} &= \frac{\cos 1^\circ + \sin 1^\circ}{\cos 1^\circ} \cdot \frac{\cos 2^\circ + \sin 2^\circ}{\cos 2^\circ} \cdots \frac{\cos 44^\circ + \sin 44^\circ}{\cos 44^\circ} \\ &= \frac{\sqrt{2}\sin(1^\circ + 45^\circ)}{\cos 1^\circ} \cdot \frac{\sqrt{2}\sin(2^\circ + 45^\circ)}{\cos 2^\circ} \cdots \frac{\sqrt{2}\sin(44^\circ + 45^\circ)}{\cos 44^\circ} \\ &= 2^{22}\frac{\cos 44^\circ}{\cos 1^\circ} \cdot \frac{\cos 43^\circ}{\cos 2^\circ} \cdots \frac{\cos 1^\circ}{\cos 44^\circ} = 2^{22}. \end{aligned}$$

It follows that $n - 1 = 22$, so $n = 23$.

2.2 2004 High School

Problem 1. How many different positive integers divide 10! ?

Answer: 270

$10! = 2^8 3^4 5^2 7$, so to choose a divisor of 10! you can choose to have from 0 to 8 factors of 2, from 0 to 4 factors of 3, from 0 to 2 factors of 5, and zero or 1 factor of 7. Thus, the number of choices for divisors of 10! is $(8+1)(4+1)(2+1)(1+1) = 270$.

Problem 2. In triangle ABC, three lines are drawn parallel to side $\overline{BC}$ dividing the altitude of the triangle into four equal parts. If the area of the second largest part is 35, what is the area of the whole triangle ABC?

Answer: 112

The diagram shows 4 similar triangles lying on top of each other, all sharing vertex A. Since their altitudes are in the ratios 1:2:3:4, their areas are in the ratios 1:4:9:16. The area of the second biggest (shaded) section is the difference in areas of the two middle sized triangles, so the ratio of the area of the smallest triangle to the area of the second biggest section is $\frac{1}{9-4}$. Since the largest triangle has an area 16 times as large as the smallest triangle, the area of the large triangle is

$$35 \cdot \frac{1}{9-4} \cdot 16 = 112.$$

Problem 3. How many real numbers are zeros of the polynomial

$$x^9 - 37x^8 - 2x^7 + 74x^6 + x^4 - 37x^3 - 2x^2 + 74x?$$

Answer: 5

$$\begin{aligned} & x^9 - 37x^8 - 2x^7 + 74x^6 + x^4 - 37x^3 - 2x^2 + 74x \\ &\quad = x(x^8 - 37x^7 - 2x^6 + 74x^5 + x^3 - 37x^2 - 2x + 74) \\ &\quad = x(x^3 - 37x^2 - 2x + 74)(x^4 + 1) \\ &\quad = x(x^2 - 2)(x - 37)(x^4 + 1) \end{aligned}$$

$x^4 + 1$ has only non-real roots, but the other factors have only real roots, so this polynomial has 5 real roots.

Problem 4. Find x so that $2^{2^{3^{2^2}}} = 4^{4^x}$.

Answer: 40

$$2^{2^{3^{2^2}}} = 2^{2^{3^4}} = 2^{2^{81}} = 4^{4^x} = (2^2)^{(2^2)^x} = 2^{2(2^{2x})} = 2^{2^{1+2x}}.$$

It follows that $81 = 1 + 2x$ and that $x = 40$.

Problem 5. The number $2.5081081081081\ldots$ can be written as $\frac{m}{n}$, where m and n are positive integers with no common factors. Find $m+n$.

Answer: 649

$$\begin{aligned} 2.5081081081081\ldots &= \frac{25.081081081\ldots}{10} = \frac{25 + \dfrac{81}{999}}{10} \\ &= \frac{5}{2} + \frac{3}{370} = \frac{5 \cdot 185 + 3}{370} = \frac{928}{370} = \frac{464}{185}. \end{aligned}$$

So the answer is $464 + 185 = 649$.

Problem 6. Evaluate the product

$$\left(1 + \frac{2}{3}\right)\left(1 + \frac{2}{4}\right)\left(1 + \frac{2}{5}\right)\cdots\left(1 + \frac{2}{98}\right).$$

Answer: 825

$$\left(1 + \frac{2}{3}\right)\left(1 + \frac{2}{4}\right)\left(1 + \frac{2}{5}\right)\cdots\left(1 + \frac{2}{98}\right) = \frac{5}{3} \cdot \frac{6}{4} \cdot \frac{7}{5} \cdots \frac{100}{98}$$

All factors cancel except for the 99 and 100 in the numerators and the 3 and 4 in the denominators yielding the answer $9900/12 = 825$.

Problem 7. How many positive integers less that 200 are relatively prime to either 15 or 24?

Answer: 120

By the principal of inclusion/exclusion, the number of positive integers less than 200 relatively prime to either 15 or 24 is the sum of the number relatively prime to 15 and the number relatively prime to 24 minus the number relatively prime to both 15 and 24. The number of positive integers less than 200 relatively prime to 15 is the number of those that are relatively prime to both 3 and 5. This yields

$$200 - \left\lfloor \frac{200}{3} \right\rfloor - \left\lfloor \frac{200}{5} \right\rfloor + \left\lfloor \frac{200}{15} \right\rfloor = 200 - 66 - 40 + 13 = 107.$$

The number of positive integers less than 200 relatively prime to 24 is the number of those that are relatively prime to both 2 and 3. This yields

$$200 - \left\lfloor \frac{200}{2} \right\rfloor - \left\lfloor \frac{200}{3} \right\rfloor + \left\lfloor \frac{200}{6} \right\rfloor = 200 - 100 - 66 + 33 = 67.$$

The number of positive integers less that 200 relatively prime to both 15 and 24 is the number of those relatively prime to 2, 3, and 5. This yields

$$200 - \left\lfloor \frac{200}{2} \right\rfloor - \left\lfloor \frac{200}{3} \right\rfloor - \left\lfloor \frac{200}{5} \right\rfloor + \left\lfloor \frac{200}{6} \right\rfloor + \left\lfloor \frac{200}{10} \right\rfloor + \left\lfloor \frac{200}{15} \right\rfloor - \left\lfloor \frac{200}{30} \right\rfloor$$
$$= 200 - 100 - 66 - 40 + 33 + 20 + 13 - 6 = 54.$$

So the number of positive integers less than 200 relatively prime to either 15 or 24 is

$$107 + 67 - 54 = 120.$$

Problem 8. One rainy afternoon you write the number 1 once, the number 2 twice, the number 3 three times, and so forth until you have written the number 99 ninety-nine times. What is the 2005th digit that you write?

Answer: 5

This is easier to count if instead of writing 1223334444... you insert zeros before all of the one-digit numbers to get

$$01020203030304040404\ldots$$

so that all the numbers you write have two digits. This adds one extra digit for each of the one-digit numbers, that is, one for the 1, two for the 2's, three for the 3's, and so forth through the nine for the 9's. Thus, you will write an extra 45 digits. So instead of looking for the 2005th digit, you look for the $2005+45 = 2050$th digit. Since each number written is two digits long, the 2050th digit is the second digit of the 1025th number. We need to know the smallest n so that

$$\frac{n(n+1)}{2} \geq 1025$$

or

$$\begin{aligned} n^2 + n &\geq 2050 \\ 44^2 + 44 &= 1980 \\ 45^2 + 45 &= 2070. \end{aligned}$$

So the 2005th digit is the second digit of 45 or 5.

Problem 9. Let M and m be the greatest and least values of x, respectively, satisfying

$$4x(x-5) \leq 375.$$

Find $M - m$.

Answer: 20

We have

$$\begin{aligned} 4x(x-5) &\leq 375 \\ 4x^2 - 20x - 375 &\leq 0 \\ (2x+15)(2x-25) &\leq 0. \end{aligned}$$

This is satisfied when x is between the two roots of $m = -\dfrac{15}{2}$ and $M = \dfrac{25}{2}$. So $M - m$ is 20.

Problem 10. Three lines are drawn parallel to each of the sides of triangle ABC so that they intersect in the interior of ABC. The resulting three smaller triangles have areas 1, 4, and 9. Find the area of triangle ABC.

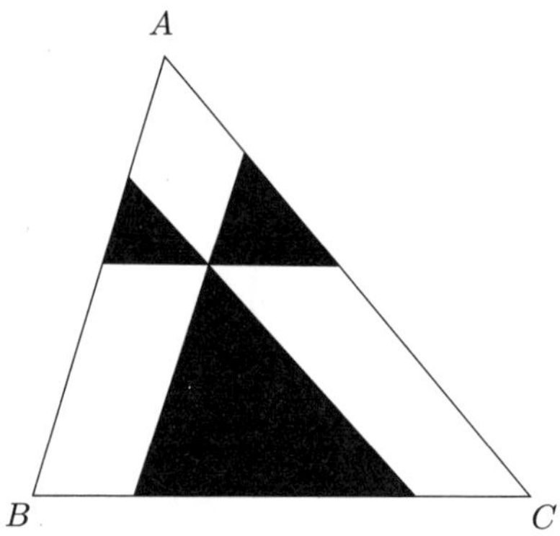

Answer: 36

Name the three intersection points along the horizontal in the center of the triangle D, E, and F, and name the two intersections along the base of the triangle G and H as shown.

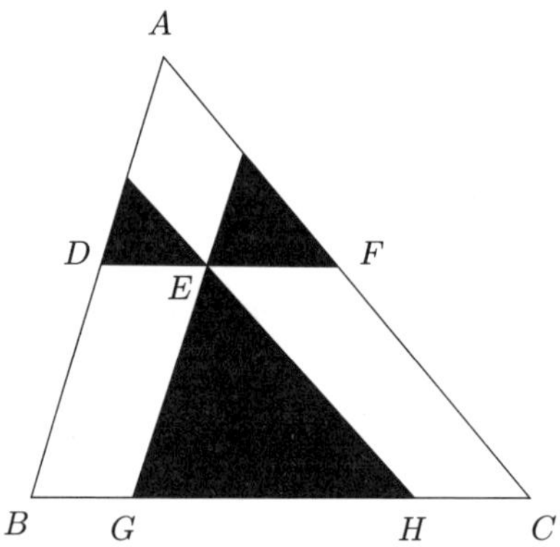

Since the areas of the shaded triangles are in the ratio 1 to 4 to 9, the sides of these similar triangle are in the ratio 1 to 2 to 3. Thus, the distance DE plus the distance EF equals the distance GH. Also, since $\overline{EF}$ is parallel to $\overline{DB}$, the distance BG = the distance DE. Similarly, the distance HC = the distance EF. This shows that the side $\overline{BC}$ is twice the length of $\overline{DF}$, and the triangle ADF has half the altitude of the triangle ABC. Thus, triangle ADF is congruent to triangle EGH, and the area of ABC has four times the area of triangle EGH. Thus, the answer is $4 \cdot 9 = 36$.

Problem 11. How far is it from the point $(9, 17)$ to its reflection across the line

$$3x + 4y = 15?$$

Answer: 32

This is twice the distance from the point $(9, 17)$ to the line $3x + 4y = 15$. That distance is given by the normal form for the equation of the line

$$\frac{3 \cdot 9 + 4 \cdot 17 - 15}{\sqrt{3^2 + 4^2}} = \frac{27 + 68 - 15}{5} = 16$$

so the final answer is 32. Alternatively, one can find the distance from $(9, 17)$ to the point (a, b) which lies on the line so that the line from $(9, 17)$ to (a, b) is perpendicular to the line $3x+4y = 15$. That is the line $3y - 4x = 3(17) - 4(9)$ or $3y - 4x = 15$. This intersects $3x + 4y = 15$ at $(a, b) = (-3/5, 21/5)$. The distance from $(9, 17)$ to (a, b) is then 16 as above.

Problem 12. The diagram shows a circle with radius 24 which contains two circles with radius 12 tangent to each other and the larger circle. The smallest circle is tangent to the three other circles. What is the radius of the smallest circle?

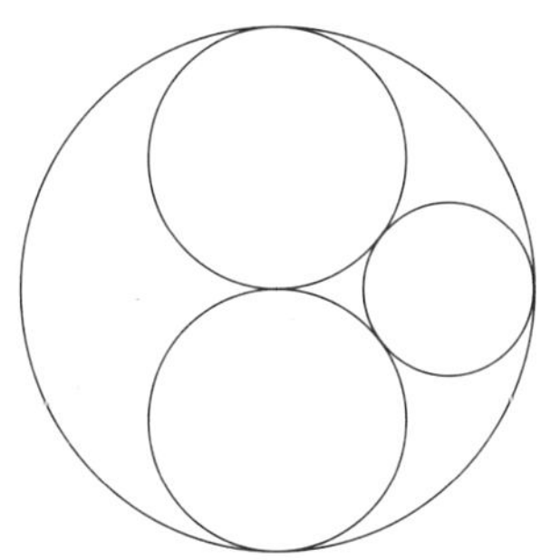

Answer: 8

Let the small circle have radius x. Draw the triangle with vertices at the centers of the three smallest circles. The base of this triangle is made of two radii of the two middle sized circles, so it has length 24. The other two sides are made up of a middle sized radius and a small radius, so they have length $12 + x$. The altitude of the triangle is given by the Pythagorean Theorem to be

$$\sqrt{(12 + x)^2 - 12^2} = \sqrt{24x + x^2}.$$

This altitude plus one small circle radius make the radius of the large circle giving

$$\begin{aligned}\sqrt{24x+x^2}+x &= 24\\ \sqrt{24x+x^2} &= 24-x\\ 24x+x^2 &= 24^2-48x+x^2\\ 3\cdot 24x &= 24^2\\ x &= 8\end{aligned}$$

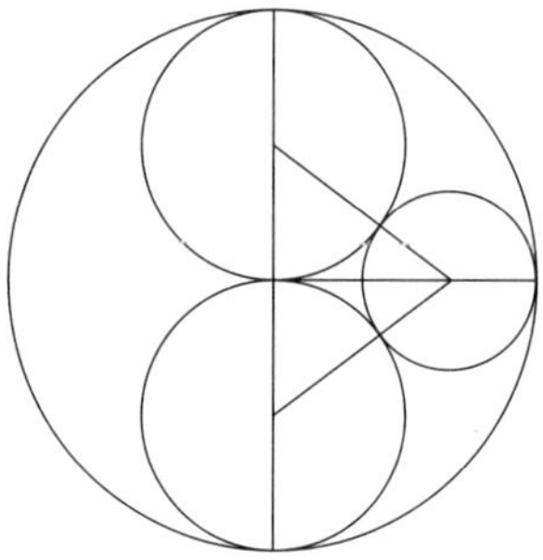

Problem 13. A cubic block with dimensions $n\times n\times n$ is made up of a collection of $1\times 1\times 1$ unit cubes. What is the least n so that if the outer two layers of unit cubes are removed from the block, more than half the original unit cubes will still remain?

Answer: 20

If the outer two layers of cubes with side 1 are removed from the cube with side n, what remains is a cube with side $n-4$. The number of cubes with side 1 in the cube with side n is n^3 while the number of cubes with side 1 in the cube with side $n-4$ is $(n-4)^3$. So we seek the smallest n satisfying $n^3 \le 2(n-4)^3$. To do this, one just has to consider the sequence of cubes 1, 8, 27, 64, 125, 216,... until the correct n is found. Alternatively, we can take a cube root and solve for n:

$$\begin{aligned}n^3 &\le 2(n-4)^3\\ n &\le \sqrt[3]{2}(n-4)\\ 4\sqrt[3]{2} &\le (\sqrt[3]{2}-1)n\\ \frac{4\sqrt[3]{2}}{\sqrt[3]{2}-1} &\le n\\ 19.389 &\le n\end{aligned}$$

So $n=20$.

Problem 14. Two circles have radii 15 and 95. If the two external tangents to the circles intersect at 60 degrees, how far apart are the centers of the circles?

Answer: 160

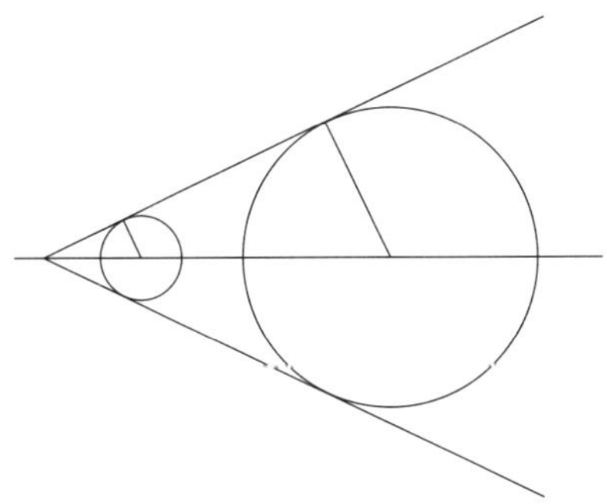

Let x be the distance from the vertex of the angle to the center of the small circle, and let y be the distance between the centers of the circles. Similar triangles gives

$$\frac{15}{x} = \frac{95}{x+y} = \sin 30^\circ = \frac{1}{2}.$$

So $x = 30$, $x + y = 190$, $y = 160$. The answer is 160.

Problem 15. Jerry purchased some stock for \$14,400 at the same time that Susan purchased a bond for \$6,250. Jerry's investment went up 20 percent the first year, fell 10 percent the second year, and rose another 20 percent the third year. Susan's investment grew at a constant rate of compound interest for three years. If both investments are worth the same after three years, what was the annual percentage increase of Susan's investment?

Answer: 44

Jeremys investment ended up being worth

$$14400(1.20)(0.9)(1.20) = 14400 \cdot \frac{6}{5} \cdot \frac{9}{10} \cdot \frac{6}{5} = \frac{2^7 3^6}{5}.$$

Susans investment is worth

$$6250(1+r)^3 = 2 \cdot 5^5 (1+r)^3.$$

Setting these equal yields

$$\begin{aligned}\frac{2^7 3^6}{5} &= 2 \cdot 5^5 (1+r)^3 \\ (1+r)^3 &= \frac{2^6 3^6}{5^6} \\ (1+r) &= \frac{2^2 3^2}{5^2} = \frac{36}{25} = \frac{144}{100} = 1.44\end{aligned}$$

So r is 44 percent.

Problem 16. Find the total length of the set of real numbers satisfying

$$\frac{x^2 - 80x + 1500}{x^2 - 55x + 700} < 0.$$

Answer: 25

$$\frac{x^2 - 80x + 1500}{x^2 - 55x + 700} = \frac{(x-50)(x-30)}{(x-20)(x-35)}.$$

This fraction is negative when an odd number of the factors are negative. This happens exactly when x is between 20 and 30 and when x is between 35 and 50. Thus, the total length of the set where the fraction is negative is

$$(30 - 20) + (50 - 35) = 25.$$

Problem 17. We want to paint some identically-sized cubes so that each face of each cube is painted a solid color and each cube is painted with six different colors. If we have seven different colors to choose from, how many distinguishable cubes can we produce?

Answer: 210

We can choose a unique way to color a cube by first choosing the one color which will not appear on a face of the cube. There are seven ways to do that. Assuming the colors are in some natural order (perhaps alphabetical) paint one side of the cube with the first of the six colors. There is only one way to do that. Then choose the color which goes on the opposite face of that first color. There are five ways to do that. Finally, choose an ordering for the four remaining colors to fit on the four remaining sides. There are three factorial ways to order four items in a circle. Thus, the total number of ways to choose how to paint a cube is $7 \cdot 5 \cdot 3! = 210$.

Problem 18. As x ranges over the interval $(0, \infty)$, the function

$$\sqrt{9x^2 + 173x + 900} - \sqrt{9x^2 + 77x + 900}$$

ranges over the interval $(0, M)$. Find M.

Answer: 16

$$\begin{aligned} &\sqrt{9x^2 + 173x + 900} - \sqrt{9x^2 + 77x + 900} \\ &= \frac{(9x^2 + 173x + 900) - (9x^2 + 77x + 900)}{\sqrt{9x^2 + 173x + 900} + \sqrt{9x^2 + 77x + 900}} \\ &= \frac{96x}{\sqrt{9x^2 + 173x + 900} + \sqrt{9x^2 + 77x + 900}} \\ &= \frac{96}{\sqrt{9 + \frac{173}{x} + \frac{900}{x^2}} + \sqrt{9 + \frac{77}{x} + \frac{900}{x^2}}} \end{aligned}$$

This last expression is clearly an increasing function of x since the denominator decreases in x. Therefore, the upper value M is obtained by letting x grow without bound allowing the fraction to increase to

$$\frac{96}{\sqrt{9} + \sqrt{9}} = \frac{96}{3 + 3} = 16.$$

Problem 19. There are three bags. One bag contains three green candies and one red candy. One bag contains two green candies and two red candies. One bag contains one green candy and three red candies. A child randomly selects one of the bags, randomly chooses a first candy from that bag, and eats the candy. If the first candy had been green, the child randomly chooses one of the other two bags and randomly selects a second candy from that bag. If the first candy had been red, the child randomly selects a second candy from the same bag as the first candy. If the probability that the second candy is green is given by the fraction $\frac{m}{n}$ in lowest terms, find $m + n$.

Answer: 217

First the child chooses a bag. The probability of choosing any one bag is $1/3$. Then the child chooses a candy. The probabilities of getting a green or red candy then depend on which bag was selected and are shown in the tree diagram below. If the first candy was green, the second candy is chosen at random from one of the eight candies in the other two bags. If the first candy was red, the second candy is chosen from the three candies remaining in the bag. The probabilities of selecting a green candy for the second candy given the results of the first two choices are shown in the diagram. The probability that the child makes three particular choices (such as bag 1, green first candy, green second candy) is the product of the three probabilities leading to that branch of the tree diagram.

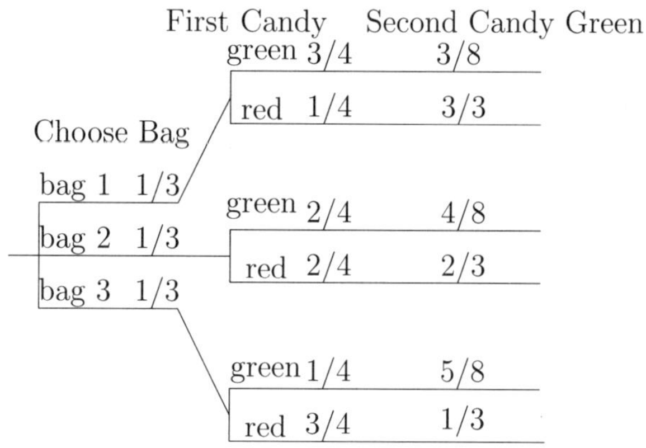

The six products are

$$\frac{1}{3}\frac{3}{4}\frac{3}{8} \quad \frac{1}{3}\frac{1}{4}\frac{3}{3} \quad \frac{1}{3}\frac{2}{4}\frac{4}{8} \quad \frac{1}{3}\frac{2}{4}\frac{2}{3} \quad \frac{1}{3}\frac{1}{4}\frac{5}{8} \quad \frac{1}{3}\frac{3}{4}\frac{1}{3}.$$

The probability that the second candy is green is the sum of these probabilities

$$\frac{1}{3}\frac{3}{4}\frac{3}{8}+\frac{1}{3}\frac{1}{4}\frac{3}{3}+\frac{1}{3}\frac{2}{4}\frac{4}{8}+\frac{1}{3}\frac{2}{4}\frac{2}{3}+\frac{1}{3}\frac{1}{4}\frac{5}{8}+\frac{1}{3}\frac{3}{4}\frac{1}{3}$$

$$\frac{9+8+5}{3\cdot 4\cdot 8}+\frac{3+4+3}{3\cdot 4\cdot 3}=\frac{11}{48}+\frac{5}{18}=\frac{73}{144}.$$

So the desired answer is $73+144=217$.

Problem 20. A 70 foot pole stands vertically in a plane supported by three 490 foot wires, all attached to the top of the pole, pulled taut, and anchored to three equally spaced points in the plane. How many feet apart are any two of those anchor points?

Answer: 840

The distance we need is the distance y in the following diagram.

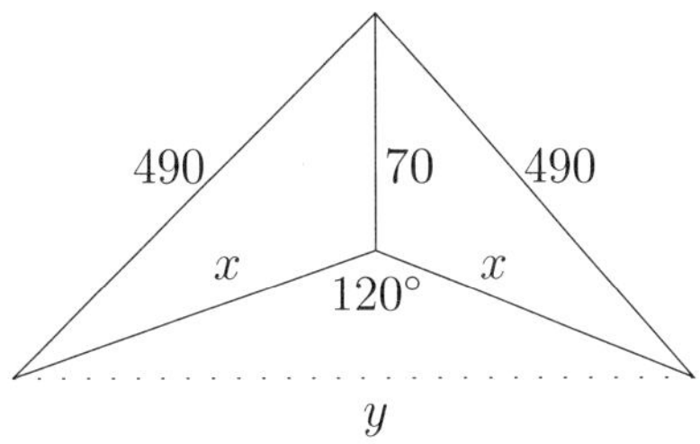

x is given by the Pythagorean Theorem as

$$\sqrt{490^2 - 70^2} = 70\sqrt{7^2 - 1^2} = 70\sqrt{48} = 70 \cdot 4\sqrt{3} = 280\sqrt{3}$$

Then y is given by the Law of Cosines as

$$\sqrt{x^2 + x^2 - 2 \cdot x \cdot x \cdot \cos 120^\circ} = \sqrt{2x^2[1 - \cos 120^\circ]}$$

$$= x\sqrt{2\left(1 + \frac{1}{2}\right)} = 280\sqrt{3}\sqrt{3} = 840.$$

Problem 21. Define $a_k = (k^2+1)k!$ and $b_k = a_1 + a_2 + \cdots + a_k$. Let

$$\frac{a_{100}}{b_{100}} = \frac{m}{n},$$

where m and n are relatively prime positive integers. Find $n - m$.

Answer: 99

Note that

$$a_k = (k^2+1)k! = [k(k+1) - (k-1)]k! = k \cdot (k+1)! - (k-1) \cdot k!$$

Thus the sum telescopes with

$$\begin{aligned} b_k &= a_1 + a_2 + a_3 + \ldots + a_k \\ &= [1 \cdot 2! - 0 \cdot 1!] + [2 \cdot 3! - 1 \cdot 2!] + [3 \cdot 4! - 2 \cdot 3!] \\ &+ \cdots + [k \cdot (k+1)! - (k-1) \cdot k!] \\ &= k \cdot (k+1)! \end{aligned}$$

So, $\dfrac{a_k}{b_k} = \dfrac{(k^2+1)k!}{k \cdot (k+1)!} = \dfrac{k^2+1}{k^2+k}$, and $\dfrac{a_{100}}{b_{100}} = \dfrac{10,001}{10,100}$. Thus the answer is $10,100 - 10,001 = 99$.

Problem 22. How many non-overlapping 2 by 2 squares fit into a circle with radius 8?

Answer: 40

You can fit 39 squares in the circle by lining up a row of squares whose centers line-up along the diameter of the circle (see diagram below). These squares reach one unit on either side of the diameter, so we can align up to $\sqrt{8^2 - 1^2} = \sqrt{63} = 3\sqrt{6} > 7$ squares. A second row of squares would extend 3 units from the diameter, so we can align up to $\sqrt{8^2 - 3^2} = \sqrt{55} > 7$ more squares on each side of the circle. A third row of squares would extend 5 units from the diameter, so we can align up to $\sqrt{8^2 - 5^2} = \sqrt{39} > 6$ more squares on each side of the circle. Finally, a fourth row of squares would extend 7 units from the diameter, so we can align up to $\sqrt{8^2 - 7^2} = \sqrt{15} > 3$ more squares on each side of the circle. This accounts for $7 + 2(7 + 6 + 3) = 39$ squares.

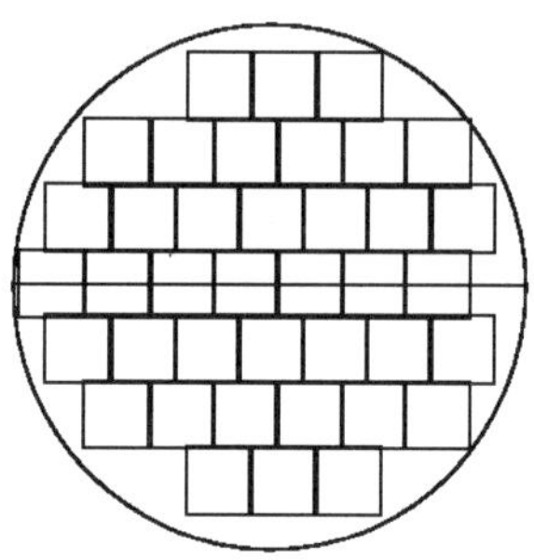

By moving this brick pattern off center, it is possible to make room for a fortieth square. Let the circle of radius 8 be centered at the origin, and take 7 vertical strips of width 2 and lengths 6, 12, 14, 14, 14, 12, and 8 centered at $(-6.08, 0)$, $(-4.08, 0)$, $(-2.08, 0)$, $(-.08, 0)$, $(1.92, 0)$, $(3.92, 0)$, and $(5.92, 0)$, respectively. It is easy to verify by the Pythagorean theorem that each of these strips is completely contained in the circle. Dividing the respective strips into 3, 6, 7, 7, 7, 6, and 4 2×2 squares gives a total of $3 + 6 + 7 + 7 + 7 + 6 + 4 = 40$ squares. Although a proof that this is optimal would be extremely difficult, it is also difficult to imagine a way to improve on this arrangement of squares.

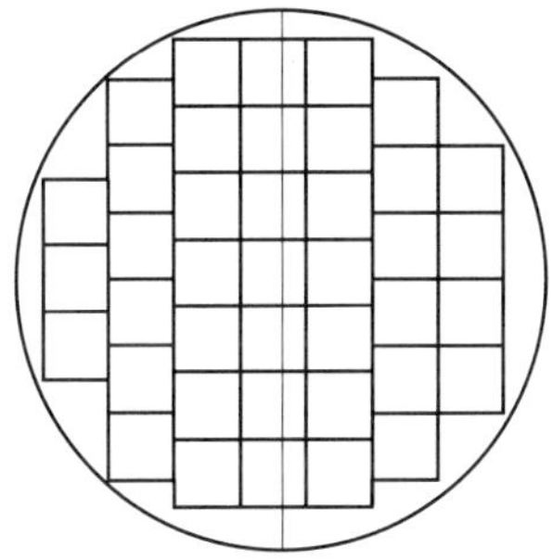

Problem 23. Let a and b be real numbers satisfying

$$a^4 + 8b = 4(a^3 - 1) - 16\sqrt{3}$$

and

$$b^4 + 8a = 4(b^3 - 1) + 16\sqrt{3}.$$

Find $a^4 + b^4$.

Answer: 56

Adding the two equations together gives

$$\begin{aligned} a^4 + 8b + b^4 + 8a &= 4(a^3 - 1) - 16\sqrt{3} + 4(b^3 - 1) + 16\sqrt{3} \\ (a^4 - 4a^3 + 8a + 4) + (b^4 - 4b^3 + 8b + 4) &= 0 \\ (a^2 - 2a - 2)^2 + (b^2 - 2b - 2)^2 &= 0 \end{aligned}$$

The only way for this sum of squares to be zero is for each square to be zero. Thus, both a and b must be roots of $x^2 - 2x - 2$. In order to satisfy the original two equations, a and b must be distinct, so a and b must be the two roots

$$\{a, b\} = \{1 + \sqrt{3}, 1 - \sqrt{3}\}.$$

These values do satisfy the original equations.

$$a^4 + b^4 = (1 + \sqrt{3})^4 + (1 - \sqrt{3})^4 = 2(1 + 6 \cdot 3 + 9) = 56.$$

Problem 24. The determinant

$$\begin{vmatrix} 3 & -2 & 5 \\ 7 & 1 & -4 \\ 5 & 2 & 3 \end{vmatrix}$$

has the same value as the determinant

$$\begin{vmatrix} x & 1+x & 2+x \\ 3 & 0 & 1 \\ 1 & 1 & 0 \end{vmatrix}$$

Find x.

Answer: 51

We need

$$\begin{vmatrix} 3 & -2 & 5 \\ 7 & 1 & -4 \\ 5 & 2 & 3 \end{vmatrix} = \begin{vmatrix} x & 1+x & 2+x \\ 3 & 0 & 1 \\ 1 & 1 & 0 \end{vmatrix} = \begin{vmatrix} x & x & x \\ 3 & 0 & 1 \\ 1 & 1 & 0 \end{vmatrix} + \begin{vmatrix} 0 & 1 & 2 \\ 3 & 0 & 1 \\ 1 & 1 & 0 \end{vmatrix}$$

$$= x \begin{vmatrix} 1 & 1 & 1 \\ 3 & 0 & 1 \\ 1 & 1 & 0 \end{vmatrix} + \begin{vmatrix} 0 & 1 & 2 \\ 3 & 0 & 1 \\ 1 & 1 & 0 \end{vmatrix}$$

Evaluating the determinants gives $160 = 3x + 7$ or $x = 51$.

Problem 25. In the addition

$$\begin{array}{cccccc} & W & H & I & T & E \\ + & W & A & T & E & R \\ \hline P & I & C & N & I & C \end{array}$$

each distinct letter represents a different digit. Find the number represented by the answer PICNIC.

Answer: 169069

There are two ways to create a solution addition problem which can be found by trial and error together with some simple inferences. The two addition problems are $83642 + 85427 = 169069$ and $85642 + 83427 = 169069$. Clearly, the solution to the problem is 169069.

2.3 2004 Middle School

Problem 1. This year February 29 fell on a Sunday. In what year will February 29 next fall on a Sunday?

Answer: 2032

Every four year period has $365 \times 4 + 1$ days. This number is

$$364 \times 4 + 5 = 7 \times 52 \times 4 + 5,$$

so when this number is divided by 7, the remainder is 5. Thus, every four years, February 29 will be five days later in the week. Since 5 and 7 are relatively prime, the first multiple of 5 which is also a multiple of 7 is 5×7, so it will take seven four year periods or 28 years before February 29 is again on a Sunday. This will be in the year $2004 + 28 = 2032$.

Problem 2. If $h(a, b, c) = \dfrac{abc}{a+b+c}$, find $h(3\sqrt{5}, 6\sqrt{5}, 9\sqrt{5})$.

Answer: 45

$$\begin{aligned} h(3\sqrt{5}, 6\sqrt{5}, 9\sqrt{5}) &= \frac{(3\sqrt{5})(6\sqrt{5})(9\sqrt{5})}{3\sqrt{5} + 6\sqrt{5} + 9\sqrt{5}} \\ &= \frac{3 \cdot 6 \cdot 9 \cdot 5 \cdot \sqrt{5}}{\sqrt{5}(3+6+9)} = \frac{3 \cdot 6 \cdot 9 \cdot 5}{18} = 45 \end{aligned}$$

Problem 3. In triangle ABC, three lines are drawn parallel to side $\overline{BC}$ dividing the altitude of the triangle into four equal parts. If the area of the second largest part is 35, what is the area of the whole triangle ABC?

Answer: 112

This is the same as High School problem number 2.

Problem 4. If the numbers $2a+2$ and $2b+2$ add up to 2004, find the sum of the numbers $\frac{a}{2}-2$ and $\frac{b}{2}-2$.

Answer: 496

$$\begin{aligned}(2a+2)+(2b+2) = 2(a+b)+4 &= 2004\\ a+b &= 1000\\ \left(\frac{a}{2}-2\right)+\left(\frac{b}{2}-2\right) &= \frac{a+b}{2}-4 = \frac{1000}{2}-4 = 496\end{aligned}$$

Problem 5. Write the number $2004_{(5)}$ [2004 base 5] as a number in base 6.

Answer: 1102

$$2004_{(5)} = 2\cdot 5^3+4 = 254 = 6(42)+2 =$$
$$6(6\cdot 7+0)+2 = 6(6(6\cdot 1+1)+0)+2 = 1102_{(6)}$$

Problem 6. How many different positive integers divide 10! ?

Answer: 270

This is the same as High School problem number 1.

Problem 7. A rectangle has area 1100. If the length is increased by ten percent and the width is decreased by ten percent, what is the area of the new rectangle?

Answer: 1089

The area of the rectangle is its length times its width. So we have

$$\begin{aligned}L\cdot W &= 1100\\ (1.1L)(0.9W) &= (1.1)(0.9)(1100) = (0.99)(1100)\\ &= 1100-11 = 1089.\end{aligned}$$

Problem 8. The number $2.5081081081081\ldots$ can be written as $\frac{m}{n}$, where m and n are positive integers with no common factors. Find $m+n$.

Answer: 649

This is the same as High School problem number 5.

Problem 9. How many positive integers less that 200 are relatively prime to either 15 or 24?

Answer: 120

This is the same as High School problem number 7.

Problem 10. One rainy afternoon you write the number 1 once, the number 2 twice, the number 3 three times, and so forth until you have written the number 99 ninety-nine times. What is the 2005th digit that you write?

Answer: 5

This is the same as High School problem number 8.

Problem 11. Find the sum of all integers x satisfying

$$1 + 8x \leq 358 - 2x \leq 6x + 94.$$

Answer: 102

$$\begin{array}{ll} 1 + 8x \leq 358 - 2x & \quad 358 - 2x \leq 6x + 94 \\ 10x \leq 357 & \quad 264 \leq 8x \\ x \leq 35.7 & \quad 33 \leq x \end{array}$$

So x can be 33, 34, or 35. The sum of these values is 102.

Problem 12. If $f(x, y) = xy + 2x + y + 1$, find $f(f(2, f(3, 4)), 5)$.

Answer: 524

We wish to calculate $f(f(2, f(3, 4)), 5)$:

$$\begin{aligned} f(3,4) &= 3 \cdot 4 + 2 \cdot 3 + 4 + 1 = 23 \\ f(2,23) &= 2 \cdot 23 + 2 \cdot 2 + 23 + 1 = 74 \\ f(74,5) &= 74 \cdot 5 + 2 \cdot 74 + 5 + 1 = 524 \end{aligned}$$

Problem 13. How many three-digit numbers are made up of three distinct digits?

Answer: 648

To form such a three-digit number, begin by choosing a first digit. There are nine possible choices (every digit except 0). Next choose a second digit. There are again nine possible choices (every digit except the one chosen first). Finally, choose the last digit. There are eight possible choices (every digit except the two already chosen). Thus, there are $9 \cdot 9 \cdot 8 = 648$.

Problem 14. A regular polygon has five times as many diagonals as it has sides. How many vertices does the polygon have?

Answer: 13

If a regular polygon has n sides, it has $\dfrac{n(n-3)}{2}$ diagonals. This can be seen by pairing each vertex with each other vertex not counting itself or its two neighbors. So we are looking for n satisfying $\dfrac{n(n-3)}{2} = 5n$, or $n(n-3) = 10n$, or $n - 3 = 10$. So $n = 13$.

Problem 15. Find the prime number p for which $p + 2500$ is a perfect square.

Answer: 101

There is an n so that $2500 + p = n^2$, so $p = n^2 - 2500 = (n-50)(n+50)$. Since p is prime, its only factors are 1 and p, and it follows that $1 = n - 50$, $n = 51$, and thus $p = 51 + 50 = 101$.

Problem 16. A week ago, Sandy's seasonal Little League batting average was 360. After five more at bats this week, Sandy's batting average improved to 400. What is the least number of hits that Sandy could have had this season?

Answer: 12

Suppose that a week ago Sandy had been at bat x times. Then Sandy would have made $(0.360)x$ hits. Suppose Sandy made y more hits in the next five times at bat. That makes a total of $(0.360)x + y$ hits after $x + 5$ times at bat giving a batting average of

$$\begin{aligned} &\frac{(0.360)x + y}{x + 5} = 0.400, \text{ so} \\ &(0.360)x + y = 0.4(x + 5) \\ &y = 0.04x + 2 \end{aligned}$$

Since x must be a positive integer, the smallest possible value for y is 3 which occurs when $x = 25$. Thus, Sandy had made $(0.360)(25) = 9$ hits up through last week and has made 3 more hits this week for a total of 12 hits.

Problem 17. Find x so that $2^{2^{3^{2^2}}} = 4^{4^x}$.

Answer: 40

This is the same as High School problem number 4.

Problem 18. Find the number of addition problems in which a two digit number is added to a second two digit number to give a two digit answer, such as in the three examples:

$$\begin{array}{r}23\\ \underline{42}\\ 65\end{array}, \quad \begin{array}{r}36\\ \underline{36}\\ 72\end{array}, \quad \begin{array}{r}42\\ \underline{23}\\ 65\end{array}.$$

Answer: 3240

The smallest two digit number is ten. You can add to ten any natural number up to 89 to get a two digit answer, so there are $89 - 9 = 80$ two digit numbers that can be added to ten to give a two digit answer. Similarly, there are $88 - 9 = 79$ numbers that can be added to eleven to give a two digit answer. There are 78 numbers that can be added to twelve, 77 that can be added to thirteen, and so forth through the one two digit number that can be added to 89. Each of these results in a new problem, so there are $80 + 79 + 78 + \ldots + 1$ distinct addition problems. This adds to $\dfrac{80(81)}{2} = 3240$.

Problem 19. Find n such that $n - 76$ and $n + 76$ are both cubes of positive integers.

Answer: 140

If $n - 76 = m^3$ and $n + 76 = k^3$, then

$(n + 76) - (n - 76) = 152 = k^3 - m^3 = (k - m)(k^2 + km + m^2)$.

So $k - m$ is a factor of 152. If $k - m = 1$, then $k = m + 1$ so

$$152 = (m + 1)^2 + (m + 1)m + m^2 = 3m^2 + 3m + 1,$$

or

$$3m^2 + 3m - 151 = 0$$

which has no integer solutions. If $k - m = 2$, then $k = m + 2$ so

$$76 = (m + 2)^2 + (m + 2)m + m^2 = 3m^2 + 6m + 4,$$

or $3m^2 + 6m - 72 = 0$ which has $m = 4$ as a solution. It follows then that $k = 6$ and $n = 140$. The other possible values for $k - m$ which are the other divisors of 152 (4, 8, 19, 38, 76, and 152) do not lead to solutions.

Problem 20. A circle of area 40 is tangent to a circle of area 10. Let R be the rectangle containing both circles as shown. The area of R is $\frac{n}{\pi}$. Find n.

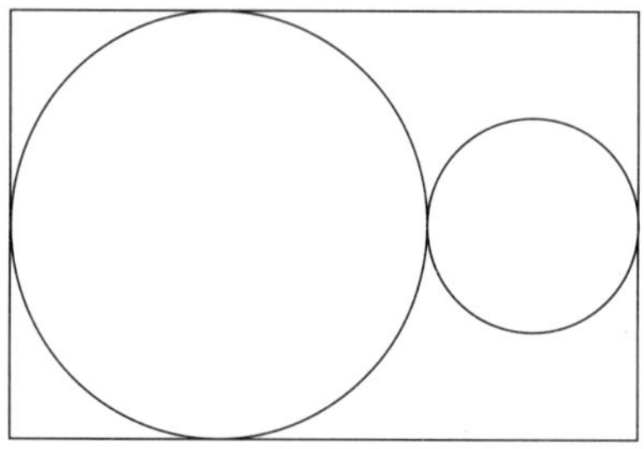

Answer: 120

The large circle has area 40, so it has radius

$$r = \sqrt{\frac{40}{\pi}} = 2\sqrt{\frac{10}{\pi}}.$$

The small circle has area 10, so it has radius

$$s = \sqrt{\frac{10}{\pi}}.$$

The rectangle has width $2r$ and length $2r + 2s$, so it has area

$$R = \left(2 \cdot 2\sqrt{\frac{10}{\pi}}\right)\left(2 \cdot 2\sqrt{\frac{10}{\pi}} + 2\sqrt{\frac{10}{\pi}}\right) = 3(4+2)\left(\sqrt{\frac{10}{\pi}}\right)^2$$
$$= 12\frac{10}{\pi} = \frac{120}{\pi}.$$

Problem 21. Find the number of different quadruples (a, b, c, d) of positive integers such that

$$ab = cd = a + b + c + d - 3.$$

Answer: 10

We have

$$ab + cd = 2(a + b + c + d - 3),$$

or

$$(a-2)(b-2) + (c-2)(d-2) = 2.$$

Assuming that a is the smallest number among a, b, c, d, we get $-1 \leq a - 2 \leq 1$.

Case 1. If $a - 2 = 1$, then $b - 2 = c - 2 = d - 2 = 1$, so $a = b = c = d = 3$.

Case 2. If $a - 2 = 0$, then either $c - 2 = 1$ and $d - 2 = 2$, or $c - 2 = 2$ and $d - 2 = 1$. It follows that $cd = 12$, $a = 2$ and $b = 6$.

Case 3. If $a - 2 = -1$, then $a = 1$ and $b + c + d - 2 = b = cd$. Hence, $c + d = 2$, implying that $c = d = 1$ and $b = 1$.

With this information we can find all the solutions to be $(3,3,3,3)$, $(2,6,3,4)$, $(6,2,3,4)$, $(2,6,4,3)$, $(6,2,4,3)$, $(3,4,2,6)$, $(3,4,6,2)$, $(4,3,2,6)$, $(4,3,6,2)$, and $(1,1,1,1)$. There are ten.

Problem 22. Two circles have radii 15 and 95. If the two external tangents to the circles intersect at 60 degrees, how far apart are the centers of the circles?

Answer: 160

This is the same as High School problem number 14.

Problem 23. A cubic block with dimensions $n \times n \times n$ is made up of a collection of $1 \times 1 \times 1$ unit cubes. What is the least n so that if the outer two layers of unit cubes are removed from the block, more than half the original unit cubes will still remain?

Answer: 20

This is the same as High School problem number 13.

Problem 24. Let a be a real number greater than 1 such that

$$\frac{20a}{a^2+1} = \sqrt{2}.$$

Find $\dfrac{14a}{a^2-1}$.

Answer: 1

We have

$$\begin{aligned}
\frac{20a}{a^2+1} &= \sqrt{2}, \text{ so } 400a^2 = 2(a^2+1)^2 \\
200a^2 &= a^4 + 2a^2 + 1 \\
196a^2 &= a^4 - 2a^2 + 1 = (a^2-1)^2 \\
14a &= a^2 - 1, \text{ so } \frac{14a}{a^2-1} = 1
\end{aligned}$$

Problem 25. In the addition

$$\begin{array}{cccccc} & W & H & I & T & E \\ + & W & A & T & E & R \\ \hline P & I & C & N & I & C \end{array}$$

each distinct letter represents a different digit. Find the number represented by the answer PICNIC.

Answer: 169069

This is the same as High School problem number 25.

2.4 2005 High School

Problem 1. The cost of producing each item is inversely proportional to the square root of the number of items produced. The cost of producing ten items is \$2100. If items sell for \$30 each, how many items need to be sold so that the producers break even?

Answer: 490

If the total cost of producing n items is x, then the statement of proportion gives

$$\frac{x/n}{2100/10} = \sqrt{\frac{10}{n}}.$$

Therefore, $x = 210\sqrt{10n}$. The income when n items are sold is $30n$. The producers break even when the total cost equals the total income or when $30n = 210\sqrt{10n}$ which happens when n is 490.

Problem 2. We glue together 990 one inch cubes into a $9 \times 10 \times 11$ inch rectangular solid. Then we paint the outside of the solid. How many of the original 990 cubes have just one of their sides painted?

Answer: 382

To have just one side painted, a cube needs to be on a face of the large rectangular solid but not on an edge of the solid. The large rectangular solid has two 9×10 faces, two 9×11 faces, and two 10×11 faces. The number of non-edge cubes forming these faces are $(9-2)(10-2)$, $(9-2)(11-2)$, and $(10-2)(11-2)$, respectively. This gives a total of $2 \cdot 7 \cdot 8 + 2 \cdot 7 \cdot 9 + 2 \cdot 8 \cdot 9 = 382$ cubes.

Problem 3. Four rectangular strips, each measuring 4 by 16 inches, are laid out with two vertical strips crossing two horizontal strips forming a single polygon which looks like a tic-tack-toe pattern. What is the perimeter of this polygon?

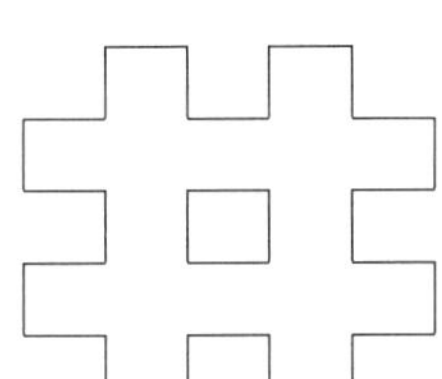

Answer: 96

Each of the strips have a perimeter of $2(4+16) = 40$, so the total perimeter of the four separate strips is $4 \cdot 40 = 160$. When laid on top of one another, each strip crossing eliminates a 4×4 square from the perimeter. There are four such crossings, each accounting for 16 units of perimeter, Therefore, the remaining perimeter is $160 - 4 \cdot 16 = 96$.

Problem 4. A palindrome is a number that reads the same forwards and backwards such as 3773 or 42924. What is the least 9-digit palindrome which is a multiple of 3 and has at least two digits that are 5's and two digits that are 7's?

Answer: 105717501

For a 9 digit palindrome to be small, it would need to start, and therefore end, with the digit 1. Also, to make the number small, the 5 and 7 digits should be as far as possible to the right with the 5 digit coming before the 7 digit. Thus, the smallest palindrome with 9 digits, two 5 digits, and two 7 digits is 105707501. But this number is not a multiple of 3 since its digits add to 26 which is one short of a multiple of 3. The smallest 9 digit palindrome of this form whose digits do add a multiple of 3 is the answer 105717501.

Problem 5. In January, Jeff's investment went up by three quarters. In February, it went down by one quarter. In March, it went up by one third. In April, it went down by one fifth. In May, it went up by one seventh. In June, Jeff's investment fell by $\frac{m}{n}$, where m and n are relatively prime positive integers. If Jeff's investment was worth the same amount at the end of June as it had been at the beginning of January, find $m+n$.

Answer: 11

The final value of Jeff's investment is

$$\begin{aligned} 1 &= \left(1+\frac{3}{4}\right)\left(1-\frac{1}{4}\right)\left(1+\frac{1}{3}\right)\left(1-\frac{1}{5}\right)\left(1+\frac{1}{7}\right)\left(1-\frac{m}{n}\right) \\ &= \frac{7}{4}\frac{3}{4}\frac{4}{3}\frac{4}{5}\frac{8}{7}\frac{n-m}{n} = \frac{8}{5}\frac{n-m}{n}. \end{aligned}$$

So $5n = 8n - 8m$ or $8m = 3n$. Thus, $m = 3$ and $n = 8$ giving $m + n = 11$.

Problem 6. Pentagon $ABCDE$ is regular. What is the degree measure of the acute angle at the intersection of line segments $\overline{AC}$ and $\overline{BD}$?

Answer: 72

Label the intersection point F as shown in the diagram. A regular polygon with n sides has internal angles given by the formula $\frac{180(n-2)}{n}$, so when n is 5, the internal angles measure 108 degrees.

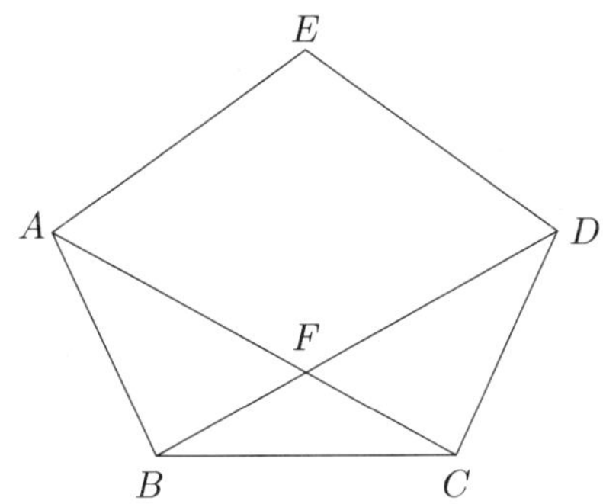

Triangle ABC is isosceles since sides $\overline{AB}$ and $\overline{BC}$ are congruent. Therefore, angle BAC has measure

$$\frac{180-108}{2} = 36.$$

Similarly, angle CBD has measure 36, and angle ABD has measure

$$108 - 36 = 72.$$

So, considering the angles in triangle ABF, one can conclude that angle AFB has measure $180 - 72 - 36 = 72$.

Problem 7. The graph of the equation $y = 5x + 24$ intersects the graph of the equation $y = x^2$ at two points. The two points are a distance $\sqrt{N}$ apart. Find N.

Answer: 3146

Substituting for y gives $x^2 = 5x + 24$ which factors as

$$x^2 - 5x - 24 = (x-8)(x+3) = 0.$$

It follows that the two intersection points are $(-3, 9)$ and $(8, 64)$. The distance between these two points is given by the distance formula

$$\sqrt{(8+3)^2 + (64-9)^2} = \sqrt{11^2 + 55^2} = 11\sqrt{1+25} = 11\sqrt{26}.$$

Hence $N = 121(26) = 3146$.

Problem 8. Find x if $\dfrac{1}{\dfrac{1}{\dfrac{1}{\frac{1}{x}+\frac{1}{2}}+\dfrac{1}{\frac{1}{x}+\frac{1}{2}}}+\dfrac{1}{\dfrac{1}{\frac{1}{x}+\frac{1}{2}}+\dfrac{1}{\frac{1}{x}+\frac{1}{2}}}} = \dfrac{x}{36}$.

Answer: 70

$$\dfrac{1}{\dfrac{1}{\dfrac{1}{\frac{1}{x}+\frac{1}{2}}+\dfrac{1}{\frac{1}{x}+\frac{1}{2}}}+\dfrac{1}{\dfrac{1}{\frac{1}{x}+\frac{1}{2}}+\dfrac{1}{\frac{1}{x}+\frac{1}{2}}}} = \dfrac{1}{\dfrac{\frac{1}{2}}{\frac{1}{x}+\frac{1}{2}}+\dfrac{\frac{1}{2}}{\frac{1}{x}+\frac{1}{2}}}$$

$$= \dfrac{1}{\dfrac{2}{\dfrac{2}{\frac{1}{x}+\frac{1}{2}}}} = \dfrac{1}{\frac{1}{x}+\frac{1}{2}} = \dfrac{2x}{2+x} = \dfrac{x}{36}.$$

Thus, since x cannot be zero, $2 \cdot 36 = 2 + x$, and $x = 70$.

Problem 9. Let T be a 30-60-90 triangle with hypotenuse of length 20. Three circles, each externally tangent to the other two, have centers at the three vertices of T. The area of the union of the circles intersected with T is $(m + n\sqrt{3})\pi$ for some rational numbers m and n. Find $m + n$.

Answer: 50

In the diagram, T is triangle ABC with right angle at C, and 60 degree angle at A. It follows that since side $\overline{AB}$ is length 20, that $\overline{AC}$ is length 10 and $\overline{BC}$ is length $10\sqrt{3}$.

Let the circles centered at vertices A, B, and C have radii a, b, and c, respectively. Then $a+b = 20$, $a+c = 10$, and $b+c = 10\sqrt{3}$. Solving this system of equations yields $a = 15-5\sqrt{3}$, $b = 5+5\sqrt{3}$, $c = 5\sqrt{3} - 5$. Since the angle at C is 90 degrees, one quarter of the area of the circle centered at C is in the triangle. Similarly, one sixth of the area of the circle centered at A is in the triangle, and one twelfth of the area of the circle centered at B is in the triangle.

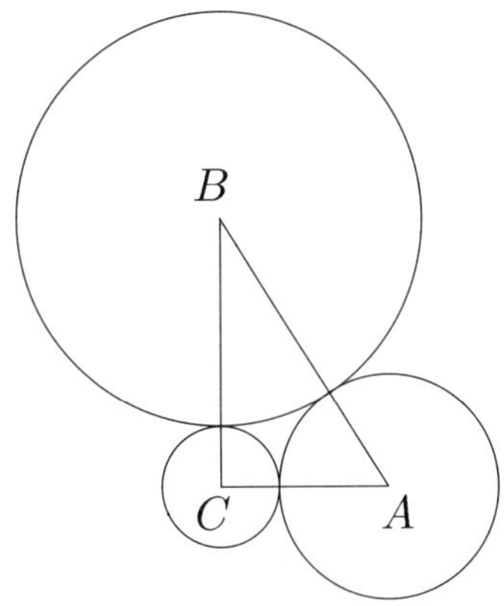

It follows that the area of the union of the circles inside the triangles is

$$\frac{\pi(5\sqrt{3}-5)^2}{4}+\frac{\pi(15-5\sqrt{3})^2}{6}+\frac{\pi(5+5\sqrt{3})^2}{12}$$
$$=\frac{25}{12}\pi[3(\sqrt{3}-1)^2+2(3-\sqrt{3})^2+(1+\sqrt{3})^2]$$
$$=\frac{25\pi}{12}[40-16\sqrt{3}]=\frac{(250-100\sqrt{3})\pi}{3}.$$

It follows that $m+n=\frac{250}{3}-\frac{100}{3}=50$.

Problem 10. A jar contains two yellow candies, four red candies, and six blue candies. Candies are randomly drawn out of the jar one-by-one and eaten. The probability that the two yellow candies will be eaten before any of the red candies are eaten is given by the fraction $\frac{m}{n}$, where m and n are relatively prime positive integers. Find $m+n$.

Answer: 16

The problem simplifies once you realize that the blue candies can be ignored completely, and that the probability is just the probability that the 2 yellow candies are selected as the first two out of the six yellow and red candies. This is just the probability that a random alignment of two yellow and 4 red candies has the two yellow candies first or one way out of the $\binom{6}{2}=15$ ways to arrange the six candies giving a probability of $\frac{1}{15}$. Alternatively, the probability that the first candy is yellow is $\frac{2}{6}$, the probability that the second candy is yellow given that the first candy is yellow is $\frac{1}{5}$, so the product $\frac{2}{6} \cdot \frac{1}{5} = \frac{1}{15}$ gives the correct answer. Thus, $1 + 15 = 16$ is the number requested.

Problem 11. The straight river is one and a half kilometers wide and has a current of 8 kilometers per hour. A boat capable of traveling 10 kilometers per hour in still water, sets out across the water. How many minutes will it take the boat to reach a point directly across from where it started?

Answer: 15

This problem is easiest when worked using vectors representing the velocity of the boat and the velocity of the river. In the diagram, the river is flowing horizontally with the horizontal vector representing a velocity of 8 kilometers per hour.

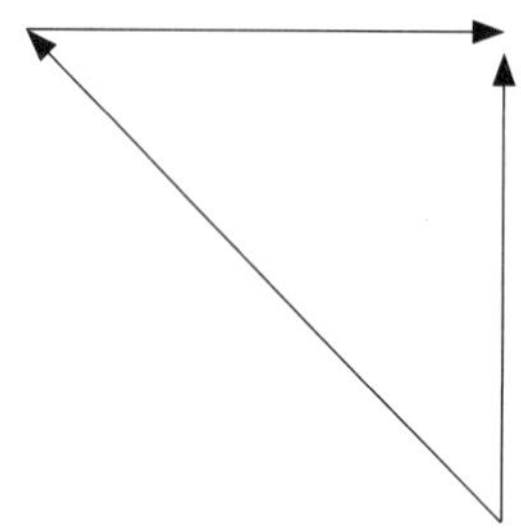

The diagonal vector represented the velocity of the boat in still water of 10 kilometers per hour. The vertical vector represented the resultant vector giving the real velocity of the boat crossing the river toward the other side. The Pythagorean Theorem shows that the resultant vector has magnitude 6 kilometers per hour. Thus, it takes the boat $\frac{1.5}{6} \cdot 60 = 15$ minutes to cross the river.

Problem 12. For how many nonnegative integers n is $(n^2 - 3n + 1)^2 + 1$ a prime number?

Answer: 4

Notice that for any integer n, the numbers n^2 and $3n$ are either both even or both odd. This means that $(n^2 - 3n + 1)^2 + 1$ is always an even number. The only even prime number is 2.

$(n^2 - 3n + 1)^2 + 1$ is equal to 2 when n is 0, 1, 2, and 3. When $n > 3$, $n^2 - 3n > 0$, so $(n^2 - 3n + 1)^2 + 1$ is bigger than 2. There are four values of n which satisfy the condition.

Problem 13. The work team was working at a rate fast enough to process 1250 items in ten hours. But after working for six hours, the team was given an additional 165 items to process. By what percent does the team need to increase its rate so that it can still complete its work within the ten hours?

Answer: 33

The work team begins work at a rate of $1250/10 = 125$ items per hour. During the last four hours the team must process an additional $165/4$ items per hour. This is an increase of $\frac{165/4}{125}100 = 33$ percent.

Problem 14. Four mathletes and two coaches sit at a circular table. How many distinct arrangements are there of these six people if the two coaches sit opposite each other?

Answer: 24

There is only one way to seat the two coaches across from each other. Each way the four mathletes now can take the remaining four seats results in a distinct arrangement, so there are $4! = 24$ ways to arrange four mathletes giving the answer to the question.

Problem 15. And it came to pass that Jeb owned over a thousand chickens. So Jeb counted his chickens. And Jeb reported the count to Hannah. And Hannah reported the count to Joshua. And Joshua reported the count to Caleb. And Caleb reported the count to Rachel. But as fate would have it, Jeb had over-counted his chickens by nine chickens. Then Hannah interchanged the last two digits of the count before reporting it to Joshua. And Joshua interchanged the first and the third digits of the number reported to him before reporting it to Caleb. Then Caleb doubled the number reported to him before reporting it to Rachel. Now it so happens that the count reported to Rachel was the correct number of chickens that Jeb owned. How many chickens was that?

Answer: 2052

Assume that Jeb's count was the four digit number $abcd$. Then Hannah would have reported $abdc$, and Joshua would have reported $dbac$. Therefore, twice $dbac$ is the correct number of chickens and is nine short of *abcd* forming the addition problem

$$\begin{array}{r} 9 \\ dbac \\ dbac \\ \hline abcd \end{array}$$

Considering the hundreds place in this addition problem, it is clear that the digit b must either be 0 or 9 since no other digit could add to itself or itself plus 1 to yield a number ending in itself. From the thousands place it is clear that the digit d cannot exceed 4. From the ones place it is clear that d must be odd, so d is either 1 or 3.

Suppose b is 9. Then the thousands place gives $2d + 1 = a$. If d is 1, then a is 3, and the ones column in the addition problem shows that c is either 1 or 6. Neither of these choices allow the tens column to work out. If d is 3, then a is 7, and the ones column in the addition problem shows that c is either 2 or 7. Again, neither of these choices allow the tens column to work out. Conclude, therefore, that b must be 0.

Reasoning as above, d must be 1 or 3 with $2d = a$. If d is 3, then a is 6, and the ones column of the addition problem shows that c is either 2 or 7. Neither of these choices allow the tens column to work out. But if d is 1, then a is 2, and the ones column of the addition problem shows that c is either 1 or 6. $c = 1$ does not work, but $c = 6$ does allowing *abcd* to be the number 2061. It follows that the correct number of chickens is $2061 - 9 = 2052$.

One does need to rule out the case of Jeb's number being more than four digits. For example, if Jeb's number is the five digit number *abcde*, it gives rise to the addition problem

$$\begin{array}{r} 9 \\ cbaed \\ cbaed \\ \hline abcde \end{array}$$

Problem 16. A tailor met a tortoise sitting under a tree. When the tortoise was the tailor's age, the tailor was only a quarter of his current age. When the tree was the tortoise's age, the tortoise was only a seventh of its current age. If the sum of their ages is now 264, how old is the tortoise?

Answer: 77

Let R be the tailor's age, S be the tortoise's age, and T be the tree's age. The tortoise was the tailor's age $S-R$ years ago. At that time the tailor was $R-(S-R)=R/4$ years old. The tree was the tortoise's age $T-S$ years ago. At that time the tortoise was $S-(T-S)=S/7$ years old. It follows from the first equation that $7R=4S$, and it follows from the second equation that $13S=7T$. Finally, $R+S+T=264$. Substituting for R and T in this last equation yields $R+S+T=\frac{4}{7}S+S+\frac{13}{7}S=\frac{24}{7}S=264$. Thus, the tortoise's age is 77. The tailor is 44, and the tree is 143.

Problem 17. Functions f and g are defined so that $f(1)=4$, $g(1)=9$, and for each integer $n\geq 1$, $f(n+1)=2f(n)+3g(n)+2n$ and $g(n+1)=2g(n)+3f(n)+5$. Find $f(2005)-g(2005)$.

Answer: 1999

Let $a_n=f(n)-g(n)$ and note that

$$\begin{aligned}a_{n+1}&=f(n+1)-g(n+1)=[2f(n)+3g(n)+2n]\\&\quad-[2g(n)+3f(n)+5]\\&=g(n)-f(n)+2n-5=2n-5-a_n.\end{aligned}$$

Iterating this formula gives

$$a_{n+2}=2(n+1)-5-a_{n+1}=2n-3-[2n-5-a_n]=2+a_n.$$

It follows that $a_{n+2k}=2k+a_n$ and

$$\begin{aligned}f(2005)-g(2005)&=a_{2005}=2004+a_1=2004+f(1)-g(1)\\&=2004+4-9=1999.\end{aligned}$$

Problem 18. The side lengths of a trapezoid are $\sqrt[4]{3}$, $\sqrt[4]{3}$, $\sqrt[4]{3}$, and $2\sqrt[4]{3}$. Its area is the ratio of two relatively prime positive integers, m and n. Find $m+n$.

Answer: 13

The diagram shows trapezoid $ABCD$ where side $\overline{AB}$ has length $2\sqrt[4]{3}$.

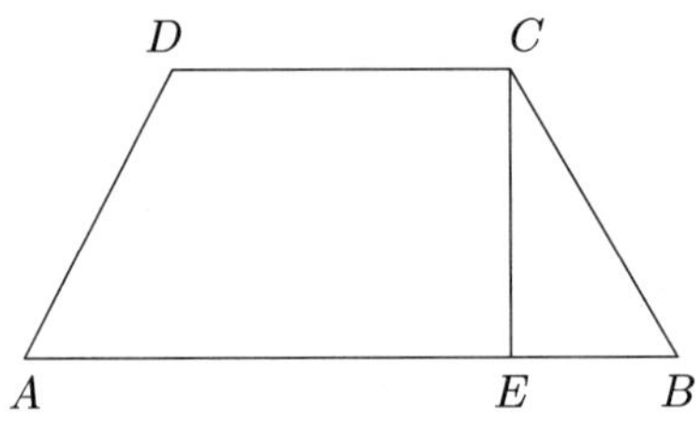

If E is the base of the altitude of the trapezoid from vertex C, then EB must have length equal to half the difference between the lengths of the two bases of the trapezoid. Thus, $\overline{EB}$ has length $\frac{\sqrt[4]{3}}{2}$, and

$$CE = \sqrt{CB^2 - EB^2} = \frac{\sqrt[4]{3}}{2} \cdot \sqrt{3}.$$

So the area of the trapezoid is

$$\frac{\sqrt[4]{3}}{2}\sqrt{3}\frac{\sqrt[4]{3} + 2\sqrt[4]{3}}{2} = \frac{\sqrt[4]{3} \cdot \sqrt{3} \cdot 3\sqrt[4]{3}}{4} = \frac{9}{4}.$$

This gives the answer $9+4 = 13$. Notice that by extending sides $\overline{AD}$ and $\overline{BC}$ until they meet, one gets an equilateral triangle with area $\frac{4}{3}$ the area of trapezoid. Since the area of an equilateral triangle with side s is $\frac{s^2\sqrt{3}}{4}$, it follows that the trapezoid has area

$$\frac{3}{4} \cdot \frac{(2\sqrt[4]{3})^2\sqrt{3}}{4} = \frac{9}{4}.$$

Problem 19. Let x and y be integers satisfying both $x^2 - 16x + 3y = 20$ and $y^2 + 4y - x = -12$. Find $x + y$.

Answer: 18

Completing the squares gives the equations $(x-8)^2 + 3y = 84$ and $(y+2)^2 - x = -8$. Then $(x-8)^2 + 3(y+2) = 90$ and $(y+2)^2 - (x-8) = 0$. So letting $a = x - 8$ and $b = y + 2$ implies $a^2 + 3b = 90$ and $b^2 - a = 0$. Substituting $a = b^2$ in the first equation gives $b^4 + 3b = 90$ which has the one positive solution of $b = 3$ and one negative irrational solution. Thus, $y = 1$, $a = 9$, $x = 17$, and $x + y = 18$.

Problem 20. If $\displaystyle\sum_{k=1}^{360} \frac{1}{k\sqrt{k+1}+(k+1)\sqrt{k}}$ is the ratio of two relatively prime positive integers m and n, find $m+n$.

Answer: 37

The series is a telescoping series as seen by

$$\sum_{k=1}^{360} \frac{1}{k\sqrt{k+1}+(k+1)\sqrt{k}} = \sum_{k=1}^{360} \frac{1}{\sqrt{k}\sqrt{k+1}} \cdot \frac{1}{\sqrt{k}+\sqrt{k+1}}$$
$$= \sum_{k=1}^{360} \frac{1}{\sqrt{k}\sqrt{k+1}} \cdot \frac{\sqrt{k+1}-\sqrt{k}}{(k+1)-k} = \sum_{k=1}^{360} \frac{(\sqrt{k+1}-\sqrt{k})}{\sqrt{k}\sqrt{k+1}}$$
$$= \sum_{k=1}^{360} \left[\frac{1}{\sqrt{k}} - \frac{1}{\sqrt{k+1}}\right] = \frac{1}{\sqrt{1}} - \frac{1}{\sqrt{361}} = \frac{18}{19}.$$

Thus, $m+n=37$.

Problem 21. In the diagram below angle CAB, angle CBD, and angle CDE are all right angles with side lengths $AC=3$, $BC=5$, $BD=12$, and $\overline{DE}=84$. The distance from point E to the line $\overline{AB}$ can be expressed as the ratio of two relatively prime positive integers, m and n. Find $m+n$.

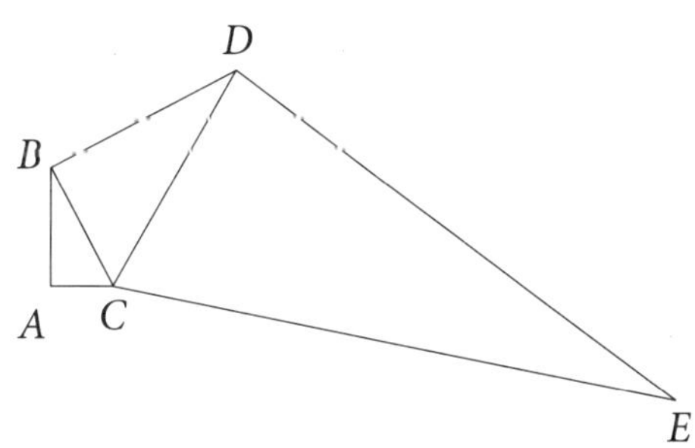

Answer: 5393

All three triangles are right triangles with Pythagorean triples for side lengths: $3-4-5$ for triangle ABC, $5-12-13$ for triangle BCD, and $13-84-85$ for triangle CDE. Construct a line through point D perpendicular to line $\overline{AB}$ which intersects line $\overline{AB}$ at point F. Construct a line through point E perpendicular to line $\overline{DF}$ which intersects line $\overline{DF}$ at point G. Construct a line through point C parallel to line $\overline{AB}$ which intersects line $\overline{FG}$ at point H.

Since angle CBD is a right angle, angle ABC and angle DBF are complementary. Thus, triangle ABC is similar to triangle FDB. From this $\dfrac{AC}{BC} = \dfrac{FB}{DB}$ so $FB = \dfrac{3}{5} \cdot 12 = \dfrac{36}{5}$. Similarly, $FD = \dfrac{4}{5} \cdot 12 = \dfrac{48}{5}$. Then $CH = AF = FB + AB = \dfrac{36}{5} + 4 = \dfrac{56}{5}$. Since angle CDE is a right angle, angle CDH and angle EDG are complementary. Thus, triangle HDC is similar to triangle GED.

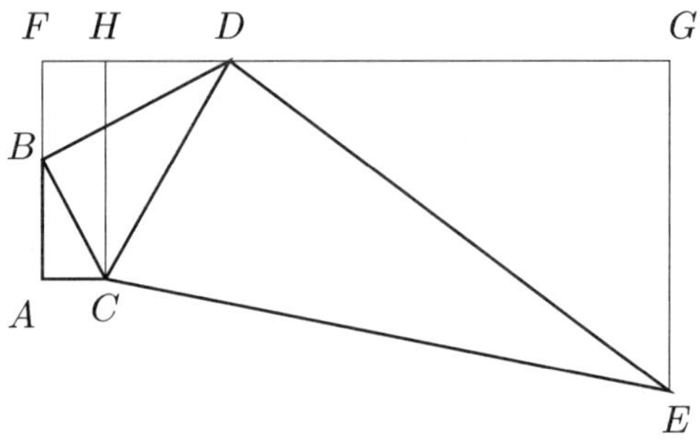

From this $\dfrac{DG}{DE} = \dfrac{CH}{CD}$ so $DG = \dfrac{\frac{56}{5}}{13} 84 = \dfrac{56 \cdot 84}{5 \cdot 13}$. The required distance from the point E to the line $\overline{AB}$ is the sum of the lengths of $\overline{FD}$ and $\overline{DG}$ which is

$$\frac{48}{5} + \frac{56 \cdot 84}{5 \cdot 13} = \frac{48}{5}\left[1 + \frac{14 \cdot 7}{13}\right] = \frac{48}{5} \cdot \frac{111}{13} = \frac{5328}{65}.$$

The requested result is $5328 + 65 = 5393$.

Problem 22. Let d_k be the greatest odd divisor of k for $k = 1, 2, 3, \ldots$. Find $d_1 + d_2 + d_3 + \cdots + d_{1024}$.

Answer: 349526

For $n \geq 0$, let $s_n = d_1 + d_2 + d_3 + \cdots + d_{2^n}$. We claim that, in general, $s_n = \dfrac{2^{2n} + 2}{3}$. Note that for any odd k, $d_k = k$, and for any k, $d_{2k} = d_k$. Thus,

$$s_{n+1} = \sum_{k=1}^{2^{n+1}} d_k = \sum_{k=1}^{2^n} d_{2k-1} + \sum_{k=1}^{2^n} d_{2k}$$
$$= \sum_{k=1}^{2^n}(2k-1) + \sum_{k-1}^{2^n} d_k = (2^n)^2 + s_n.$$

Hence,

$$s_n = s_0 + (2^0)^2 + (2^1)^2 + \cdots + (2^{n-1})^2 = 1 + 1 + 4 + \cdots + 4^{n-1}$$
$$= 1 + \frac{4^n - 1}{4 - 1} = \frac{2^{2n} + 2}{3}.$$

In our special case, $s_{10} = \dfrac{2^{20} + 2}{3} = 349526.$

Problem 23. Let $a = \sqrt[401]{4} - 1$ and for each $n \geq 2$ let

$$b_n = \binom{n}{1} + \binom{n}{2}a + \cdots + \binom{n}{n}a^{n-1}.$$

Find $b_{2006} - b_{2005}$.

Answer: 1024

The binomial theorem gives

$$\binom{n}{0} + \binom{n}{1}a + \binom{n}{2}a^2 + \cdots + \binom{n}{n}a^n = (1+a)^n.$$

So,

$$b_n = \frac{(1+a)^n - 1}{a}$$

and

$$b_{2006} - b_{2005} = \frac{[(1+a)^{2006} - 1] - [(1+a)^{2005} - 1]}{a}$$
$$= (1+a)^{2005}\frac{(1+a) - 1}{a} = (1+a)^{2005}$$
$$= \sqrt[401]{4^{2005}} = 4^5 = 1024.$$

Problem 24. Triangle ABC has area 240. Points X, Y, Z lie on sides $\overline{AB}$, $\overline{BC}$, $\overline{CA}$, respectively. Given that $\dfrac{AX}{BX} = 3$, $\dfrac{BY}{CY} = 4$, $\dfrac{CZ}{AZ} = 5$, find the area of triangle XYZ.

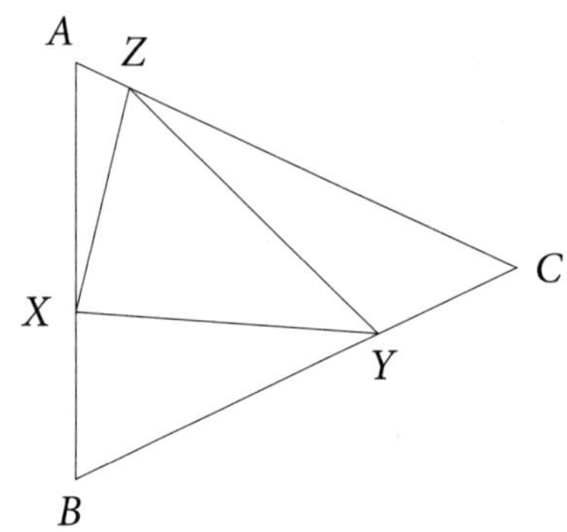

Answer: 122

Triangle XBC has the same altitude from vertex C as does triangle ABC. $\dfrac{AX}{BX} = 3$ implies that the length of $\overline{BX}$ is only one quarter the length of $\overline{AB}$, so the area of XBC is only one quarter the area of ABC.

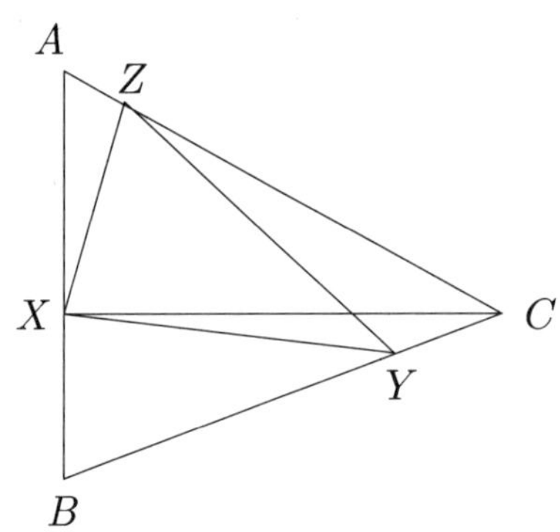

A similar argument shows that triangle BXY has four fifths the area of triangle XBC. Therefore, triangle BXY has area $240 \cdot \dfrac{1}{4} \cdot \dfrac{4}{5} = 48$. Repeat this argument to show that triangle CYZ has area 40, and triangle AXZ has area 30. It follows that triangle XYZ has area $240 - (48 + 40 + 30) = 122$.

Problem 25. Find the number of quadruples (a, b, c, d) of non-zero integers satisfying both

$$\frac{1}{a} + \frac{1}{b} + \frac{1}{c} + \frac{1}{d} = \frac{1}{2}$$

and

$$2(a+b+c+d) = ab + cd + (a+b)(c+d) + 1.$$

Answer: 48

Multiply the first equation by $2abcd$ and then rewrite it as

$$abcd - 2(bcd + acd + abd + abc) = 0.$$

Rewrite the second equation to get

$$2(a+b+c+d) - (ab + cd + ac + ad + bc + bd) = 1.$$

Now note that

$$\begin{aligned}(a-2)(b-2)(c-2)(d-2) &= abcd - 2(bcd + acd + abd + abc)\\ &+ 4(ab + cd + ac + ad + bc + bd) - 8(a+b+c+d) + 16\\ &= [abcd - 2(bcd + acd + abd + abc)]\\ &- 4[2(a+b+c+d) - (ab + cd + ac + ad + bc + bd)] + 16\\ &= [0] + 4[1] + 16 = 12.\end{aligned}$$

There are four ways to factor 12 as the product of four positive integers: $12{\cdot}1{\cdot}1{\cdot}1$, $6{\cdot}2{\cdot}1{\cdot}1$, $4{\cdot}3{\cdot}1{\cdot}1$, and $3{\cdot}2{\cdot}2{\cdot}1$. If one takes one of these products and assigns the first factor to be $a-2$, the second factor to be $b-2$, the third to be $c-2$, and the last to be $d-2$, the assignment does not result in an assignment of values for a, b, c, and d which satisfy the original two equations. On the other hand, if two or four of the four factors in some of the products are made negative, then some solutions can be found. There are three families of such solutions resulting from the products $(-3)(2)(2)(-1)$, $(-4)(-3)(-1)(-1)$, and $(4)(-3)(-1)(1)$. For example, if $a-2=4$, $b-2=-3$, $c-2=-1$, and $d-2=1$, then $(a,b,c,d) = (6,-1,1,3)$ which does satisfy the original two equations. Of course, the assignment of factors $4, -3, -1$, and 1 to the expressions $a-2$, $b-2$, $c-2$, and $d-2$ is arbitrary, and there are $4! = 24$ equivalent assignments resulting from taking the variables in all of their possible permutations. Both of the other two products, $(-3)(2)(2)(-1)$ and $(-4)(-3)(-1)(-1)$, have a repeated factor, so there are only $4!/2! = 12$ members of their families. This accounts for a total of $24 + 12 + 12 = 48$ possible assignment of value to (a,b,c,d) which satisfy the two given equations.

2.5 2005 Middle School

Problem 1. A cubic inch of the newly discovered material madelbromium weighs 5 ounces. How many pounds will a cubic yard of madelbromium weigh?

Answer: 14580

We have

$$\begin{aligned} 5\frac{\text{ounce}}{\text{inch}^3} &= 5\frac{\text{ounce}}{\text{inch}^3}\cdot\frac{1\text{pound}}{16\text{ounce}}\left(\frac{36\text{inch}}{1\text{yard}}\right)^3 \\ &= \frac{5\cdot 36\cdot 36\cdot 36}{16} = 5\cdot 9\cdot 9\cdot 36 = 14580. \end{aligned}$$

Problem 2. Jerry is mowing a rectangular lawn which is 77 feet north to south by 83 feet east to west. His lawn mower cuts a path 18 inches wide. Jerry mows the grass by cutting a path from west to east across the north side of the lawn and then making a right turn cutting a path along the east side of the lawn. When he completes mowing each side of the lawn, he continues by making right turns to mow a path along the next side. How many right turns will he make?

Answer: 102

Each time Jerry mows all four sides of the rectangle, he makes four right turns and reduces both the width and the length of the rectangle by 3 feet. After doing this 25 times, the unmoved rectangle will be 75 feet narrower and 75 feet shorter than it began, that is, 2 feet by 8 feet. Jerry will mow this remaining part by removing 18 inches off the north side, turning right, removing 18 inches off the east side, turning right, and finally mowing the last 6 inch strip. Thus, Jerry will make three more right turns after completing 25 full circuits with $25\cdot 4 = 100$ right turns. Jerry will have made 102 right turns in all.

Problem 3. Four rectangular strips each measuring 4 by 16 inches are laid out with two vertical strips crossing two horizontal strips forming a single polygon which looks like a tic-tack-toe pattern. What is the perimeter of this polygon?

Answer: 96

This is the same as High School problem number 3.

Problem 4. Fill in numbers in the boxes below so that the sum of the entries in each three consecutive boxes is 2005. What is the number that goes into the leftmost box?

		999					888	

Answer: 118

Suppose that four adjacent boxes contain the numbers a, b, c, and d as read from left to right. Then $a + b + c = 2005$ and $b + c + d = 2005$, so it follows that $a = d$. In other words, the numbers in the boxes repeat every three boxes as shown here.

x	y	999	x	y	z	x	888	z

We can conclude that $y = 888$ and $z = 999$.

Since $x + y + z = 2005$, $x = 118$.

Problem 5. A palindrome is a number that reads the same forwards and backwards such as 3773 or 42924. Find the sum of the twelve smallest five digit palindromes.

Answer: 126632

The ten smallest five digit palindromes are simple enough to find; they are 10001, 10101, 10201,. . . , 10901. The next two are 11011 and 11111. The sum of all of these numbers is 126632.

Problem 6. We glue together 990 one inch cubes into a $9 \times 10 \times 11$ inch rectangular solid. Then we paint the outside of the solid. How many of the original 990 cubes have just one of their sides painted?

Answer: 382

This is the same as High School problem number 2.

Problem 7. Bill's age is one third larger than Tracy's age. In 30 years Bill's age will be one eighth larger than Tracy's age. How many years old is Bill?

Answer: 24

Let Bill's current age be b and Tracy's current age be t. Then the relationship between their current ages is $b = t\frac{4}{3}$. In 30 years their ages will be $b+30$ and $t+30$ which will satisfy the relationship

$$(b+30) = (t+30)\frac{9}{8}.$$

These two equations simplify to $3b = 4t$ and $8b+240 = 9t+270$ or $8b = 9t+30$. Replacing t by $\frac{3b}{4}$ in this last equations yields

$$8b = 9\frac{3b}{4} + 30$$

and

$$30 = 8b - 9\frac{3b}{4} = \frac{32-27}{4}b = \frac{5}{4}b.$$

Thus,

$$b = 30 \cdot \frac{4}{5} = 24.$$

Problem 8. The number 1 is special. The number 2 is special because it is relatively prime to 1. The number 3 is not special because it is not relatively prime to the sum of the special numbers less than it, $1+2$. The number 4 is special because it is relatively prime to the sum of the special numbers less than it. So, a number bigger than 1 is special only if it is relatively prime to the sum of the special numbers less than it. Find the twentieth special number.

Answer: 29

There seems to be no way to solve this which is faster than just writing out each special number and the sum of the special numbers up to that value as in

Number	1	2	4	5	7	8	10	11	13	14
Sum	1	3	7	12	19	27	37	48	61	75

Number	16	17	19	20	22	23	25	26	28	29
Sum	91	108	127	147	169	192	217	243	271	300

Problem 9. For how many nonnegative integers n is $(n^2-3n+1)^2+1$ a prime number?

Answer: 4

This is the same as High School problem number 12.

Problem 10. What is the 1000th digit to the right of the decimal point in the decimal representation of $\dfrac{37}{5500}$?

Answer: 7

We have

$$\begin{aligned}\frac{37}{5500} &= \frac{37}{11\cdot 5\cdot 100} = \frac{37\cdot 2}{11\cdot 1000} = \frac{\frac{74}{11}}{1000} = \frac{6+\frac{8}{11}}{1000}\\ &= \frac{6+\frac{8\cdot 9}{11\cdot 9}}{1000} = \frac{6+\frac{72}{99}}{1000} = \frac{6.7272\ldots}{1000}\\ &= 0.0067272\ldots\end{aligned}$$

so after the third digit to the right of the decimal point, the even numbered positions contain the digit 7.

Problem 11. The work team was working at a rate fast enough to process 1250 items in ten hours. But after working for six hours, the team was given an additional 150 items to process. By what percent does the team need to increase its rate so that it can still complete its work within the ten hours?

Answer: 30

The work team begins work at a rate of $1250/10 = 125$ items per hour. During the last four hours the team must process an additional $150/4$ items per hour. This is an increase of $\dfrac{150/4}{125}100 = 30$ percent.

Problem 12. Four mathletes and two coaches sit at a circular table. How many distinct arrangements are there of these six people if the two coaches sit opposite each other?

Answer: 24

This is the same as High School problem number 14.

Problem 13. Find x such that $\dfrac{\dfrac{5}{x-50}+\dfrac{7}{x+25}}{\dfrac{2}{x-50}-\dfrac{3}{x+25}} = 17.$

Answer: 125

Multiplying numerator and denominator by the $(x-50)(x+25)$

gives

$$\frac{\frac{5}{x-50}+\frac{7}{x+25}}{\frac{2}{x-50}-\frac{3}{x+25}}\cdot\frac{(x-50)(x+25)}{(x-50)(x+25)}=\frac{5(x+25)+7(x-50)}{2(x+25)-3(x-50)}$$
$$=\frac{12x-225}{200-x}=17.$$

Now multiplying by $200-x$ yields

$$12x-225=17(200-x)=3400-17x \text{ and } 29x=3625.$$

Dividing by 29 gives $x=125$.

Problem 14. Eight identical $1\times1\times1$ cubes each have the numbers 1 through 6 written on their faces with the number 1 written on the face opposite number 2, number 3 written on the face opposite number 5, and number 4 written on the face opposite number 6. The eight cubes are stacked into a single $2\times2\times2$ cube. Add all of the numbers appearing on the outer surface of the new cube. Let M be the maximum possible value for this sum, and N be the minimum possible value for this sum. Find $M-N$.

Answer: 40

Each cube has eight corners. Adding the numbers on the faces adjacent to each corner yield the eight possible sums of

$$(1+3+4)=8,\ (1+3+6)=10,$$
$$(1+4+5)=10,\ (1+5+6)=12,$$
$$(2+3+4)=9,\ (2+3+6)=11,$$
$$(2+4+5)=11, \text{ and } (2+5+6)=13.$$

So, if each cube is oriented so that its faces numbered 1, 3, and 4 face outward, then the sum of the faces will be smallest so $N=8\cdot8=64$. Similarly, if each cube is oriented so that its faces numbered 2, 5, and 6 face outward, then the sum of the faces will be largest so $M=8\cdot13=104$. Thus, $M-N=40$.

Problem 15. And it came to pass that Jeb owned over a thousand chickens. So Jeb counted his chickens. And Jeb reported the count to Hannah. And Hannah reported the count to Joshua. And Joshua reported the count to Caleb. And Caleb reported the count to Rachel. But as fate would have it, Jeb had over-counted his chickens by nine chickens. Then Hannah interchanged the last two digits of the count before reporting it to Joshua. And Joshua interchanged the first and the third digits of the number reported to him before reporting it to Caleb. Then Caleb doubled the number reported to him before reporting it to Rachel. Now it so happens that the count reported to Rachel was the correct number of chickens that Jeb owned. How many chickens was that?

Answer: 2052

This is the same as High School problem number 15.

2.6 2006 High School

Problem 1. The sizes of the freshmen class and the sophomore class are in the ratio 5:4. The sizes of the sophomore class and the junior class are in the ratio 7:8. The sizes of the junior class and the senior class are in the ratio 9:7. If these four classes together have a total of 2158 students, how many of the students are freshmen?

Answer: 630

Let the sizes of the freshman, sophomore, junior, and senior classes be w, x, y, and z, respectively. Then $4w = 5x$, $8x = 7y$, $7y = 9z$, and $w + x + y + z = 2158$. Thus

$$x = \frac{4}{5}w, \; y = \frac{8}{7}x = \frac{8 \cdot 4}{7 \cdot 5}w, \text{ and } z = \frac{7}{9}y = \frac{8 \cdot 4 \cdot 7}{7 \cdot 5 \cdot 9}w.$$

This shows that

$$w + x + y + z = w + \frac{4}{5}w + \frac{32}{35}w + \frac{32}{45}w = \frac{1079}{315}w = 2158.$$

This gives $w = 630$.

Problem 2. At a movie theater tickets for adults cost 4 dollars more than tickets for children. One afternoon the theater sold 100 more child tickets than adult tickets for a total sales amount of 1475 dollars. How many dollars would the theater have taken in if the same tickets were sold, but the costs of the child tickets and adult tickets were reversed?

Answer: 1875

Let x be the cost of a child ticket, and let a be the number of adult tickets sold. Then the cost of an adult ticket is $x + 4$ and the number of child tickets sold is $A + 100$. This gives $(x + 4)A + x(100 + A) = 1475$. If the costs of the tickets were reversed, the theater would have taken in

$$\begin{aligned} xA + (x + 4)(100 + A) &= (x + 4)A + x(100 + A) + 400 \\ &= 1475 + 400 = 1875. \end{aligned}$$

Problem 3. Point P lies outside a circle and two rays drawn from P intersect the circle as shown. One ray intersects the circle at points A and B while the other ray intersects the circle at M and N. Lines

$\overline{AN}$ and $\overline{MB}$ intersect at X. Given that $\angle AXB$ measures $127°$ and the minor arc $\overset{\frown}{AM}$ measures $14°$, find the measure of the angle at P.

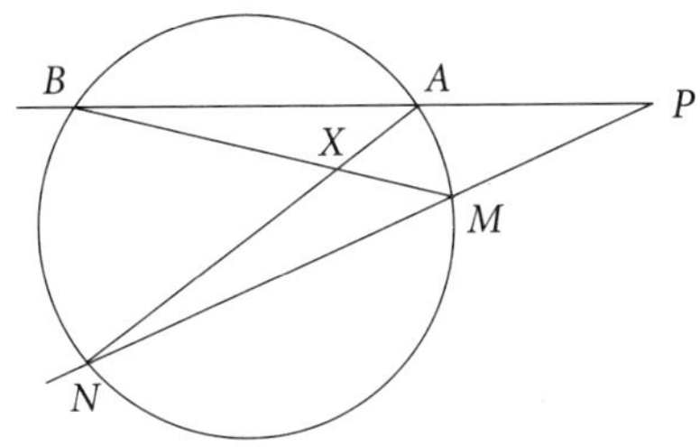

Answer: 39

$\angle AXM$ and $\angle AXB$ are supplementary, so $\angle AXM$ measures $53°$. This means the measures of arc AM and BN average to be $53°$, so arc BN measures $92°$. Then the angle at P is

$$\frac{BN - AM}{2} = \frac{92 - 14}{2} = 39.$$

Problem 4. A rogue spaceship escapes, and 54 minutes later the police leave in a spaceship in hot pursuit. If the police spaceship travels 12% faster than the rogue spaceship along the same route, how many minutes will it take for the police to catch up with the rogues?

Answer: 450

Let x be the number of minutes it will take the police to catch the rogues, and let r be the rate at which the rogues are traveling. Then the police are traveling at a rate of $1.12r$, and the rogues will have traveled for $54 + x$ minutes.

Since distance = (rate × time), it follows that

$$(54 + x)r = x(1.12)r, \text{ and } 54 = (.12)x, \text{ so } x = \frac{54}{.12} = 450.$$

Problem 5. Find the sum of all positive integers less than 2006 which are both multiples of six and one more than a multiple of seven.

Answer: 47,094

If the integer n is a multiple of six and one more than a multiple of seven, then there is an integer m so that $n = 6m \equiv 1 \pmod{7}$. Multiplying by 6 yields $36m \equiv 6 \pmod{7}$ which is equivalent to

$m \equiv 6 \pmod 7$. Thus, the numbers $6m$ which are one more than a multiple of seven are those where m is 6 more than a multiple of 7. The correct sum is then

$$6 \cdot (6 + 13 + 20 + 27 + \cdots + 328) = 6 \cdot \frac{6 + 328}{2} \cdot \frac{329}{7} = 47,094.$$

Problem 6. The positive integers v, w, x, y, and z satisfy the equation

$$v + \cfrac{1}{w + \cfrac{1}{x + \cfrac{1}{y + \cfrac{1}{z}}}} = \frac{222}{155}.$$

Compute $10^4 v + 10^3 w + 10^2 x + 10^1 y + 10^0 z$.

Answer: 12,354

Since v, w, x, y, and z are all positive integers, the denominators of each of the unit fractions on the left hand side of the equation is greater than or equal to 1. It follows that the value of each of these fractions is between 0 and 1. Therefore we can surmise that v must equal $\left\lfloor \frac{222}{155} \right\rfloor = 1$. Then the denominator of the fraction on the left must equal $\frac{155}{67}$, and w must equal $\left\lfloor \frac{222}{67} \right\rfloor =$ 2. Similarly, we find that $x = 3$, $y = 5$, and $z = 4$ giving a final answer of 12,354.

Problem 7. Heather and Kyle need to mow a lawn and paint a room. If Heather does both jobs by herself, it will take her a total of nine hours. If Heather mows the lawn and, after she finishes, Kyle paints the room, it will take them a total of eight hours. If Kyle mows the lawn and, after he finishes, Heather paints the room, it will take them a total of seven hours. If Kyle does both jobs by himself, it will take him a total of six hours. It takes Kyle twice as long to paint the room as it does for him to mow the lawn. The number of hours it would take the two of them to complete the two tasks if they worked together to mow the lawn and then worked together to paint the room is $\frac{m}{n}$, where m and n are relatively prime positive integers. Find $m + n$.

Answer: 41

Suppose Heather can mow the lawn in w hours and paint the room in x hours. Suppose Kyle can mow the lawn in y hours and paint the room in z hours. Then the facts in the problem state that

$$w+x=9, \quad w+z=8, \quad y+x=7, \quad y+z=6, \quad \text{and} \quad z=2y.$$

The first equation gives $x = 9 - w$. The second gives $z = 8 - w$. Subtracting the third from the first gives $y = w - 2$. Then the last equation implies that $8-w = 2(w-2)$, so solving this yields $w = 4$, $x = 5$, $y = 2$, and $z = 4$. Heather mows the lawn at the rate of one lawn every $\frac{1}{w}$ hours while Kyle mows the lawn at a rate of one lawn every $\frac{1}{y}$ hours. So, working together, they can mow the lawn in

$$\frac{1}{\frac{1}{w}+\frac{1}{y}} = \frac{wy}{w+y} = \frac{4\cdot 2}{6} = \frac{4}{3} \text{ hours.}$$

Similarly, working together, they can paint the room in

$$\frac{1}{\frac{1}{x}+\frac{1}{z}} = \frac{xz}{x+z} = \frac{5\cdot 4}{9} = \frac{20}{9} \text{ hours.}$$

The total time is then

$$\frac{4}{3}+\frac{20}{9} = \frac{12}{9}+\frac{20}{9} = \frac{32}{9}$$

so $m+n = 32+9 = 41$.

Problem 8. Evaluate $\dfrac{2i}{\dfrac{1}{30-5i}-\dfrac{1}{30+5i}}$ where $i^2 = -1$.

Answer: 185

We have

$$\begin{aligned}
\frac{2i}{\frac{1}{30-5i}-\frac{1}{30+5i}} &= \frac{2i}{\frac{1}{30-5i}\cdot\frac{30+5i}{30+5i}+\frac{1}{30+5i}\cdot\frac{30-5i}{30-5i}} \\
&= \frac{2i}{\frac{30+5i}{30^2+5^2}-\frac{30-5i}{30^2+5^2}} = \frac{2i(30^2+5^2)}{(30+5i)-(30-5i)} \\
&= \frac{2i(925)}{10i} = 185.
\end{aligned}$$

Problem 9. How many rectangles are there in the diagram below such that the sum of the numbers within the rectangle is a multiple of 7?

43	44	45	46	47	48	49
36	37	38	39	40	41	42
29	30	31	32	33	34	35
22	23	24	25	26	27	28
15	16	17	18	19	20	21
8	9	10	11	12	13	14
1	2	3	4	5	6	7

Answer: 163

It is sufficient to consider the numbers in the table when the numbers are replaced by their remainders when divided by 7 (that is, the numbers mod 7) since the extra 7's will not affect whether a sum is divisible by 7:

1	2	3	4	5	6	0
1	2	3	4	5	6	0
1	2	3	4	5	6	0
1	2	3	4	5	6	0
1	2	3	4	5	6	0
1	2	3	4	5	6	0
1	2	3	4	5	6	0

Every rectangle which contains seven rows of numbers will contain a sum which is a multiple of 7 since the sum will be 7 times the sum of the numbers in the bottom row of the rectangle. There are 8 vertical lines in the table. By selecting any 2 of the vertical lines, we get a unique rectangle of height 7. There are, therefore, $\binom{8}{2} = 28$ such rectangles. If a rectangle does not contain 7 rows of numbers, then the sum of the numbers in the rectangle will be a multiple of the sum of the numbers in the bottom row of the rectangle. Since the number of rows is less than seven, the sum will be a multiple of 7 only if the sum in the bottom row of the rectangle is a multiple of seven. The only possible choices for columns of numbers which make the sum of the numbers in the bottom row a multiple of seven are columns 1 to 6, columns 1 to 7, columns 2 to 5, columns 3 and 4, and column 7 by itself. For each of these five choices of column ranges, there are again 28 possible rectangles which span those columns, although, in each case, the rectangle of height 7 was already counted in the original list of 28 rectangles of height 7. Therefore, the total number of rectangles with the correct sum is $28 + 5(27) = 163$.

Problem 10. An equilateral triangle with side length 6 has a square of side length 6 attached to each of its edges as shown. The distance between the two farthest vertices of two squares (marked A and B in the figure) can be written as $m + \sqrt{n}$ where m and n are positive integers. Find $m + n$.

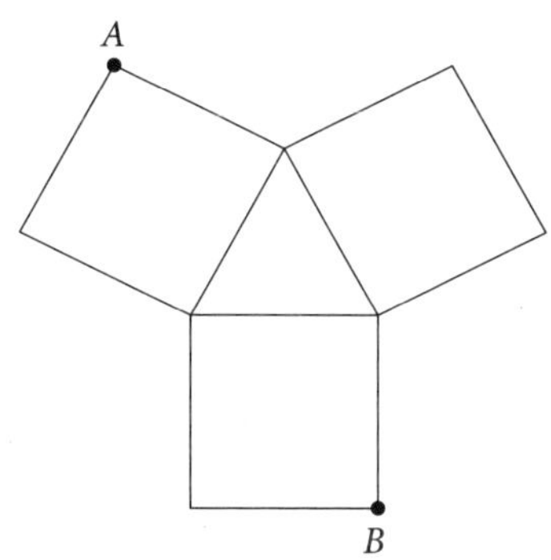

Answer: 114

To find the distance between the vertex A of one square to vertex B of the other square, let C be the vertex that the two squares have in common. Then both $\overline{CA}$ and $\overline{CB}$ are diagonals of squares with side length 6, so both have length $6\sqrt{2}$. In addition, angle ACB is made up of two 45 degree angles and a 60 degree angle, so its measure is 150 degrees. By the Law of Cosines

$$\begin{aligned}(AB)^2 &= (6\sqrt{2})^2 + (6\sqrt{2})^2 - 2(6\sqrt{2})(6\sqrt{2})\cos 150^\circ \\ &= 36(4)(1 - \cos 150^\circ) = 36(4)(1 + \cos 30^\circ) \\ &= 36(4 + 2\sqrt{3}).\end{aligned}$$

So the desired distance is

$$AB = 6\sqrt{4 + 2\sqrt{3}}.$$

Alternatively, consider the diagram with a (horizontal) line drawn through point A and a perpendicular (vertical) line which bisects the equilateral triangle as shown. Let G be the intersection of the two new lines, D be the vertex of the triangle on the vertical line, E be the intersection of the vertical line and the base of the triangle, and F be the intersection of the vertical line and the bottom side of the square with point B.

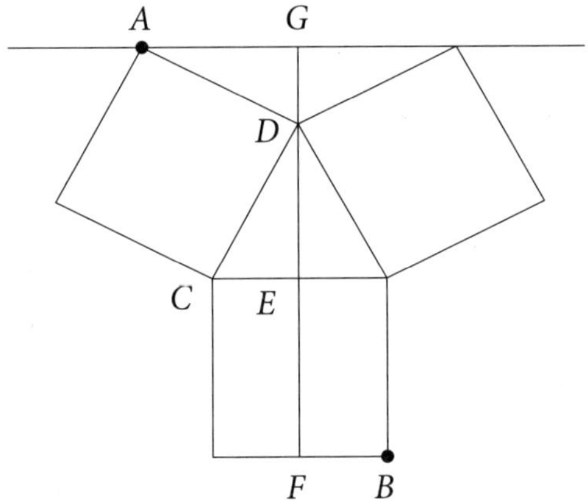

Then triangle AGD is a $30-60-90$ triangle congruent to half the equilateral triangle. It follows that both segments $\overline{AG}$ and $\overline{DE}$ are length $3\sqrt{3}$, the segments $\overline{GD}$ and $\overline{FB}$ are both length 3, and the segment $\overline{EF}$ is length 6. Then the horizontal distance between A and B is $3\sqrt{3}+3$, and the vertical distance is

$$3+3\sqrt{3}+6=9+2\sqrt{3}.$$

So by the Pythagorean Theorem, the distance from A to B is

$$\sqrt{(3+3\sqrt{3})^2+(9+3\sqrt{3})^2}=\sqrt{144+36\sqrt{3}}=6\sqrt{4+\sqrt{3}}.$$

Now suppose that $\sqrt{4+2\sqrt{3}}=m+\sqrt{n}$ for some integers m and n. Then squaring both sides yields

$$4+\sqrt{12}=m^2+n+2m\sqrt{n}=m^2+n+\sqrt{4m^2n}.$$

This has solution $m=1$ and $n=3$. Thus,

$$AB=6(1+\sqrt{3})=6+\sqrt{108}.$$

This gives the answer $108+6=114$.

Problem 11. Let k be the product of every third positive integer from 2 to 2006, that is

$$k = 2 \cdot 5 \cdot 8 \cdot 11 \cdot \dots \cdot 2006.$$

Find the number of trailing zeros that k has.

Answer: 168

The product includes many factors of 2 since alternate factors are all even. Thus the number of zeros at the right end of the k is equal to the number of factors of 5 which appear in the product. The factors of the product which contain at least one prime factor of five are $5, 20, 35, 50, \dots, 2000$. Dividing each of these factors by 5 gives the sequence

$$1, 4, 7, 10, 13, 16, 19, 22, 25, \dots, 400.$$

There are 134 numbers in this sequence accounting for 134 factors of 5 in k. In addition, the numbers $10, 25, 40, 55, \dots, 400$ contain an additional factor of 5. Dividing these numbers by 5 gives the sequence $2, 5, 8, 11, \dots, 80$ representing an additional 27 factors of five in k. But this sequence still contains multiples of 5, namely 5, 20, 35, 50, 65, 80. Each of these numbers contains exactly one factor of 5 except for 50 which contains two factors of 5. This gives an additional 7 factors of 5 for a total number of factors of 5 in k of $134 + 27 + 7 = 168$.

Problem 12. We draw a triangle inside of a circle with one vertex at the center of the circle and the other two vertices on the circumference of the circle. The angle at the center of the circle measures 75 degrees. We draw a second triangle, congruent to the first, also with one vertex at the center of the circle and the other vertices on the circumference of the circle rotated 75 degrees clockwise from the first triangle so that it shares a side with the first triangle. We draw a third, fourth, and fifth such triangle each rotated 75 degrees clockwise from the previous triangle. The base of the fifth triangle will intersect the base of the first triangle. What is the degree measure of the obtuse angle formed by the intersection?

Answer: 120

Let the center of the circle be C, the other vertices of the first triangle be A and B, and the other vertices of the fifth triangle be D and E as shown. Let F be the intersection of the bases of the two triangles. $5(75) = 375$, so $\angle ACE = 15°$.

$$\angle CAB = \angle CBA = \angle CED = \angle CDE = \frac{180 - 75}{2} = 52\frac{1}{2}.$$
$$\angle FCE = \frac{1}{2}\angle ACE = 7\frac{1}{2}.$$

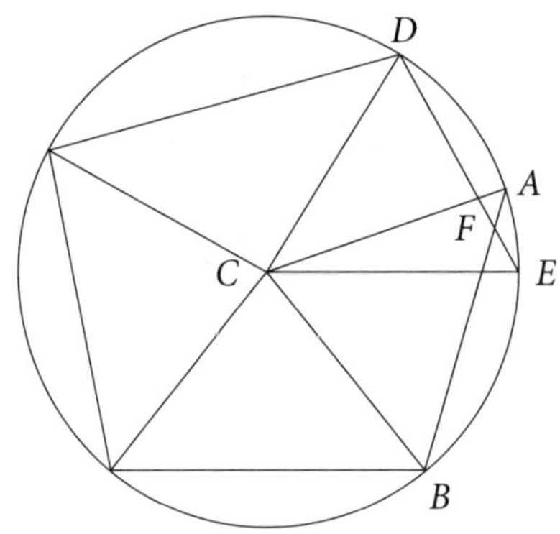

Then

$$\angle DCF = \angle DCE - \angle FCE = 67\frac{1}{2}.$$

It follows that the desired angle

$$\angle DFB = 2(\angle DFC) = 2(180 - \angle DCF - \angle CDE) = 120.$$

Problem 13. Twelve students need to form five study groups. They will form three study groups with two students each and two study groups with three students each. In how many ways can these groups be formed?

Answer: 138,600

To select students for their study groups you could

- choose which six students form the two groups of 3 ($\binom{12}{6}$ ways)

- select any one student of those six (for example, the shortest) (1 way)

- choose two students of the remaining five to study with that selected student ($\binom{5}{2}$ ways)

- select any one of six students to be in a group of two (1 way)

- choose one student to study with that selected student

(5 ways)

- choose one of the remaining four students to be in a group

(1 way)

- choose one of the remaining three students to study with that student (3 ways).

This determines the study groups, and the number of ways of doing this is

$$\begin{aligned}\binom{12}{6}\cdot 1\cdot\binom{5}{2}\cdot 1\cdot 5\cdot 1\cdot 3 &= \frac{12\cdot 11\cdot 10\cdot 9\cdot 8\cdot 7}{6\cdot 5\cdot 4\cdot 3\cdot 2\cdot 1}\cdot\frac{5\cdot 4}{2\cdot 1}\cdot 5\cdot 3\\ &= 11\cdot 10\cdot 9\cdot 8\cdot 7\cdot 5 = 138,600.\end{aligned}$$

Alternatively, number the groups from 1 to 5 with groups 1 and 2 having three students each. There are $\binom{12}{3}$ ways to select 3 students out of 12 students for group 1, $\binom{9}{3}$ ways to select 3 more students out of the remaining 9 students for group 2, $\binom{6}{2}$ ways to select 2 more students out of the remaining 6 students for group 3, $\binom{4}{2}$ ways to select 2 more students out of the remaining 4 students for group 4, and, $\binom{2}{2}$ ways to select 2 more students out of the remaining 2 students for group 5. Then the number of ways to make all of these designations is $\binom{12}{3}\binom{9}{3}\binom{6}{2}\binom{4}{2}\binom{2}{2}$ which is also the multinomial coefficient $\binom{12}{3\ \ 3\ \ 2\ \ 2\ \ 2}$. But selecting one group of students to be in group 1 and a second group of students to be in group 2 results in the same groupings as putting the first three students into group 2 and the second three students into group 1. Thus these groupings have been over-counted by a factor of 2. Similarly, each set of groupings for groups 3, 4, and 5 have been over-counted by a factor of $3! = 6$. Therefore the correct answer is

$$\frac{\binom{12}{3}\binom{9}{3}\binom{6}{2}\binom{4}{2}\binom{2}{2}}{2\cdot 6} = \frac{12!}{3!\cdot 3!\cdot 2!\cdot 2!\cdot 2!\cdot 2\cdot 6} = 138,600.$$

Problem 14. Consider all ordered pairs (m, n) of positive integers satisfying $59m - 68n = mn$. Find the sum of all the possible values of n in these ordered pairs.

Answer: 237

Suppose that $59m - 68n = mn$. Then $59m = (68 + m)n$. Since 59 is prime, it must either divide n or $68 + m$. If 59 divides n, then $n = 59k$ for some positive integer k. It would then follow that $59m = (68 + m)59k$ and $m = (68 + m)k$ which cannot happen for positive integers m and k. Thus 59 must divide $68 + m$. Then there is a positive integer k such that $68 + m = 59k$. Substituting this into the original equation gives $59\,(59k - 68) - 68n = (59k - 68)n$ which simplifies to $59k - 68 = kn$. This implies that $(59 - n)k = 68$. It follows that $59 - n$ must divide 68, and therefore must be one of 1, 2, 4, 17, 34, or 68. $59 - n$ cannot be 68, but the other values give solutions. The requested sum of possible values for n is

$$(59 - 1) + (59 - 2) + (59 - 4) + (59 - 17) + (59 - 34) = 237.$$

The ordered pairs (m, n) satisfying the equation are

$$(3944, 58),\ (1938, 57),\ (935, 55),\ (168, 42),\ \text{and } (50, 25).$$

Problem 15. A snowman is built on a level plane by placing a ball radius 6 on top of a ball radius 8 on top of a ball radius 10 as shown. If the average height above the plane of a point in the snowman is $\frac{m}{n}$, where m and n are relatively prime positive integers, find $m + n$.

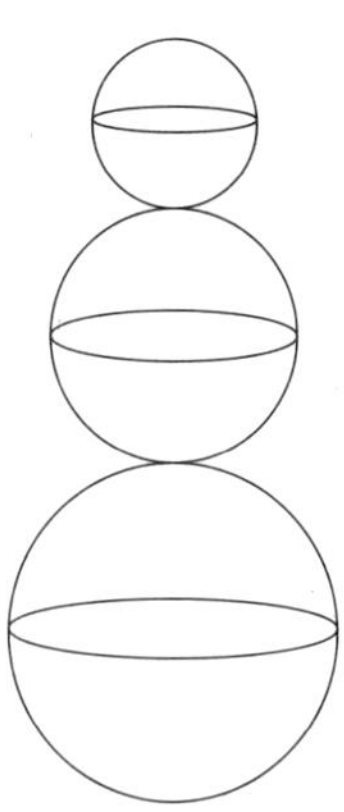

Answer: 61

The bottom ball has volume $\frac{4}{3}\pi 10^3$, and points in the bottom ball are an average of 10 units above the plane. The middle ball has volume $\frac{4}{3}\pi 8^3$, and points in the middle ball are an average of $2 \cdot 10 + 8 = 28$ units above the plane. The bottom ball has volume $\frac{4}{3}\pi 6^3$, and points in the bottom ball are an average of $2 \cdot 10 + 2 \cdot 8 + 6 = 42$ units above the plane. Then the average height of points above the plane is

$$\frac{\frac{4}{3}\pi 10^3 \cdot 10 + \frac{4}{3}\pi 8^3 \cdot 28 + \frac{4}{3}\pi 6^3 \cdot 42}{\frac{4}{3}\pi 10^3 + \frac{4}{3}\pi 8^3 + \frac{4}{3}\pi 6^3} = \frac{5^3 \cdot 10 + 4^3 \cdot 28 + 3^3 \cdot 42}{5^3 + 4^3 + 3^3} = \frac{4176}{216} = \frac{58}{3}.$$

This gives the answer $58 + 3 = 61$.

Problem 16. Let $f(x)$ and $g(x)$ be linear functions such that for all x

$$f(g(x)) = g(f(x)) = x.$$

If $f(0) = 4$ and $g(5) = 17$, compute $f(2006)$.

Answer: 122

Since $f(x)$ and $g(x)$ are linear, let

$$f(x) = ax + b, \text{ and } g(x) = cx + d.$$

Since $f(g(x)) = x$ holds for all x, we have that $a(cx + d) + b = x$ for all x. Then $acx + (ad + b) = x$ for all x. So $ac = 1$ and $ad + b = 0$. Then $c = \frac{1}{a}$ and $d = -\frac{b}{a}$. Thus $g(x) = \frac{x - b}{a}$. $f(0) = 4$ implies that $b = 4$. Then $g(5) = 17$ implies that $17 = \frac{5 - 4}{a}$ and $a = \frac{1}{17}$. Finally

$$f(x) = \frac{x}{17} + 4, \quad \text{so} \quad f(2006) = \frac{2006}{17} + 4 = 122.$$

Problem 17. A concrete sewer pipe fitting is shaped like a cylinder with diameter 48 with a cone on top. A cylindrical hole of diameter 30 is bored all the way through the center of the fitting as shown. The cylindrical portion has height 60 while the conical top portion has height 20. Find n such that the volume of the concrete is $n\pi$.

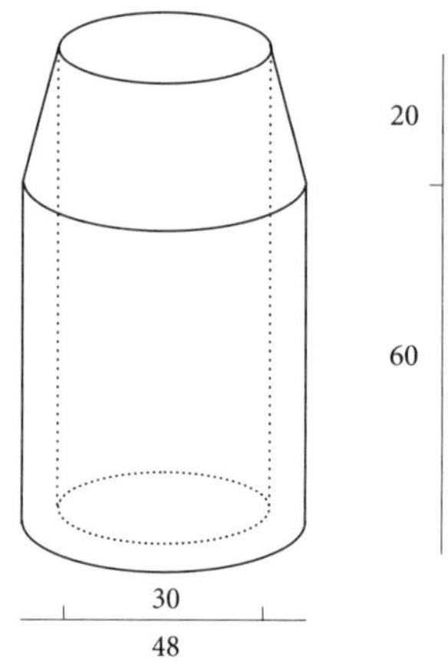

Answer: 24,300

The conical portion of the pipe fitting drops from a radius of 24 to a radius of 15 over its height of 20, or a decrease of 9 over a height of 20. If the cone is completed so that it comes to a point, its radius would need to drop an additional 15, or 5/3 as much as the original 9. So the completed cone would add an additional $20 \cdot (5/3) = 100/3$ in height. If the fitting were solid with no hole, its volume would be the volume of the cylinder radius 24 and height 60, plus the volume of the cone radius 24 and height $(20 + 100/3)$, minus the volume of the cone radius 15 and height 100/3. This gives a volume of

$$\begin{aligned}\pi \cdot 24^2 \cdot 60 + \frac{\pi \cdot 24^2 \cdot \left(20 + \dfrac{100}{3}\right)}{3} - \frac{\pi \cdot 15^2 \cdot \dfrac{100}{3}}{3} \\ = \pi[24^2 \cdot 60 + 8^2 \cdot 160 - 5^2 \cdot 100] \\ = 42,300\pi.\end{aligned}$$

The hole in the fitting is a cylinder with radius 15 and height 80, so its volume is

$$\pi 225 \cdot 80 = 18,000\pi.$$

This gives a final volume of the fitting of

$$(42,300 - 18,000)\pi = 24,300\pi.$$

Problem 18. In how many ways can 100 be written as the sum of three positive integers x, y, z satisfying $x < y < z$?

Answer: 784

Let $a = x - 1$, $b = y - x - 1$, and $c = z - y - 1$. Then a, b, and c can be any non-negative integers satisfying

$$(a+1) + (a+b+2) + (c+b+a+3) = x + y + z = 100.$$

Thus

$$3a + 2b + c = 94.$$

Since there is a one-to-one relationship between the solutions in x, y, z and the solutions in a, b, c, one only needs to count the solutions in a, b, c. Clearly, a can be any integer between 0 and 31 inclusive. For each value of a, the condition requires that $2b + c = 94 - 3a$. There are $\left\lfloor \frac{94-3a}{2} + 1 \right\rfloor$ possible solutions for b and c for a given value of a. So the total number of solutions is

$$\begin{aligned}\sum_{a=0}^{31} \left\lfloor \frac{94-3a}{2} + 1 \right\rfloor &= \left(\sum_{a=0}^{31} \frac{94-3a}{2} + 1 \right) - \frac{32}{2} \cdot \frac{1}{2} \\ &= \sum_{a=0}^{31} 48 - \frac{3}{2} \sum_{a=0}^{31} a - 8 \\ &= 32 \cdot 48 - \frac{3}{2} \cdot \frac{31 \cdot 32}{2} - 8 \\ &= 784.\end{aligned}$$

Problem 19. There is a popular race course where runners frequently go for a daily run. Assume that all runners randomly select a start time, a starting position on the course, and a direction to run. Also assume that all runners make exactly one complete circuit of the race course, all runners run at the same speed, and all runners complete the circuit in one hour. Suppose that one afternoon you go for a run on this race course, and you count 300 runners who pass you going in the opposite direction, although some of those runners you count twice since they pass you twice. What is the expected value of the number of different runners that you pass not counting duplicates?

Answer: 225

Consider a graph of possible starting times and starting locations of people running the opposite direction that you run. Measure time so that time zero is the time you begin running, and measure starting locations as distance along your route. The shaded region in the following graph shows those start times and locations for runners that you will see at least one time.

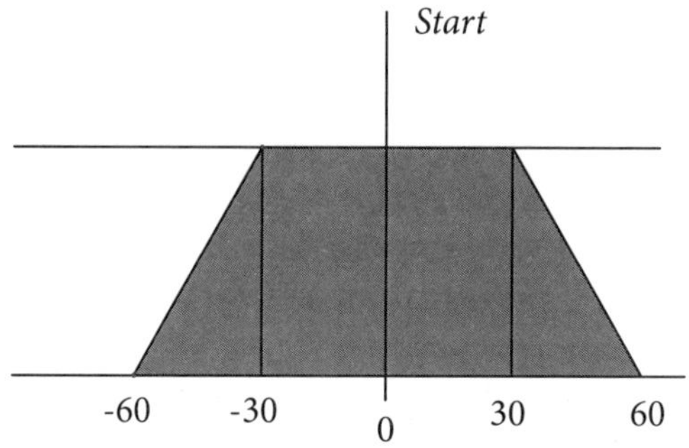

This reflects that fact that if a runner begins running within 30 minutes of your start time, then you will pass this runner no matter where the runner starts. When the runner's start time is more than 30 minutes away from your start time, then you will pass the runner only when the runner's start location is close to your start location. The shaded region in the following graph shows the times and locations of the runners who you will pass twice.

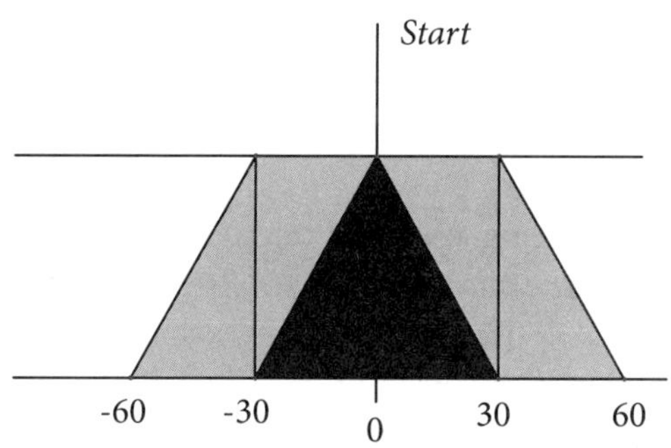

Each starting location and time indicated by a point on this graph is equally likely to occur. This shows that of the runners you pass, the probability that you will pass the runner twice is one third. Therefore, it is expected that one third of the different runners you pass, you will pass twice. On average you pass $4/3$ as many runners as the number of different runners you pass. So, if you pass 300 runners, the expected number of different runners is $300(3/4) = 225$.

Problem 20. Find the sum of all positive integers having at most three, not necessarily distinct, prime factors where the primes are from the set $\{2, 3, 5, 7\}$.

Answer: 1932

The integers in question will have either 0, 1, 2, or 3 prime factors. The sum of the numbers with 0 or 1 prime factor is $1+2+3+5+7 = 18$. The sum of the numbers with two prime factors is given by

$$\frac{(2+3+5+7)^2 + (2^2+3^2+5^2+7^2)}{2} = \frac{17^2+87}{2} = 188.$$

The sum of the numbers with three prime factors is given by

$$\begin{aligned}&(2^2+3^2+5^2+7^2)(2+3+5+7)\\ &\quad + (2\cdot3\cdot5 + 2\cdot3\cdot7 + 2\cdot5\cdot7 + 3\cdot5\cdot7)\\ &\quad = (87)(17) + (30+42+70+105) = 1726.\end{aligned}$$

This gives a final answer of $18 + 188 + 1726 = 1932$.

Problem 21. In triangle ABC, $AB = 52$, $BC = 56$, $CA = 60$. Let D be the foot of the altitude from A, and let E be the intersection of the internal angle bisector of $\angle BAC$ with $\overline{BC}$. Find DE.

Answer: 6

Let the altitude $\overline{AD}$ be h, $\overline{BD}$ be x, and $\overline{EC}$ be y. Then

$$h^2 + x^2 = 52^2 \quad \text{and} \quad h^2 + (56-x)^2 = 60^2.$$

Subtracting these two gives

$$60^2 - 52^2 = 56^2 - 2(56)x$$

which yields $x = 20$ and $h = 48$. The angle bisector theorem says that

$$\frac{y}{60} = \frac{56-y}{52}$$

which yields $y = 30$.

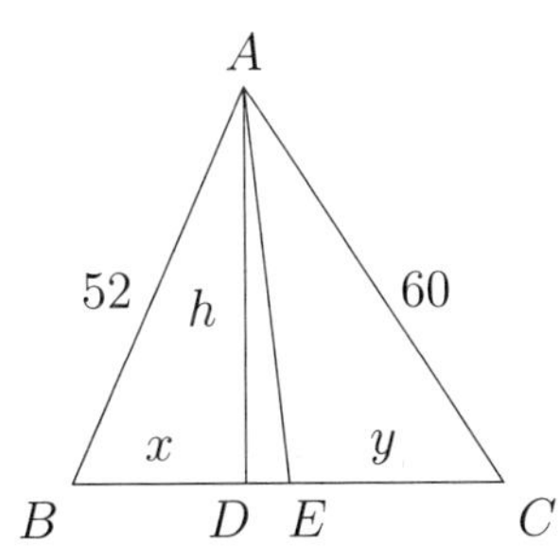

It follows that $DE = 56 - 20 - 30 = 6$.

Problem 22. Let $F_0 = 0$, $F_1 = 1$, and for $n \geq 1$, $F_{n+1} = F_n + F_{n-1}$. Define

$$a_n = \left(\frac{1+\sqrt{5}}{2}\right)^n F_n.$$

Then there are rational numbers A and B such that

$$\frac{a_{30} + a_{29}}{a_{26} + a_{25}} = A + B\sqrt{5}.$$

Find $A + B$.

Answer: 34

Let $\gamma = \dfrac{1+\sqrt{5}}{2}$. Then for any $n > 0$

$$a_{n+1} = \gamma^{n+1} F_{n+1} = \gamma^{n+1}(F_n + F_{n-1}) = \gamma a_n + \gamma^2 a_{n-1}.$$

From this we get

$$a_{n+1} + a_n = (\gamma + 1)a_n + \gamma^2 a_{n-1}.$$

Note that

$$\gamma^2 = \left(\frac{1+\sqrt{5}}{2}\right)^2 = \frac{3+\sqrt{5}}{2} = 1 + \gamma,$$

so

$$a_{n+1} = (\gamma + 1)[a_n + a_{n-1}].$$

It follows that

$$\begin{aligned}
\frac{a_{30} + a_{29}}{a_{26} + a_{25}} &= (\gamma + 1)^4 = \left(\frac{3+\sqrt{5}}{2}\right)^4 \\
&= \left(\frac{9 + 6\sqrt{5} + 5}{4}\right)^2 = \left(\frac{7 + 3\sqrt{5}}{2}\right)^2 \\
&= \left(\frac{49 + 42\sqrt{5} + 45}{4}\right) = \frac{47 + 21\sqrt{5}}{2}.
\end{aligned}$$

Therefore $A + B = 34$.

Problem 23. We have two positive integers both less than 1000. The arithmetic mean and the geometric mean of these numbers are consecutive odd integers. Find the maximum possible value of the difference of the two integers we have.

Answer: 120

Let the two numbers be m and n with $m > n$. Then, since the geometric mean of two different positive numbers is less than their arithmetic mean,

$$\frac{m+n}{2} = \sqrt{mn} + 2.$$

Then

$$\begin{aligned}
&m + n - 4 = 2\sqrt{mn} \\
&m^2 + n^2 + 16 + 2mn - 8m - 8n = 4mn \\
&(m-n)^2 = 8(m+n-2)
\end{aligned}$$

or equivalently

$$\begin{aligned}
&(m+n-4)^2 = 4mn \\
&(m-n)^2 + 16 - 8(m+n) = 0 \\
&m - n = 2\sqrt{2m+2n-4}.
\end{aligned}$$

Clearly, $2m + 2n - 4$ must be an even perfect square less than 4000. If $2m+2n-4 = 62^2$, then $m+n = 1924$ and $m-n = 124$ implying that $m = 1024$ which is too large. But if $2m+2n-4 = 60^2$, then $m+n = 1802$ and $m-n = 120$ implying that $m = 961$ and $n = 841$ which satisfies all the necessary conditions.

Alternatively, from $\dfrac{m+n}{2} = \sqrt{mn} + 2$, calculate that

$$\begin{array}{lll}
m+n = 2\sqrt{mn}+4 & m - 2\sqrt{mn} + n = 4 & (\sqrt{m}-\sqrt{n})^2 = 4 \\
\sqrt{m} - \sqrt{n} = 2 & \sqrt{m} + \sqrt{n} = \dfrac{m-n}{\sqrt{m}-\sqrt{n}} = \dfrac{m-n}{2}. &
\end{array}$$

It follows that both $\sqrt{m} - \sqrt{n}$ and $\sqrt{m} + \sqrt{n}$ are rational numbers. Thus both $\sqrt{m}$ and $\sqrt{n}$ are rational numbers and are, therefore, integers. Conclude that m and n are squares of two integers which differ by two. The largest difference between m and n for m and n less than 1000 occurs when $m = 31^2$ and $n = 29^2$ giving $m - n = 961 - 841 = 120$.

Problem 24. A semicircle with diameter of length 16 contains a circle of radius 3 tangent both to the inside of the semicircle and its diameter as shown. A second larger circle is tangent to the inside of the semicircle, the outside of the first circle, and the diameter of the semicircle. The diameter of the second circle can be written as $\frac{n+k\sqrt{2}}{m}$, where m, n, k are positive integers and m and n have no factors in common. Find $m+n+k$.

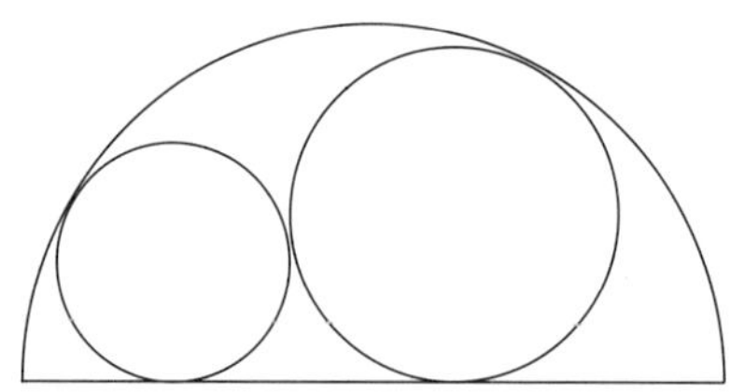

Answer: 361

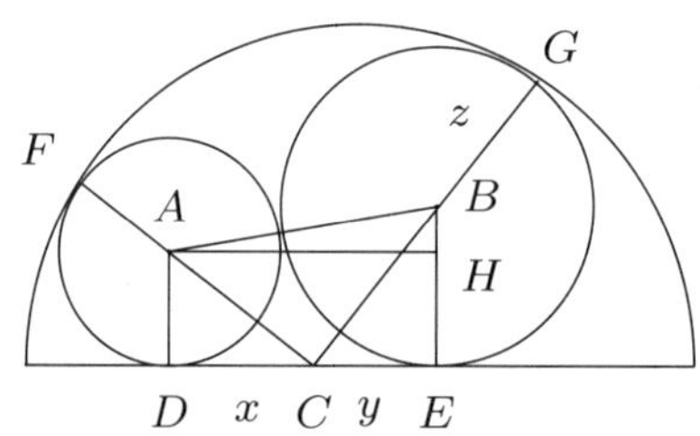

Let A be the center of the circle radius 3, B be the center of the second circle, and C be the center of the semicircle as shown. Let D and E be the points where the circles are tangent to the diameter of the semicircle, and let F and G be the points where the circles are tangent to the circumference of the semicircle. Locate point H on $\overline{BE}$ so that $\overline{AH}$ is parallel to the diameter of the semicircle. Let $\overline{CD}$ have length x, $\overline{CE}$ have length y, and the second circle have radius z. Then $\overline{AC}$ has length $\sqrt{3^2+x^2}$. Since $\overline{FC}$ is a radius of the semicircle, it follows that

$$3+\sqrt{3^2+x^2}=8.$$

Solving this for x shows that $x=4$. Similarly, since $\overline{CG}$ is a radius of the semicircle, it follows that $z+\sqrt{y^2+z^2}=8$. Solving for the radical and squaring yields $y^2=64-16z$ and $y=4\sqrt{4-z}$. Since $\overline{AH}$ has length $x+y$, $\overline{HB}$ has length $z-3$, and $\overline{AB}$ has length $z+3$, it follows that

$$\sqrt{(4+y)^2+(z-3)^2}=z+3.$$

Squaring yields

$$(4 + y)^2 = 12z.$$

Substituting for y gives

$$(4 + 4\sqrt{4 - z})^2 = 12z$$

implying

$$16 + 32\sqrt{4 - z} + 16(4 - z) = 12z$$

and

$$8\sqrt{4 - z} = 7z - 20.$$

Again, squaring and simplifying yields

$$0 = 49z^2 - 216z + 144.$$

This gives

$$z = \frac{108 \pm \sqrt{108^2 - 49 \cdot 144}}{49} = \frac{108 \pm 12\sqrt{9^2 - 49}}{49} = \frac{108 \pm 48\sqrt{2}}{49}.$$

The requested diameter is then gives $\frac{216 + 96\sqrt{2}}{49}$ and the desired answer is $216 + 98 + 49 = 361$. The extraneous solution $\frac{108 - 48\sqrt{2}}{49}$ gives the radius of the small circle shown here.

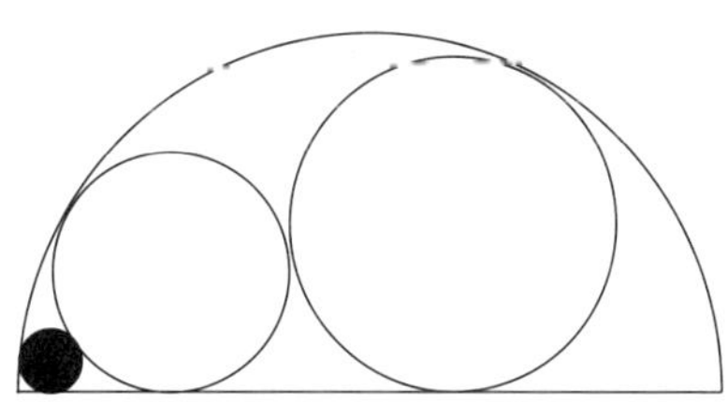

Problem 25. Let x and y be real numbers such that

$$2 \sin x \sin y + 3 \cos y + 6 \cos x \sin y = 7.$$

Find $\tan^2 x + 2 \tan^2 y$.

Answer: 9

Notice that and that $2^2 + 3^2 + 6^2 = 7^2$ and that

$$(\sin x \sin y)^2 + (\cos y)^2 + (\cos x \sin y)^2 = 1^2.$$

The Cauchy-Schwarz Inequality states that

$$\left|\sum a_j b_j\right|^2 \le \sum a_j^2 \sum b_j^2$$

with equality only when there is a constant c such that $a_j = cb_j$ for each j. Equivalently, this states that the absolute value of the dot product of two vectors is less than or equal to the product of the lengths of the two vectors with equality only when the two vectors point in the same direction. The equation in the problem states that if

$$A = \langle a_1, a_2, a_3 \rangle = \langle 2, 3, 6 \rangle$$

and

$$B = \langle b_1, b_2, b_3 \rangle = \langle \sin x \sin y, \cos y, \cos x \sin y \rangle,$$

then

$$A \cdot B = |A| \cdot |B|.$$

It follows that the vector A is a multiple of the vector B, so that

$$\frac{\sin x \sin y}{2} = \frac{\cos y}{3} = \frac{\cos x \sin y}{6}$$
$$\frac{\sin x \sin y}{2} = \frac{\cos x \sin y}{6}$$

which implies that

$$\tan x = \frac{1}{3}$$

which means that

$$\sin x = \frac{\pm 1}{\sqrt{10}} \cdot \frac{\sin x \sin y}{2} = \frac{\cos y}{3}$$

and implies that

$$\tan y = \frac{2}{3 \sin x} = \frac{\pm 2\sqrt{10}}{3}.$$

Thus

$$\tan^2 x + 2\tan^2 y = \left(\frac{1}{3}\right)^2 + 2\left(\frac{2\sqrt{10}}{3}\right)^2 = \frac{1}{9} + \frac{80}{9} = 9.$$

2.7 2006 Middle School

Problem 1. Michael is celebrating his fifteenth birthday today. How many Sundays have there been in his lifetime?

[This problem was originally stated for the contest held on Wednesday and Thursday, April 26 & 27, 2006.]

Answer: 783

Michael was born in $2006 - 15 = 1991$. Each year since then has had 52 full weeks plus one extra day except for the four leap years (1992, 1996, 2000, 2004) which had 2 extra days. So, there have been $52(7)(15) + 15 + 4 + 1 = 52(7)(15) + 20$ days in Michael's lifetime. This gives $52(15) + 2$ full weeks plus 6 extra days which accounts for 783 Sundays.

Problem 2. Evaluate $\dfrac{\dfrac{1}{\frac{1}{10}-\frac{1}{12}}}{\dfrac{1}{\frac{1}{8}-\frac{1}{6}}+\dfrac{1}{\frac{1}{5}-\frac{1}{6}}}$.

Answer: 10

$$\frac{\dfrac{1}{\frac{1}{10}-\frac{1}{12}}}{\dfrac{1}{\frac{1}{8}-\frac{1}{6}}+\dfrac{1}{\frac{1}{5}-\frac{1}{6}}} = \frac{\dfrac{1}{\frac{6}{60}-\frac{5}{60}}}{\dfrac{1}{\frac{3}{24}-\frac{4}{24}}+\dfrac{1}{\frac{6}{30}-\frac{5}{30}}} = \frac{\dfrac{1}{\frac{1}{60}}}{\dfrac{1}{-\frac{1}{24}}+\dfrac{1}{\frac{1}{30}}}$$
$$= \frac{60}{6} = 10.$$

Problem 3. Find the sum of all prime numbers less than 100 that are one more than a multiple of six.

Answer: 527

The numbers which are one more than a multiple of six are 1, 7, 13, 19, 25, 31, 37, 43, 49, 55, 61, 67, 73, 79, 85, 91, 97. The sum of the prime numbers is

$$7 + 13 + 19 + 31 + 37 + 43 + 61 + 67 + 73 + 79 + 97 = 527.$$

Problem 4. At the beginning of each hour from 1 o'clock AM to 12 NOON and from 1 o'clock PM to 12 MIDNIGHT a coo-coo clock's coo-coo bird coo-coos the number of times equal to the number of the hour. In addition, the coo-coo clock's coo-coo bird coo-coos a single time at 30 minutes past each hour. How many times does the coo-coo bird coo-coo from 12:42 PM on Monday until 3:42 AM on Wednesday?

Answer: 279

In one twelve hour period, the coo-coo coo-coos $1+2+3+4+5+\ldots+12$ times on the hour and a total of 12 times 30 minutes past each hour for a total of $\frac{12 \cdot 13}{2} + 12 = 90$ times. So for the three twelve hour periods from 12:42 PM on Monday until 12:42 AM on Wednesday, the coocoo coo-coos 270 times. Then it will coo-coo 1 time at 1:00, 1 time at 1:30, 2 times at 2:00, 1 time at 2:30, 3 times at 3:00, and 1 time at 3:30 for a total of 9 more times. This gives a final answer of 279.

Problem 5. The sizes of the freshmen class and the sophomore class are in the ratio 5:4. The sizes of the sophomore class and the junior class are in the ratio 7:8. The sizes of the junior class and the senior class are in the ratio 9:7. If these four classes together have a total of 2158 students, how many of the students are freshmen?

Answer: 630

This is the same as High School problem number 1.

Problem 6. We draw a radius of a circle. We draw a second radius 23 degrees clockwise from the first radius. We draw a third radius 23 degrees clockwise from the second. This continues until we have drawn 40 radii each 23 degrees clockwise from the one before it. What is the measure in degrees of the smallest angle between any two of these 40 radii?

Answer: 7

15 times 23 is 345, so the 16th radius drawn is 345 degrees from the first. Thus, the 17th radius drawn is $345 + 23 - 360 = 8$ degrees from the first radius drawn. The following 15 radii drawn will also be 8 degrees from another radius drawn. Similarly, the 33rd radius drawn will be 8 degrees from the 17th radius and, therefore, 16 degrees from the first radius. This places it 7 degrees from the second radius drawn. The remaining seven radii are similarly spaced 7 degrees from the nearest drawn radii.

Problem 7. At a movie theater tickets for adults cost 4 dollars more than tickets for children. One afternoon the theater sold 100 more child tickets than adult tickets for a total sales amount of 1475 dollars. How much money would the theater have taken in if the same tickets were sold, but the costs of the child tickets and adult tickets were reversed?

Answer: 1875

This is the same as High School problem number 2.

Problem 8. A rogue spaceship escapes, and 54 minutes later the police leave in a spaceship in hot pursuit. If the police spaceship travels 12% faster than the rogue spaceship along the same route, how many minutes will it take for the police to catch up with the rogues?

Answer: 450

This is the same as High School problem number 4.

Problem 9. Moving horizontally and vertically from point to point along the lines in the diagram below, how many routes are there from point A to point B consisting of six horizontal moves and six vertical moves?

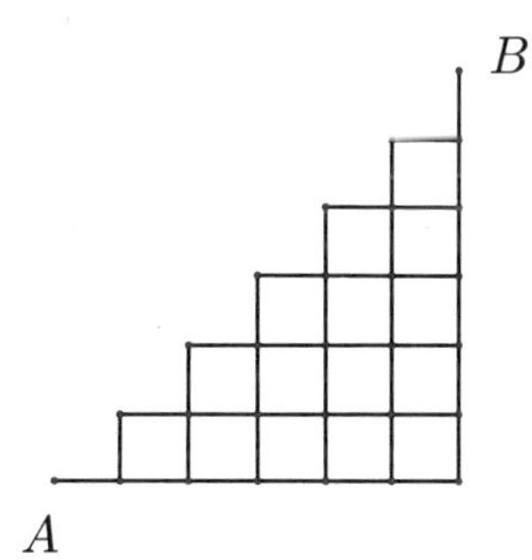

Answer: 132

Label point A and the other points on the bottom row with the number 1. For each other point, label the point with a number which is the sum of the numbers on the labels of the points immediately to the left and immediately below the point since this will give a count of the number of paths from A to that point.

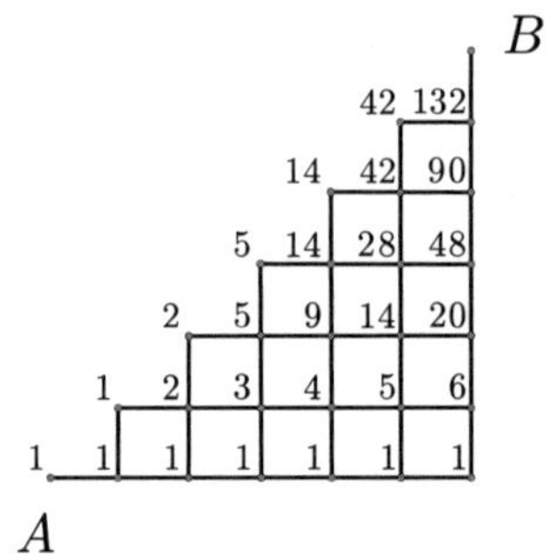

Continuing this until point B is labeled, gives the number of paths from A to B which is 132. The answer is also given by the Catalan number

$$C_6 = \frac{1}{6+1}\binom{2\cdot 6}{6} = \frac{1}{7}\binom{12}{6} = 132.$$

Problem 10. How many rectangles are there in the diagram below such that the sum of the numbers within the rectangle is a multiple of 7?

43	44	45	46	47	48	49
36	37	38	39	40	41	42
29	30	31	32	33	34	35
22	23	24	25	26	27	28
15	16	17	18	19	20	21
8	9	10	11	12	13	14
1	2	3	4	5	6	7

Answer: 163

This is the same as High School problem number 9.

Problem 11. Consider the polynomials

$$P(x) = (x+\sqrt{2})(x^2-2x+2)$$

$$Q(x) = (x-\sqrt{2})(x^2+2x+2)$$

$$R(x) = (x^2+2)(x^8+16).$$

Find the coefficient of x^4 in $P(x)\cdot Q(x)\cdot R(x)$.

Answer: 0

$$\begin{aligned}P(x)\cdot Q(x)\cdot R(x) &= (x-\sqrt{2})(x+\sqrt{2})(x^2+2)[(x^2+2)-2x]\\&\quad\cdot[(x^2+2)+2x](x^8+16)\\&=(x^2-2)(x^2+2)[(x^2+2)^2-4x^2](x^8+16)\\&=(x^4-4)(x^4+4)(x^8+16)\\&=(x^8-16)(x^8+16)=x^{16}-256\end{aligned}$$

which contains no x^4 term.

Problem 12. How many positive integers divide the number

$$10! = 1\times 2\times 3\times 4\times 5\times 6\times 7\times 8\times 9\times 10?$$

Answer: 270

$$10! = 10\cdot 9\cdot 8\cdot 7\cdot 6\cdot 5\cdot 4\cdot 3\cdot 2\cdot 1 = 2^8\cdot 3^4\cdot 5^2\cdot 7^1.$$

Thus a number divides 10! if and only if its prime factorization has no more than eight factors of 2, four factors of 3, two factors of five, and one factor of 7. Therefore to get a number that divides 10!, select from 0 to 8 factors of two $(8+1$ ways), select from 0 to 4 factors of 3 $(3+1$ ways), select 0 to 2 factors of 5 $(2+1$ ways), and select 0 or one factor or 7 $(1+1$ ways) and take the product of all the selected factors. Thus the number of ways to select such a positive integer divisor of 10! is given by

$$(8+1)\cdot(4+1)\cdot(2+1)\cdot(1+1)=270.$$

Problem 13. An equilateral triangle with side length 6 has a square of side length 6 attached to each of its edges as shown. The distance between the two farthest vertices of this figure (marked A and B in the figure) can be written as $m+\sqrt{n}$ where m and n are positive integers. Find $m+n$.

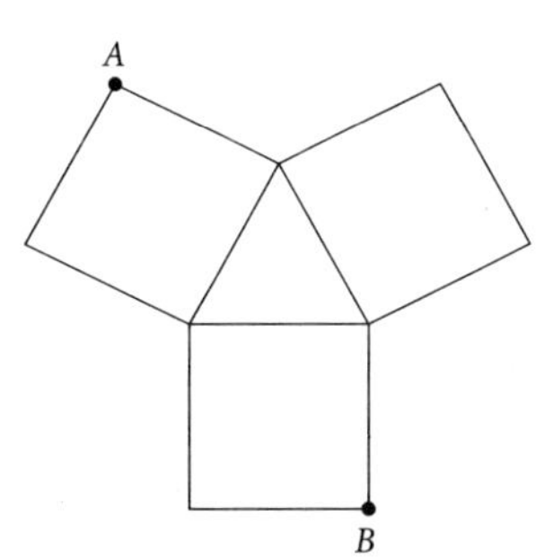

Answer: 114

This is the same as High School problem number 10.

Problem 14. The rodent control task force went into the woods one day and caught 200 rabbits and 18 squirrels. The next day they went into the woods and caught three fewer rabbits and two more squirrels than the day before. Each day they went into the woods and caught three fewer rabbits and two more squirrels than the day before. This continued through the day when they caught more squirrels than rabbits. Up through that day how many rabbits did they catch in all?

Answer: 5491

The number of rabbits caught on the nth day is $203 - 3n$. The number of squirrels caught on the nth day is $16 + 2n$. The task force kept on catching rodents up through the first day when $203 - 3n < 16 + 2n$ or $187 < 5n$. This happens when $n = 38$. The number of rabbits caught is then the sum of the arithmetic series

$$(203 - 3) + (203 - 6) + (203 - 9) + \ldots + (203 - 38 \cdot 3)$$

$$= 38 \cdot \frac{(203 - 3) + (203 - 38 \cdot 3)}{2} = 19(200 + 89) = 5491.$$

Alternatively,

$$(203 - 3) + (203 - 6) + (203 - 9) + \ldots + (203 - 38 \cdot 3)$$

$$= 203 \cdot 38 - (3 + 6 + 9 + \ldots + 38 \cdot 3) = 203 \cdot 38 - 3 \cdot (1 + 2 + 3 + \ldots + 38)$$

$$= 7714 - 3 \cdot \frac{38 \cdot 39}{2} = 5491.$$

Problem 15. A concrete sewer pipe fitting is shaped like a cylinder with diameter 48 with a cone on top. A cylindrical hole of diameter 30 is bored all the way through the center of the fitting as shown. The cylindrical portion has height 60, while the conical top portion has height 20. Find n such that the volume of the concrete is $n\pi$.

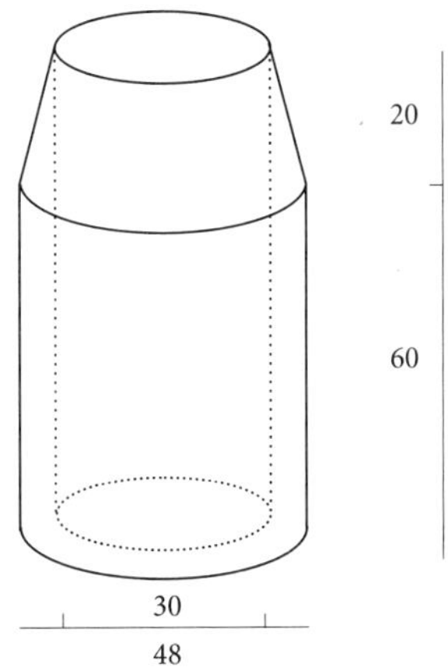

Answer: 24,300

This is the same as High School problem number 17.

2.8 2007 High School

Problem 1. The sum of nine consecutive odd integers is 2007. Find the greatest of these nine integers.

Answer: 231

Suppose that the first of the sequence of numbers is n. Then the greatest of these numbers is $n + 16$. The arithmetic mean of these two values is $\dfrac{n + (n + 16)}{2} = n + 8$ so the sum of the nine numbers must be 9 times this mean or $9n + 72 = 2007$. Solving for n gives $n = 215$, so the greatest of the nine numbers is $215 + 16 = 231$.

Problem 2. A positive number $\dfrac{m}{n}$ has the property that it is equal to the ratio of 7 plus the number's reciprocal to 65 minus the number's reciprocal. Given that m and n are relatively prime positive integers, find $2m + n$.

Answer: 7

The conditions state that

$$x = \frac{7 + \dfrac{1}{x}}{65 - \dfrac{1}{x}} = \frac{7x + 1}{65x - 1}.$$

Thus, which

$$x(65x - 1) = 7x + 1$$

which reduces to

$$65x^2 - 8x - 1 = 0.$$

Since

$$65x^2 - 8x - 1 = (13x + 1)(5x - 1),$$

there are two possible values for x, $-\dfrac{1}{13}$ and $\dfrac{1}{5}$. Thus, x is $\dfrac{1}{5}$, and

$$2m + n = 2 \cdot 1 + 5 = 7.$$

Problem 3. A bowl contained 10% blue candies and 25% red candies. A bag containing three quarters red candies and one quarter blue candies was added to the bowl. Now the bowl has 16% blue candies. What percentage of the candies in the bowl are now red?

Answer: 45

Suppose that the bowl originally had x candies. Then it use to contain $.10x$ blue candies and $.25x$ red candies. If y candies were added, then $.25y$ blue candies were added making the proportion of blue candies in the bowl

$$\frac{.10x + .25y}{x+y} = .16.$$

Thus $.10x + .25y = .16x + .16y$ and $6x = 9y$ or $2x = 3y$. The proportion of red candies in the bowl is now

$$\frac{.25x + .75y}{x+y} = .25\frac{x+3y}{x+y} = .25\frac{3x+9y}{3x+3y} = .25\frac{3x+6x}{3x+2x} = .45$$

and the answer is 45%.

Problem 4. To the nearest degree, find the measure of the largest angle in a triangle with side lengths 3, 5, and 7.

Answer: 120

The Law of Cosines states that if a triangle has sides with lengths a, b, and c and angle θ between the sides with lengths a and b, then

$$a^2 + b^2 - 2ab\cos\theta = c^2.$$

Applying this to the triangle with sides $a = 3$, $b = 5$, and $c = 7$ to get

$$3^2 + 5^2 - 2 \cdot 3 \cdot 5 \cdot \cos\theta = 7^2$$

which shows

$$\cos\theta = \frac{7^2 - 3^2 - 5^2}{-2 \cdot 3 \cdot 5} = -\frac{15}{30} = -\frac{1}{2}.$$

This means that the angle θ is 120 degrees.

Problem 5. Let $F(0) = 3$ and $F(n) = F(n-1) + 4$ when n is positive. Find $F(F(F(5)))$.

Answer: 383

Since $F(n) - F(n-1) = 4$, $F(n)$ increases by 4 whenever n increases by 1, so it follows that $F(n) = 3 + 4n$. Thus,

$$F(5) = 3 + 4 \cdot 5 = 23, \quad F(23) = 3 + 4 \cdot 23 = 95,$$

and

$$F(95) = 3 + 4 \cdot 95 = 383.$$

Problem 6. Find the sum of all positive integers that are divisors of either 96 or 180.

Answer: 770

$96 = 2^5 \cdot 3^1$ and $180 = 2^2 \cdot 3^2 \cdot 5^1$ so the greatest common divisor of 96 and 180 is $12 = 2^2 \cdot 3^1$. The sum of the divisors of 96 is

$$(2^0 + 2^1 + 2^2 + 2^3 + 2^4 + 2^5)(3^0 + 3^1) = 63 \cdot 4 = 252.$$

The sum of the divisors of 180 is

$$(2^0 + 2^1 + 2^2)(3^0 + 3^1 + 3^2)(5^0 + 5^1) = 7 \cdot 13 \cdot 6 = 546.$$

The sum of the number which divide both 96 and 180 is the sum of the numbers which divide 12 which is

$$(2^0 + 2^1 + 2^2)(3^0 + 3^1) = 7 \cdot 4 = 28.$$

So, the required sum is $252 + 546 - 28 = 770$.

Problem 7. If x is a real number, what is the greatest value that can be attained by the function

$$25 \sin 4x - 60 \cos 4x?$$

Answer: 65

For each value of x, there is an angle θ so that

$$\begin{aligned} 25 \sin 4x - 60 \cos 4x &= \sqrt{25^2 + 60^2} \left[\frac{25 \sin 4x - 60 \cos 4x}{\sqrt{25^2 + 60^2}} \right] \\ &= 65 \left[\frac{25}{65} \sin 4x - \frac{60}{65} \cos 4x \right] \\ &= 65[\cos \theta \sin 4x - \sin \theta \cos 4x] \\ &= 65 \sin(4x - \theta). \end{aligned}$$

This has a maximum value when the sine function is equal to 1, and the maximum is 65.

Problem 8. You know that the Jones family has five children, and the Smith family has three children. Of the eight children you know that there are five girls and three boys. Let $\frac{m}{n}$ be the probability that at least one of the families has only girls for children. Given that m and n are relatively prime positive integers, find $m + n$.

Answer: 67

Since there are eight children with three of them in the Smith family, there are $\binom{8}{3} = 56$ equally likely collections of children for the Smith family. There are $\binom{5}{3} = 10$ ways for the Smith family to have three girls. There is only one way for the Smith family to have three boys so that the Jones children are all girls. Thus, there are $10 + 1 = 11$ ways for one family to have only girls for children. This gives a probability of $\frac{11}{56}$ so that

$$m + n = 11 + 56 = 67.$$

Problem 9. Sets A, B, C, D each have 400 elements. The intersection of any two of the sets has 115 elements. The intersection of any three of the sets has 53 elements. The intersection of all four sets has 28 elements. How many elements are there in the union of the four sets?

Answer: 1094

The inclusion-exclusion principal says that the number of elements in a union of four sets is the sum of the numbers of elements in each of the four sets minus the numbers of elements in each of the intersections of each pair of sets plus the numbers of elements in each of the intersections of each trio of sets minus the number of elements in the intersection of all four sets. Since there are $\binom{4}{2} = 6$ pairs of sets and $\binom{4}{3} = 4$ trios of sets, the required number of elements in the union of these four sets is

$$4 \cdot 400 - 6 \cdot 115 + 4 \cdot 53 - 28 = 1094.$$

Problem 10. For each value of the angle θ we take the product of the two complex numbers

$$(8 + i)\sin\theta + (7 + 4i)\cos\theta \quad \text{and} \quad (1 + 8i)\sin\theta + (4 + 7i)\cos\theta$$

to get a complex number in the form $a + bi$, where a and b are real numbers. Find the greatest possible value for $a + b$.

Answer: 125

Note that

$$(8 + i)(1 + 8i) = (7 + 4i)(4 + 7i) = 65i$$

and

$$(8+i)(4+7i)+(7+4i)(1+8i)=120i.$$

The desired product is then

$$\begin{aligned}&[(8+i)\sin\theta+(7+4i)\cos\theta][(1+8i)\sin\theta+(4+7i)\cos\theta]\\&=65i\cdot\sin^2\theta+65i\cdot\cos^2\theta+120i\cdot\sin\theta\cdot\cos\theta\\&=65i+60i\cdot\sin(2\theta)\\&=i\cdot[65+60\cdot\sin(2\theta)].\end{aligned}$$

So for every value of θ, $a=0$ and b has a maximum value when $\sin(2\theta)=1$. This gives a maximum for $a+b=65+60=125$.

Problem 11. A dart board consists of three concentric circles with radii of 4, 6, and 8. Three darts are thrown at the board so that they stick at three random locations on the board. The probability that one dart sticks in each of the three regions of the dart board is $\frac{m}{n}$, where m and n are relatively prime positive integers. Find $m+n$.

Answer: 617

Since the darts land randomly, the probability that a dart hits one of the region is proportional to area of that region. The area of the entire dart board is $8^2\pi=64\pi$. The inside region has area $4^2\pi=16\pi$, the middle region has area $(6^2-4^2)\pi=20\pi$, and the outer region has area $(8^2-6^2)\pi=28\pi$. Since the three darts can land in the three regions in $3!=6$ orders, the probability is

$$6\cdot\frac{16\pi}{64\pi}\cdot\frac{20\pi}{64\pi}\cdot\frac{28\pi}{64\pi}=\frac{105}{512}$$

giving a final answer of $105+512=617$.

Problem 12. Find the maximum possible value of

$$8\cdot 27^{\log_6 x}+27\cdot 8^{\log_6 x}-x^3$$

as x varies over the positive real numbers.

Answer: 216

Note that

$$8\cdot 27^{\log_6 x}+27\cdot 8^{\log_6 x}-x^3=8\cdot 27-[8-8^{\log_6 x}]\cdot[27-27^{\log_6 x}].$$

The product of differences is the product of two positive numbers when $\log_6 x<1$, the product of two negative numbers when

$\log_6 x > 1$, or the product of two zeros when $\log_6 x = 1$. The expression is, therefore, at a maximum when $\log_6 x = 1$, and the maximum is $8 \cdot 27 = 216$.

Problem 13. Find the circumradius of the triangle with side lengths 104, 112, and 120.

Answer: 65

Since all the sides of the triangle are multiples of 8, it is sufficient to find the circumradius of a triangle with side lengths 13, 14, and 15, and then multiple the answer by 8. The semiperimeter of this triangle is

$$s = \frac{13 + 14 + 15}{2} = 21.$$

Then a well known formula for circumradius gives

$$R = \frac{abc}{4\sqrt{s(a+b-s)(a+c-s)(b+c-s)}} = \frac{13(14)15}{4\sqrt{21(6)(7)(8)}}$$

$$= \frac{13(14)15}{4(4)(3)(7)} = \frac{65}{8}.$$

This gives a final answer of 65.

Problem 14. Square $ABCD$ has side length 1000. Squares with side lengths 400 are cut from each of the corners C and D as shown. If m and n are relatively prime positive integers such that the centroid (the center of mass or balance point) of the resulting figure is at P at a distance $\frac{m}{n}$ from side $\overline{AB}$, find $m + n$.

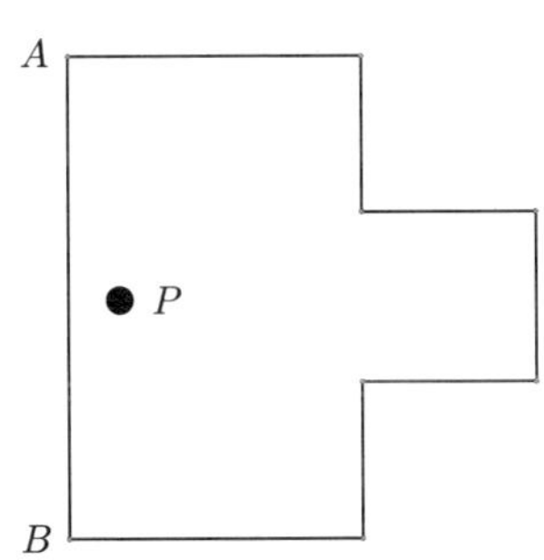

Answer: 6117

For purposes of balance, objects act as though their entire mass (area in the case of plane figures) were concentrated at their centroid (center of mass). We know that the centroid of the entire square $ABCD$ is a point a distance 500 from side $\overline{AB}$. Similarly, the two squares cut from the corners each have centroids a distance 800 from side $\overline{AB}$. This means that

$$\frac{m}{n}[1000^2 - 2(400^2)] + 800[2(400^2)] = 500[1000^2].$$

This yields

$$\frac{m}{n} = \frac{500(100) - 800(32)}{100 - 32} = \frac{50000 - 25600}{68} = \frac{24400}{68} = \frac{6100}{17}$$

giving a final answer of $6100 + 17 = 6117$.

Problem 15. The alphabet in its natural order

$$ABCDEFGHIJKLMNOPQRSTUVWXYZ$$

is T_0. We apply a permutation to T_0 to get T_1 which is

$$JQOWIPANTZRCVMYEGSHUFDKBLX.$$

If we apply the same permutation to T_1, we get T_2 which is

$$ZGYKTEJMUXSODVLIAHNFPWRQCB.$$

We continually apply this permutation to each T_m to get T_{m+1}. Find the least positive integer n so that $T_n = T_0$.

Answer: 252

The permutation sends

$$A \to J \to Z \to X \to B \to Q \to G \to A$$

which is a cycle of length 7

$$C \to O \to Y \to L \to C$$

which is a cycle of length 4

$$D \to W \to K \to R \to S \to H \to N \to M \to V \to D$$

which is a cycle of length 9

$$E \to I \to T \to U \to F \to P \to E$$

which is a cycle of length 6.

If we repeatedly apply the permutation, these cycles return their letters to the starting position every 7, 4, 9, and 6 repetitions, respectively. Therefore, the entire permutation returns to the starting position when the number of applications of the permutations is a multiple of 7, 4, 9, and 6. This first happens at the least common multiple of these four numbers which is $4 \cdot 7 \cdot 9 = 252$.

Problem 16. We have some identical paper squares which are black on one side of the sheet and white on the other side. We can join nine squares together to make a 3 by 3 sheet of squares by placing each of the nine squares either white side up or black side up. Two of these 3 by 3 sheets are distinguishable if neither can be made to look like the other by rotating the sheet or by turning it over. How many distinguishable 3 by 3 squares can we form?

Answer: 70

First consider how many distinguishable patterns can appear on one side of a 3 by 3 sheet. There are three cases to consider.

Case 1. Patterns that look like themselves when rotated 90 degrees.

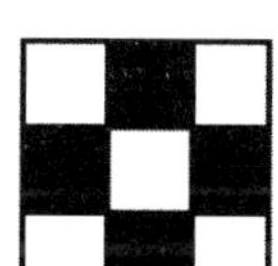

In these patterns the four edge squares are all the same color, and the four corner squares are all the same color. Since there are 2 choices of color for the center square, 2 choices of color for the edge squares, and 2 choices of color for the corner squares, there are $2 \times 2 \times 2 = 8$ such patterns.

Case 2. Patterns that look like themselves when rotated 180 degrees but not when rotated 90 degrees.

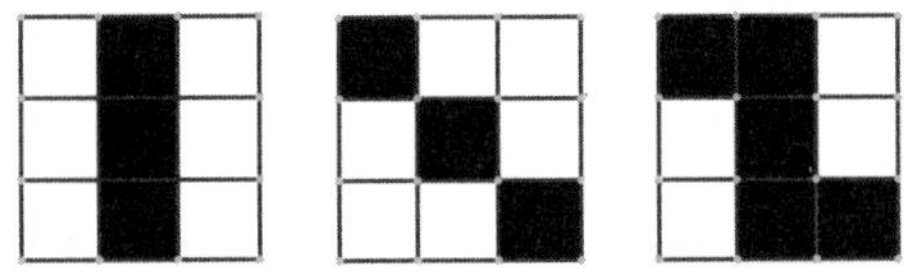

In these patterns opposite corner squares are the same color, and opposite edge squares are the same color, but either the 2 pair of opposite edge colors differ, or the 2 pair of opposite corner colors differ, or both.

If all the corner colors are the same, there are 2 choices of color for the center square and 2 choices of color for the corner squares giving $2 \times 2 = 4$ patterns. Similarly, there are 4 patterns if all the edge colors are the same. If both the edges and corners use 2 colors, then there are 2 choices of color for the center square and 2 orientations of black squares on the edges and corners (whether the corner black squares are adjacent to the edge black squares clockwise or counterclockwise) for $2 \times 2 = 4$ patterns. This gives a total of 12 patterns.

Case 3. Patterns that look different under all rotations.

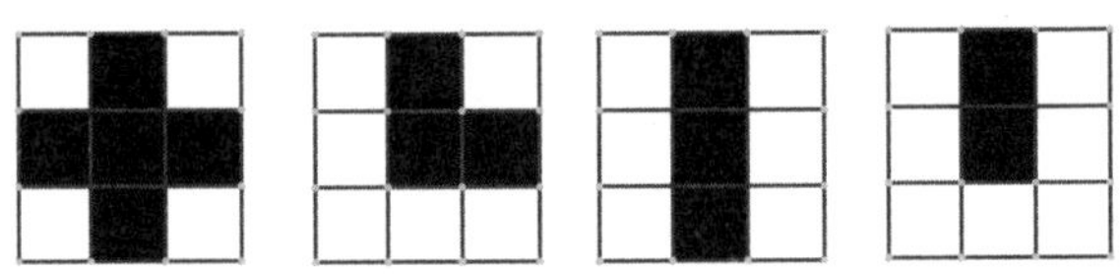

Consider the 4 subcases of A) all edge squares have the same color, B) 2 edge squares of each color where the 2 opposite edges are different, C) 2 edge squares of each color where the 2 opposite edges are the same, D) 3 edge squares of the same color.

In case A, at least one pair of opposite corners must be different in color, so there are 2 choices of color for the center square, 2 choices of color for the edge squares, and 3 possible patterns for the corners (1 white, 2 white, or 3 white). This gives $2 \times 2 \times 3 = 12$ patterns. In case B, all rotations are distinguishable, so there are 2 choices of color for the center square and 2 choices of color for each of the 4 corner squares. This gives $2^5 = 32$ patterns. In case C, at least one pair of opposite corners must be different in color, so there are 2 choices of color for the center square and 6 choices of colors for the corner squares (2 each with 1 white, 2 whites, and 3 whites). This gives $2 \times 6 = 12$ patterns. Finally, in case D, all rotations are distinguishable, so there are 2 choices of color for the center square, 2 choices of color for the edge squares (1 white or 3 white), and 2 choices of color for each of the 4 corner squares. This gives $2^6 = 64$ patterns.

The total number of distinguishable patterns for case 3 is, therefore, $12+32+12+64 = 120$. The total number of distinguishable patterns on one side of a sheet is given by the sum of the totals for cases 1, 2, and 3 which is $8 + 12 + 120 = 140$. If any one of these 3 by 3 sheets is turned over, the reverse side will be a different pattern from the first side since the numbers of white and black squares on the reverse side must be different from the numbers of white and black squares on the first side. This means that each of the 3 by 3 sheets has two distinct patterns, one on each side. It follows that the number of distinguishable 3 by 3 sheets is only half the number of distinguishable patterns on one side of a sheet. This gives the final answer of 70.

Note that the number of patterns on one side of a sheet can be counted very quickly using the method of Polya counting. In that method one would construct the polynomial

$$F(x, y) = \frac{(x+y)^9 + 2(x^4+y^4)^2(x+y) + (x^2+y^2)^4(x+y)}{4}.$$

Then the number of distinguishable patterns is given by $F(1, 1) = 140$.

Problem 17. A circle with diameter 20 has points A, B, C, D, E, and F equally spaced along its circumference. A second circle is tangent to the lines $\overline{AB}$ and $\overline{AF}$ and internally tangent to the first circle. If the second circle has diameter $\sqrt{m}+n$ for some integers m and n, find $m+n$.

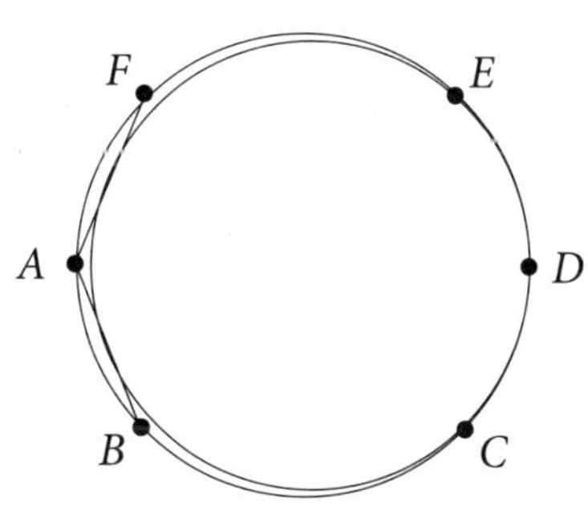

Answer: 19080

Let P be the center of the smaller circle with radius r. P lies on the diameter $\overline{AD}$ of the larger circle. Let G be the point on the line $\overline{AF}$ where the smaller circle is tangent to line $\overline{AF}$. Then triangle AGP has a right angle at G and a 60 degree angle at A. Thus, the segment $\overline{AP}$ has length $20 - r$ and we have

$$\frac{r}{20-r} = \sin(60) = \frac{\sqrt{3}}{2}.$$

Solving for r gives

$$2r = (20-r)\sqrt{3} = 20\sqrt{3} - r\sqrt{3}$$

so

$$r = \frac{20\sqrt{3}}{2+\sqrt{3}} = \frac{20\sqrt{3}}{2+\sqrt{3}} \cdot \frac{2-\sqrt{3}}{2-\sqrt{3}} = \frac{40\sqrt{3}-60}{4-3} = 40\sqrt{3} - 60.$$

The smaller circle has diameter $80\sqrt{3} - 120 = \sqrt{19200} - 120$ giving a final answer to the problem $19200 - 120 = 19{,}080$. The smaller circle's diameter is approximately 18.564.

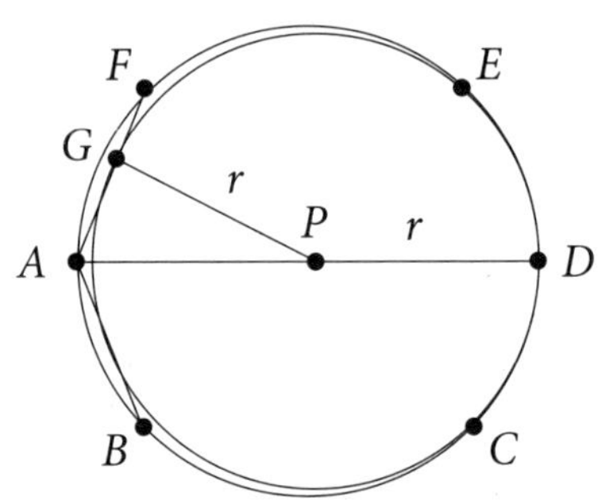

Problem 18. Let S be the graph of $y = x^3$, and T be the graph of $y = \sqrt[3]{x}$. Let S^* be S rotated around the origin 15 degrees clockwise, and T^* be T rotated around the origin 45 degrees counterclockwise. Curves S^* and T^* intersect at a point in the first quadrant a distance $m + \sqrt{n}$ from the origin, where m and n are positive integers. Find $m + n$.

Answer: 16

Since we are only interested in how far the intersection of S^* and T^* is from the origin, we can assume that $S^* = S$ and that T^*

is T rotated 60 degrees counterclockwise. S and T are graphs of functions which are inverses of each other, so S and T are reflections of each other across the line $x = y$. When the rotations are performed, S and T^* will be reflections of each other across the line $x = y$ rotated $\frac{60}{2} = 30$ degrees counterclockwise. The line $x = y$ makes a 45 degree angle with the positive x axis, so after it is rotated 30 degrees, it will make a 75 degree angle with the positive x axis. The rotated line will have equation

$$y = (\tan 75)x$$

$$\tan 75 = \tan(45 + 30) = \frac{\tan 45 + \tan 30}{1 - (\tan 45)(\tan 30)}$$

$$= \frac{1 + \frac{1}{\sqrt{3}}}{1 - \frac{1}{\sqrt{3}}} = \frac{1 + \frac{1}{\sqrt{3}}}{1 - \frac{1}{\sqrt{3}}} \cdot \frac{1 + \frac{1}{\sqrt{3}}}{1 + \frac{1}{\sqrt{3}}} = 2 + \sqrt{3}$$

so the equation of the line is

$$y = (2 + \sqrt{3})x.$$

This line intersects S exactly where T^* intersect S or at a value x such that

$$x^3 = (2 + \sqrt{3})x.$$

It follows that

$$x^2 = 2 + \sqrt{3} \quad \text{and} \quad y^2 = (2 + \sqrt{3})^2 x^2 = (2 + \sqrt{3})^3 = 26 + 15\sqrt{3}.$$

Then

$$(m + \sqrt{n})^2 = x^2 + y^2 = 28 + 16\sqrt{3}.$$

This implies that $m^2 + n = 28$ and $2m\sqrt{n} = 16\sqrt{3}$. The only integer solutions for m and n are $m = 4$ and $n = 12$ giving a final answer of $4 + 12 = 16$.

Problem 19. Six chairs sit in a row. Six people randomly seat themselves in the chairs. Each person randomly chooses either to set their feet on the floor, to cross their legs to the right, or to cross their legs to the left. There is a problem only if two people sitting next to each other have the person on the right crossing their legs to the left and the person on the left crossing their legs to the right. The probability that this will **not** happen is given by $\frac{m}{n}$, where m and n are relatively prime positive integers. Find $m + n$.

Answer: 1106

Let the six people select their chairs. Then each person randomly selects one of three alternative ways to position their legs, so there are $3^6 = 729$ equally likely ways to decide. Let the chairs be numbered from 1 to 6 starting at the left most chair. Let the set A_j be the set of ways that people can cross their legs so that the person in chair j has their legs crossed to the right and person in chair $j+1$ has their legs crossed to the left. Then the inclusion-exclusion principle says that the number of ways two people can sit next to each other while bumping feet is

$$\begin{aligned}&\#(A_1 \cup A_2 \cup A_3 \cup A_4 \cup A_5 \cup A_6)\\&= \sum_{j=1}^{6} \#(A_j) - \sum_{i<j} \#(A_i \cap A_j) + \sum_{i<j<k} \#(A_i \cap A_j \cap A_k).\end{aligned}$$

$\#(A_j) = 3^4$ for $j < 6$ and 0 when $j = 6$, $\#(A_i \cap A_j) = 3^2$ when $i+1 < j$ (that is, six times when (i,j) is $(1,3)$, $(1,4)$, $(1,5)$, $(2,4)$, $(2,5)$, and $(3,5)$) and 0 otherwise, and $\#(A_i \cap A_j \cap A_k) = 1$ when $i = 1$, $j = 3$, and $k = 5$ and 0 otherwise. It follows that

$$\begin{aligned}&\sum_{j=1}^{6} \#(A_j) - \sum_{i<j} \#(A_i \cap A_j) + \sum_{i<j<k} \#(A_i \cap A_j \cap A_k)\\&= 5 \cdot 3^4 - 6 \cdot 3^2 + 1 = 352.\end{aligned}$$

Thus, the probability that no two people will bump feet is

$$\frac{3^6 - 352}{3^6} = \frac{377}{729}.$$

The requested answer is $377 + 729 = 1106$.

Problem 20. Three congruent ellipses are mutually tangent. Their major axes are parallel. Two of the ellipses are tangent at the end points of their minor axes as shown. The distance between the centers of these two ellipses is 4. The distances from those two centers to the center of the third ellipse are both 14. There are positive integers m and n so that the area between the three ellipses is $\sqrt{n} - m\pi$. Find $m + n$.

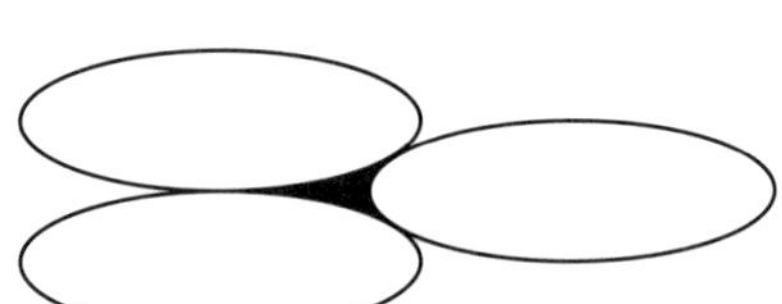

Answer: 776

Let A, B, and C be the centers of the ellipses as shown. The height of the triangle from base AB to vertex C is given by

$$\sqrt{14^2 - 2^2} = \sqrt{192} = 8\sqrt{3}.$$

The height of an equilateral triangle with side length 4 is $2\sqrt{3}$. The problem simplifies if we transform the plane by contracting it in the horizontal direction by a factor of 4. This is equivalent to contracting the coordinate plane so that every point (x, y) gets mapped to the point $\left(\frac{x}{4}, y\right)$. If this transformation is performed, the triangle ABC will be contracted to an equilateral triangle with side length 4, and the three ellipses will be transformed into three new ellipses. Since the distance between the centers of the ellipses will now all be equal to 4, the ellipses will all be circles giving the second drawing below. The area bounded by these circles is the area of the equilateral triangle minus one sixth of the areas of each of the circles. This gives an area of

$$4\sqrt{3} - 3\frac{4\pi}{6} = 4\sqrt{3} - 2\pi.$$

This area is one quarter the area bounded by the original three ellipses, so the original area is

$$16\sqrt{3} - 8\pi = \sqrt{768} - 8\pi.$$

Thus, the requested answer is $768 + 8 = 776$.

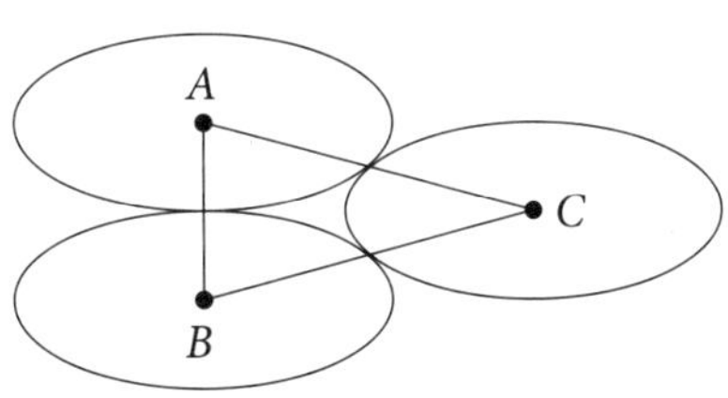

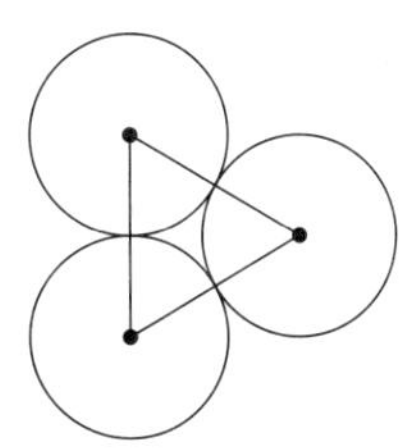

Problem 21. What is the greatest positive integer m such that

$$n^2(1 + n^2 - n^4) \equiv 1 \pmod{2^m}$$

for all odd integers n?

Answer: 7

The question is equivalent to finding the largest power of 2 which divides $n^6 - n^4 - n^2 + 1$ for every odd integer n. This expression factors as $(n^2+1)(n-1)^2(n+1)^2$. When n is odd, both $n+1$ and $n-1$ are divisible by 2 and exactly one of them is divisible by 4. Therefore, $(n-1)(n+1)$ is always divisible by 8, and $(n-1)^2(n+1)^2$ is always divisible by 64. When n is odd, n^2+1 is divisible by 2. It follows that $(n^2+1)(n-1)^2(n+1)^2$ is always divisible by $64 \cdot 2 = 2^7$ for every odd number n. On the other hand, when $n = 3$, $n^6 - n^4 - n^2 + 1$ is 640 which is not divisible by 2^8. Thus, the correct answer is 7.

Problem 22. Let $a = 3^{1/223}$ and for all $n \geq 3$ let

$$f(n) = \binom{n}{0} a^{n-1} - \binom{n}{1} a^{n-2} + \binom{n}{2} a^{n-3} - \cdots + (-1)^{n-1} \binom{n}{n-1} a^0.$$

Find $f(2007) + f(2008)$.

Answer: 19683

By the Binomial Theorem

$$(a-1)^n = \binom{n}{0} a^n - \binom{n}{1} a^{n-1} + \binom{n}{2} a^{n-2} - \ldots - \binom{n}{n} a^0,$$

so it follows that

$$f(n) = \frac{(a-1)^n - (-1)^n}{a} = \frac{3^{n/223} - (-1)^n}{a}.$$

Thus,

$$f(2007) + f(2008) = \frac{3^{2007/223} + 3^{2008/223}}{a} = 3^{2007/223} \frac{1 + 3^{1/223}}{a}$$

$$= 3^{2007/223} = 3^9 = 19683.$$

Problem 23. Two circles with radius 2 and radius 4 have a common center at P. Points A, B, and C on the larger circle are the vertices of an equilateral triangle. Point D is the intersection of the smaller circle and the line segment $\overline{PB}$. Find the square of the area of triangle ADC.

Answer: 192

Angle APC is $\frac{360°}{3} = 120°$, so the Law of Cosines gives the length AC to be

$$\sqrt{4^2 + 4^2 - 2 \cdot 4 \cdot 4 \cdot \cos(120)} = \sqrt{4^2 + 4^2 + 4 \cdot 4} = 4\sqrt{3}.$$

The Law of Cosines applied to triangle APD gives the length of $\overline{AD}$ to be

$$\sqrt{4^2 + 2^2 - 2 \cdot 4 \cdot 2 \cdot \cos(120)} = \sqrt{4^2 + 2^2 + 4 \cdot 2} = 2\sqrt{7}.$$

By symmetry the length of $\overline{DC}$ is also $2\sqrt{7}$. By Hero's formula, the square of the area of a triangle with sides $2\sqrt{7}$, $2\sqrt{7}$ and $4\sqrt{3}$ is

$$[2(\sqrt{7} + \sqrt{3})][2\sqrt{3}][2\sqrt{3}][2(\sqrt{7} - \sqrt{3})] = 16 \cdot 3 \cdot 4 = 192.$$

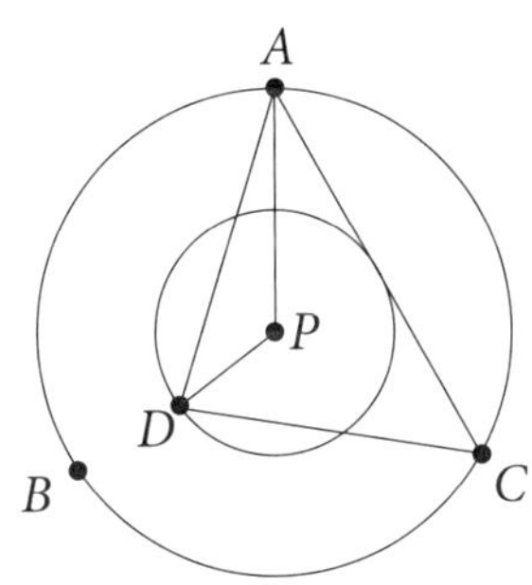

Alternatively, as above the length of side $\overline{AC}$ is $4\sqrt{3}$. Thus, the height of equilateral triangle ABC from B to side $\overline{AC}$ is $AC\frac{\sqrt{3}}{2} = 6$. Since P is the centroid of triangle ABC, and the centroid of an equilateral triangle is one third of the way from the center of a side to the opposite vertex, the point P is a distance 2 from side $\overline{AC}$. Since P lies on the altitude to vertex D in triangle ADC, it follows that the height of triangle ADC to vertex D is $2 + 2 = 4$. The area of triangle ADC is then one half its base times its height or

$$\frac{4\sqrt{3} \cdot 4}{2} = 8\sqrt{3} = \sqrt{192}.$$

Problem 24. Starting with a sequence of n 1's, you can insert plus signs to get various sums. For example, when $n = 10$, you can get the sum

$1+1+1+11+11+111 = 136$ and the sum $1+1+11+111+111 = 235$. Find the number of values of n so that the sum of 1111 is possible.

Answer: 121

First note that if you can write 1111 using a sequence of n 1's, then n must be congruent to 4 mod 9. Indeed, if

$$1111 = a \cdot 1 + b \cdot 11 + c \cdot 111 + d \cdot 1111$$

when $n = a + 2 \cdot b + 3 \cdot c + 4 \cdot d$, then subtracting these two equations yields

$$1111 - n = b \cdot 9 + c \cdot 108 + d \cdot 1107$$

which is divisible by 9. Clearly, 1111 can be written with a sequence of 4 ones. The next shortest sequence of ones uses ten copies of 111 and one single 1 to get

$$1111 = 1 + 111 + 111 + \ldots + 111$$

using a sequence of 31 ones. The next shortest sequence of ones is obtained by replacing one of the copies of 111 with ten copies of 11 plus one 1 to get

$$1111 = 1 + 1 + 11 + 11 + \ldots + 11 + 111 + 111 + \ldots + 111$$

using a sequence of 49 ones.

It is now claimed that we can write 1111 using a sequence of n ones for every n congruent to 4 mod 9 where $49 \le n \le 1111$. To see this, begin with the representation of 1111 using 31 ones which adds one 1 and ten copies of 111. By successively replacing one copy of 111 with one 1 and ten copies of 11, we get representations of 1111 using $31 + 18k$ ones for all k between 1 and 10. Each of these representation includes at least ten copies of the number 11. To get representations of 1111 using $31 + 9p$ for odd values of p, take the representation using $31 + 18(p-1)$ ones and replace one of the copies of 11 with eleven ones. Note that our representation of $31 + 18 \cdot 10 = 211$ uses 100 copies of 11 and 11 copies of 1. By successively replacing one copy of 11 with eleven ones, we get representations using n ones for every n congruent to 4 mod 9 between 211 and 1111. This proves the claim.

Thus, the answer is the number of numbers n congruent to 4 mod 9 between 49 and 1111 plus 2 (for 4 and 31) which is

$$2 + \frac{1111 - 49}{9} + 1 = 3 + \frac{1062}{9} = 121.$$

Problem 25. Let x be a positive integer less than 200 and let y be obtained by writing the base 10 digits of x in reverse order. Given that x and y satisfy $11x^2 + 363y = 7xy + 6571$, find x.

Answer: 133

Reducing the given equation modulo 11 gives that $0 \equiv 7xy + 4 \pmod{11}$ which is equivalent to $xy \equiv 1 \pmod{11}$. If x were a two-digit number, then $x + y$ would be a multiple of 11, and y would be equivalent to $-x$. That would imply that $x^2 \equiv -1 \pmod{11}$ which has no solutions. Thus, x must have either one digit or three digits. If x has just one digit, $x = y$ and $11x^2 + 363x = 7x^2 + 6571$, so $4x^2 + 363x = 6571$. This cannot be satisfied by any one digit number x, for then $4x^2 + 363x < 4 \cdot 10^2 + 363 \cdot 10 = 4030 < 6571$. It follows that x must be a three-digit number. This implies that x and y are equivalent modulo 11, and $0 \equiv 7xy + 4 \pmod{11}$ implies that $7x^2 + 4 \equiv 0 \pmod{11}$ or that $x^2 \equiv 1 \pmod{11}$. Thus, x must be equivalent to either 1 or 10 modulo 11.

Reducing the given equation modulo 9 gives that $2x^2 + 3y \equiv 7xy+1 \pmod 9$. Since x and y have the same three digits, x and y must be equivalent modulo 9, so $2x^2 + 3x \equiv 7x^2 + 1 \pmod 9$ and $5x^2 - 3x + 1 \equiv 0 \pmod 9$. Multiplying by 2 shows that this is equivalent to $x^2 - 6x + 2 \equiv 0 \pmod 9$ and to $x^2 + 3x + 2 \equiv (x + 1)(x + 2) \equiv 0 \pmod 9$. Since $x + 1$ and $x + 2$ cannot both be multiples of 3, it follows that either $x + 1$ or $x + 2$ must be a multiple of 9, or x is equivalent to either 8 or 9 modulo 9.

Reducing the given equation modulo 5 gives that $x^2 + 3y \equiv 2xy + 1 \pmod 5$. Since x is in the range from 100 to 199, y must be equivalent to 1 modulo 5, so $x^2 + 3 \equiv 2x + 1 \pmod 5$ and $x^2 - 2x + 2 \equiv 0 \pmod 5$. This is equivalent to $x^2 + 3x + 2 \equiv (x+1)(x+2) \equiv 0 \pmod 5$. It follows that x must be equivalent to either 3 or 4 modulo 5.

Knowing that x is either 1 or 10 modulo 11 and either 8 or 9 modulo 9 allows us to find possible values for x modulo $11 \times 9 = 99$. Those values are 34, 43, 89, and 98. The possible values of x in the range from 100 to 199 are then $34 + 99 = 133$,

$43 + 99 = 142$, $89 + 99 = 188$, and $98 + 99 = 197$. Only 133 and 188 are equivalent to either 3 or 4 modulo 5. If $x = 188$, then

$$11x^2 + 363y - 7xy < 11 \cdot 200^2 + 400 \cdot 900 - 7 \cdot 150 \cdot 800$$

$$= 440,000 + 360,000 - 840,000 = -40,000 < 6571.$$

It follows that x can only be 133 which does satisfy all the conditions.

2.9 2007 Middle School

Problem 1. Last Sunday at noon the date on the calendar was 15 (April 15, 2007). What will be the date on the calendar one million minutes after that time?

Answer: 9

Each hour contains 60 minutes, and each day contains 24 hours. So each day contains $60 \cdot 24 = 1440$ minutes. One million divided by 1440 is $694\frac{4}{9}$. Since $24 \cdot \frac{4}{9} < 24 \cdot \frac{1}{2} = 12$, one million minutes after noon on April 15 is on the day 694 after April 15. Since 2008 is a leap year, April 15, 2009 is $366 + 365 = 731$ days after April 15, 2007. 694 days after April 15, 2007 must be $731 - 694 = 37$ days before April 15. Since March has 31 days, 37 days before April 15, 2009 will be March 9, 2009.

Problem 2. How many numbers n have the property that both $\frac{n}{2}$ and $2n$ are four-digit whole numbers?

Answer: 1500

For both $\frac{n}{2}$ and $2n$ to be four digit numbers, they must satisfy $\frac{n}{2} \geq 1000$ and $2n < 10,000$. This means $2000 \leq n < 5000$. There are exactly 3000 whole numbers satisfying this. Exactly half of them are even whole numbers. Since $\frac{n}{2}$ must be a whole number, n must be even. It follows that there are $3000/2 = 1500$ possible values for n.

Problem 3. Square $ABCD$ has side length 36. Point E is on side $\overline{AB}$ a distance 12 from B, point F is the midpoint of side $\overline{BC}$, and point G is on side $\overline{CD}$ a distance 12 from C. Find the area of the region that lies inside triangle EFG and outside triangle AFD.

Answer: 144

Let points H and J be the intersections of the sides of triangle AFD with the line $\overline{EG}$ as shown. Since the length EB is one-third the length AB, the height of triangle HFJ is one-third the height of triangle AFD. Since the triangles HFJ and AFD are similar, it follows that the length of $\overline{HJ}$ is one-third of the length of $\overline{AD}$. Thus, triangle HFJ has area $\frac{1}{2} \cdot \frac{36}{3} \cdot 12 = 72$. Triangle EFG has area $\frac{1}{2} \cdot 36 \cdot 12 = 216$. The area of the region in question is given by the difference between these two areas or $216 - 72 = 144$.

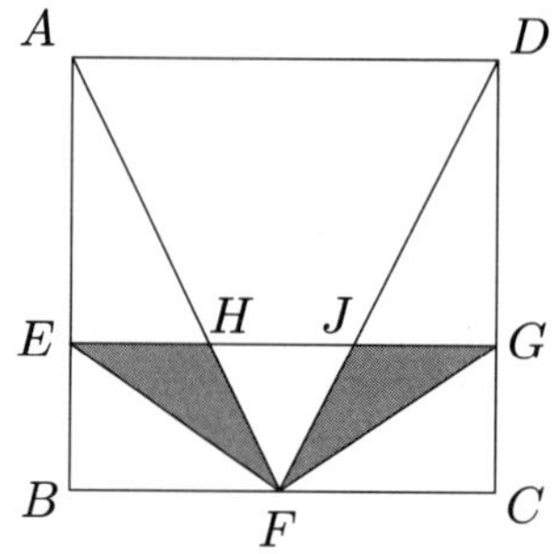

Problem 4. Terry drove along a scenic road using 9 gallons of gasoline. Then Terry went onto the freeway and used 17 gallons of gasoline. Assuming that Terry gets 6.5 miles per gallon better gas mileage on the freeway than on the scenic road, and Terry's average gas mileage for the entire trip was 30 miles per gallon, find the number of miles Terry drove.

Answer: 780

Terry got an average of 30 miles per gallon of gas and used $9+17 = 26$ gallons of gas. This means that Terry drove $26 \cdot 30 = 780$ miles.

Problem 5. The repeating decimal $0.328181818181\ldots$ can equivalently be expressed as $\frac{m}{n}$, where m and n are relatively prime positive integers. Find $m + n$.

Answer: 1461

$$\begin{aligned} 0.328181818181\ldots = .32 + .008181818181\ldots &= \frac{32}{100} + \frac{0.818181}{100} \\ &= \frac{8}{25} + \frac{81/99}{100} = \frac{8}{25} + \frac{9}{1100} \\ &= \frac{44 \cdot 8 + 9}{1100} = \frac{361}{1100} \end{aligned}$$

giving the answer $1100 + 361 = 1461$.

Problem 6. The product of two positive real numbers is equal to 50 times their sum and 75 times their difference. Find their sum.

Answer: 360

Let the two numbers be x and y. Then the information states that $xy = 50(x+y) = 75(x-y)$. It follows that $125y = 25x$ and $5y = x$. Thus, $5y \cdot y = 50(5y+y)$ or $5y^2 = 300y$. Dividing by $5y$ gives $y = 60$ and, thus, $x = 300$. The answer is $300 + 60 = 360$.

Problem 7. There is an interval $[a, b]$ that is the solution to the inequality

$$|3x - 80| \leq |2x - 105|.$$

Find $a + b$.

Answer: 12

First consider the case when $2x - 105 \geq 0$. Then, $-(2x-105) \leq 3x - 80 \leq 2x - 105$. This implies that $-2x + 105 \leq 3x - 80$ and $3x-80 \leq 2x-105$. These two inequalities have the solution $37 \leq x$ and $x \leq -25$. These conditions are never met. Alternatively, consider the case when $2x-105 < 0$. Then, $2x-105 \leq 3x-80 \leq -(2x - 105)$. These two inequalities have the solution $-25 \leq x$ and $x \leq 37$. Note that for these values of x, the assumption that $2x - 105 < 0$ or $x < 52.5$ is satisfied. Therefore, the required interval is $[-25, 37]$ and the required answer is $-25 + 37 = 12$.

Problem 8. Penelope plays a game where she adds 25 points to her score each time she wins a game and deducts 13 points from her score each time she loses a game. Starting with a score of zero, Penelope plays m games and has a total score of 2007 points. What is the least possible value for m?

Answer: 97

Penelope gets 2000 points for winning 80 games, so she will need to win more than 80 games to accumulate 2007 points. To find the minimum number of games over 80 that she needs to win, find the smallest number n so that if she wins n games, she will accumulate a number of points which is a multiple of 13 more than 7. That is, find n so that $25n \equiv 7 \pmod{13}$. But this is equivalent to $-n \equiv 7 \pmod{13}$ or $n \equiv 6 \pmod{13}$. Thus, $6 \cdot 25 - 11 \cdot 13 = 7$, and the fastest way for Penelope to get 2007 points is for her to win $80 + 6$ games and lose 11 games for a total of $80 + 6 + 11 = 97$ games.

Problem 9. Purple College keeps a careful count of its students as they progress each year from the freshman class to the sophomore class to the junior class and, finally, to the senior class. Each year at the college one third of the freshman class drops out of school, 40 students in the sophomore class drop out of school, and one tenth of the junior class drops out of school. Given that the college only admits new freshman students, and that it wants to begin each school year with 3400 students enrolled, how many students does it need to admit into the freshman class each year?

Answer: 1185

Let the number of students in each new freshman class be N. Since one third of the freshmen drop out, each sophomore class will begin with $\frac{2}{3}N$ students. Since 40 sophomores drop out, each junior class will begin with $\frac{2}{3}N - 40$ students. Since one tenth of the junior class drops out, each senior class will begin with

$$\frac{9}{10}\left(\frac{2}{3}N - 40\right) = \frac{3}{5}N - 36 \text{ students.}$$

Then each fall the college will have

$$3400 = N + \left(\frac{2}{3}N\right) + \left(\frac{2}{3}N - 40\right) + \left(\frac{3}{5}N - 36\right) \text{ students.}$$

This simplifies to

$$3400 = \frac{44}{15}N - 76 \quad \text{so} \quad N = 3476 \cdot \frac{15}{44} = 1185.$$

Problem 10. Tom can run to Beth's house in 63 minutes. Beth can run to Tom's house in 84 minutes. At noon, Tom starts running from his house toward Beth's house while at the same time Beth starts running from her house toward Tom's house. When they meet, they both run at Beth's speed back to Beth's house. At how many minutes after noon will they arrive at Beth's house?

Answer: 72

Let the distance between Tom's house and Beth's house be 1 unit. Then Tom runs at the rate $\frac{1}{63}$ units per minute and Beth runs at the rate $\frac{1}{84}$ units per minute. They cover the distance between their houses at the rate

$$\frac{1}{63} + \frac{1}{84} = \frac{7}{252} = \frac{1}{36}.$$

Thus, they meet after running for 36 minutes. If Beth then returns to her house at the same speed that she has run to meet Tom, she will run for an additional 36 minutes to arrive home after 72 minutes.

Problem 11. The alphabet in its natural order

$$ABCDEFGHIJKLMNOPQRSTUVWXYZ$$

is T_0. We apply a permutation to T_0 to get T_1 which is

$$JQOWIPANTZRCVMYEGSHUFDKBLX.$$

If we apply the same permutation to T_1, we get T_2 which is

$$ZGYKTEJMUXSODVLIAHNFPWRQCB.$$

We continually apply this permutation to each T_m to get T_{m+1}. Find the least positive integer n so that $T_n = T_0$.

Answer: 252

This is the same as High School problem number 15.

Problem 12. If you alphabetize all of the distinguishable rearrangements of the letters in the word *PURPLE*, find the number n such that the word *PURPLE* is the n-th item in the list.

Answer: 240

The word *PURPLE* has six letters with two of the letters repeated. Thus, there are $\frac{6!}{2!} = 360$ permutations of these letters. If a permutation begins with E, there are five other letters with one repeat showing that there are $\frac{5!}{2!} = 60$ permutations which begin with E. Similarly, there are 60 permutations which begin with L. All 120 of these permutations appear in the alphabet before the word *PURPLE*. If a permutation begins with P, there are five other letters with no repeats. Thus, there are $5! = 120$ permutations which begin with P. The letters *URPLE* by coincidence appear in reverse alphabetical order, so the word *PURPLE* is the last word which appears in the list of permutations beginning with P. It follows that *PURPLE* appears as entry number $60 + 60 + 120 = 240$.

Problem 13. Evaluate the sum

$$1^2 + 2^2 - 3^2 - 4^2 + 5^2 + 6^2 - 7^2 - 8^2 + \cdots - 1000^2 + 1001^2.$$

Answer: 1001

Consider four consecutive terms of this series:

$$\begin{aligned}(4n-3)^2 + (4n-2)^2 - (4n-1)^2 - (4n)^2 \\ &= [32n^2 - 40n + 13] - [32n^2 - 8n + 1] \\ &= 12 - 32n.\end{aligned}$$

The desired sum adds these foursomes from $n = 1$ through $n = 250$ before adding on the final term 1001^2.

Thus, the sum is

$$\begin{aligned}1001^2 + \sum_{n=1}^{250}(12 - 32n) &= 1001^2 + 12(250) - 32\sum_{n=1}^{250} n \\ &= 1001^2 + 3000 - 32\frac{250(251)}{2} \\ &= 1000(1001) + 1001 + 3000 - 4000(251) \\ &= 1000[1001 + 1 + 3 - 1004] + 1 = 1001.\end{aligned}$$

Problem 14. A rectangular storage bin measures 10 feet by 12 feet, is 3 feet tall, and sits on a flat plane. A pile of dirt is pushed up against the outside of the storage bin so that it slants down from the top of the storage bin to points on the ground 4 feet away from the base of the storage bin, as shown. The number of cubic feet of dirt needed to form the pile can be written as $m + n\pi$ where m and n are positive integers. Find $m + n$.

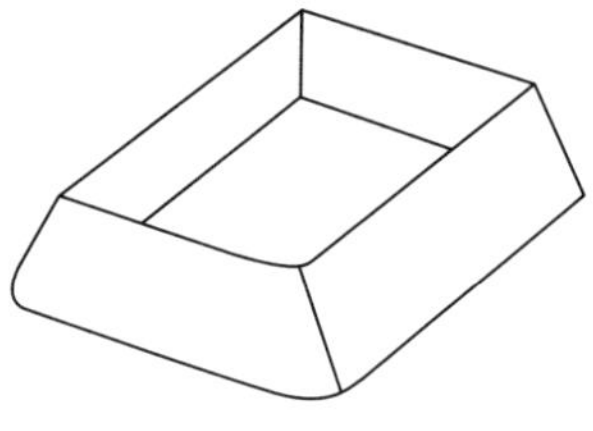

Answer: 280

The dirt pile is made up of four triangular prism sections and four corner sections each the shape of a quarter of a right circular cone. The triangular prism sections have triangular bases which are right triangles with legs of length 3 and 4, so the base area is 6. The total length of the triangular prism sections is the perimeter of the storage bin which is $10 + 12 + 10 + 12 = 44$.

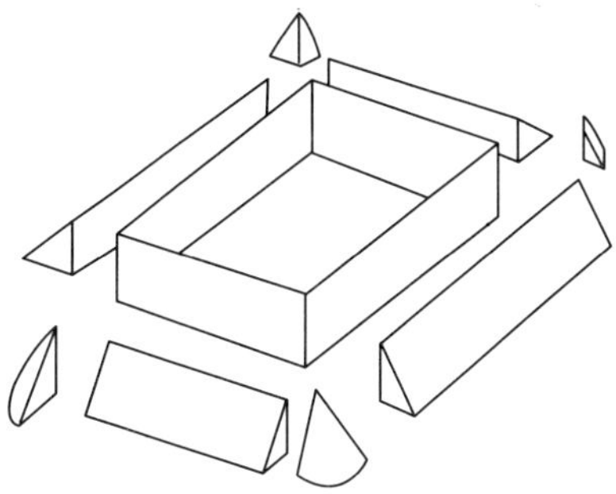

The volume of the triangular prism sections of the pile is, therefore, $6 \cdot 44 = 264$. The four quarter cone sections of the pile can be put together to get one cone with base radius 4 and height 3 having a volume of $\frac{\pi 4^2 3}{3} = 16\pi$. Thus, the requested answer is $264 + 16 = 280$.

Problem 15. We have some identical paper squares which are black on one side of the sheet and white on the other side. We can join nine squares together to make a 3 by 3 sheet of squares by placing each of the nine squares either white side up or black side up. Two of these 3 by 3 sheets are distinguishable if neither can be made to look like the other by rotating the sheet or by turning it over. How many distinguishable 3 by 3 squares can we form?

Answer: 70

This is the same as High School problem number 16.

2.10 2008 High School

Problem 1. What is the least positive integer with the property that the product of its digits is 9! ?

Answer: 2578899

$$9! = 9 \cdot 8 \cdot 7 \cdot 6 \cdot 5 \cdot 4 \cdot 3 \cdot 2 \cdot 1 = 2^7 \cdot 3^4 \cdot 5^1 \cdot 7^1.$$

To write this value as a product of digits, there clearly must be one 5 digit and one 7 digit. The products of 2's and 3's can be done in several ways such as

$$2 \cdot 2 \cdot 2 \cdot 2 \cdot 2 \cdot 2 \cdot 2 \cdot 3 \cdot 3 \cdot 3 \cdot 3 \cdot 3 = 2 \cdot 4 \cdot 4 \cdot 4 \cdot 9 \cdot 9 = 2 \cdot 8 \cdot 8 \cdot 9 \cdot 9.$$

This cannot be done with fewer than 5 digits since

$$9^4 = 6561 < 2^7 \cdot 3^4 = 10368.$$

To do this with 5 digits where at least one digit is a 2 requires the product $2 \cdot 8 \cdot 8 \cdot 9 \cdot 9$. Adding in the 5 and 7 digits so that the digits form the least possible integer gives the answer 2578899.

Problem 2. A canister contains two and a half cups of flour. Greg and Sally have a brownie recipe which calls for one and one third cups of flour. Greg and Sally want to make one and a half recipes of brownies. To the nearest whole percent, what percent of the flour in the canister would they use?

Answer: 80

Greg and Sally will use one and one half times one and one third cups out of the available two and a half cups flour. The percentage is, therefore,

$$\frac{\frac{3}{2} \cdot 1\frac{1}{3}}{2\frac{1}{2}} \cdot 100\% = \frac{\frac{3}{2} \cdot \frac{4}{3}}{\frac{5}{2}} \cdot 100\% = \frac{4}{5} \cdot 100\% = 80\%.$$

Problem 3. Find the least integer n greater than 345 such that

$$\frac{3n+4}{5}, \quad \frac{4n+5}{3} \text{ and } \frac{5n+3}{4}$$

and are all integers.

Answer: 397

The sum of the three given fractions is

$$\begin{aligned}\frac{3n+4}{5}+\frac{4n+5}{3}+\frac{5n+3}{4} &= \frac{12(3n+4)+20(4n+5)+15(5n+3)}{60}\\ &= \frac{191n+193}{60} = 3n+3+\frac{11n+13}{60}.\end{aligned}$$

This sum is an integer exactly when $11n+13$ is a multiple of 60. If $11n+13$ is a multiple of 60, so is

$$11\cdot(11n+13) = 121n+143 = 60(2n+2)+(n+23).$$

This can be a multiple of 60 only when $n+23$ is a multiple of 60 or, equivalently when n is 37 more than a multiple of 60. The least such number greater than 345 is $360+37=397$ which does satisfy the conditions in the problem.

Problem 4. The student population at one high school consists of freshmen, sophomores, juniors, and seniors. There are 25 percent more freshmen than juniors, 10 percent fewer sophomores than freshmen, and 20 percent of the students are seniors. If there are 144 sophomores, how many students attend the school?

Answer: 540

Since there are 10 percent fewer sophomores than freshman, there are $\frac{1}{1-0.1}=\frac{10}{9}$ as many freshmen as sophomores or $144\cdot\frac{10}{9}=160$ freshmen. Since there are 25 percent more freshmen than juniors, there are $\frac{1}{1+0.25}=\frac{4}{5}$ as many juniors as freshmen or $160\cdot\frac{4}{5}=128$ juniors. Since there are $160+144+128=432$ students who are freshmen, sophomores, or juniors, and these account for 80 percent of the school, the school must have $432\cdot\frac{100}{80}=432\cdot\frac{5}{4}=540$ students.

Problem 5. Find the sum of all digits in the decimal representations of all positive integers less than 1000.

Answer: 13500

Write each number from 0 to 999 as a three digit number, perhaps with leading zeros as in 000, 001, 002, 003, ..., 999. Note that there are 3 times 1000 digits written and that each of the ten possible digits appears the same number of times. Thus, the sum of all of the digits is

$$300 \cdot (0+1+2+3+4+5+6+7+8+9) = 300 \cdot 45 = 13500.$$

Problem 6. Three friends who live in the same building leave the building at the same time to go rock climbing on a cliff nearby. Susan walks to the cliff and climbs to the top of the cliff. Fred runs to the cliff twice as fast as Susan walks and climbs the top of the cliff at a rate that is only two-thirds as fast as Susan climbs. Ralph bikes to the cliff at a speed twice as fast as Fred runs and takes two hours longer to climb to the top of the cliff than Susan does. If all three friends reach the top of the cliff at the same time, how many minutes after they left home is that?

Answer: 320

Let x be the amount of time it takes Susan to walk to the cliff, and y be the amount of time it takes Susan to climb the cliff. Then the problem states that the time taken by each climber is

$$x+y = \frac{x}{2} + \frac{y}{2/3} = \frac{x}{4} + (y+120).$$

Subtracting each of these quantities from $x+y$ yields $\frac{x}{2} - \frac{y}{2} = 0$ and $\frac{3x}{4} = 120$. Thus, $x = y$, and $x = 160$, so the total time is $x+y = 320$.

Problem 7. A line through the origin passes through the curve whose equation is $y = 2x^2 - 9x + 10$ at two points whose x-coordinates add up to 77. Find the slope of the line.

Answer: 29

Let the slope of the line be m so that the equation of the line is $y = mx$. Then the x coordinates at the places where the line intersects the curve satisfy

$$mx = 2x^2 - 9x + 10$$

or

$$0 = 2x^2 - (9+5m)x + 10.$$

The sum of the roots of the polynomial $ax^2 + bx + c$ is given by $\frac{b}{a}$ which give us that $77 = \frac{9+5m}{2}$. Solving this for m gives $m = 29$.

Problem 8. The shape of a container is a square-based pyramid where the base has side length 23 centimeters and the height is 120 centimeters. The container is open at the base of the pyramid and stands in an open field with its vertex pointing down. One afternoon 5 centimeters of rain falls in the open field partially filling the previously empty container. Find the depth in centimeters of the rainwater in the bottom of the container after the rain.

Answer: 60

Since the area of the opening at the top of the container is $(23)^2$, the volume of rainwater entering the container is $5 \cdot (23)^2$. Suppose the water fills the container to a height h forming a square pyramid with base side x as shown. Then by similar triangles,

$$\frac{x}{23} = \frac{h}{120} \quad \text{and} \quad x = \frac{23h}{120}.$$

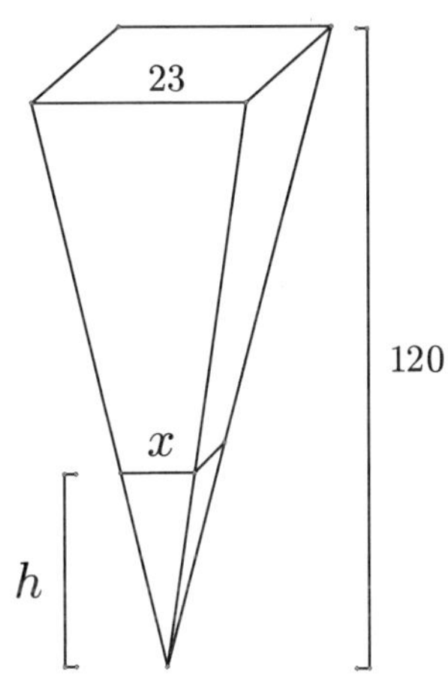

Thus, since the volume of a pyramid is equal to one third the product of the area of the base times the height, the volume of the water is given by

$$\frac{hx^2}{3} = \frac{h\left(\frac{23h}{120}\right)^2}{3} = \frac{(23)^2h^3}{3 \cdot 120^2} = 5(23)^2.$$

It follows that

$$h^3 = 5 \cdot 3 \cdot (120)^2 = 5 \cdot 3 \cdot (8 \cdot 15)^2 = 4^3 \cdot (15)^3 = (60)^3$$

so $h = 60$. Note that the answer would be the same for a pyramid with any shape base (or cone) and height 120.

Problem 9. Find the sum of all the integers $N > 1$ with the properties that each prime factor of N is either 2, 3, or 5, and N is not divisible by any perfect cube greater than 1.

Answer: 2820

The prime factorization of N must be $2^a 3^b 5^c$ where each a, b, and c is either 0, 1, or 2 and at least one of a, b, and c is positive. It follows that the sum of all such N is given by

$$\begin{aligned}&(2^0+2^1+2^2)(3^0+3^1+3^2)(5^0+5^1+5^2)-1\\&\quad=(1+2+4)(1+3+9)(1+5+25)-1=7\cdot 13\cdot 31-1\\&\quad=2820.\end{aligned}$$

Problem 10. There are three men and eleven women taking a dance class. In how many different ways can each man be paired with a woman partner and then have the eight remaining women be paired into four pairs of two?

Answer: 103950

To choose a pairing, first choose a woman partner for each of the three men. There are ${}_{11}P_3 = 11 \cdot 10 \cdot 9 = 990$ ways to do this. Then choose a way to pair the remaining eight women by lining up the women and choosing a partner for the first woman in one of 7 ways, then a partner for the next unpaired woman in one of 5 ways, and finally a partner for the next unpaired woman in one of 3 ways to give $7(5)3 = 105$ ways. Alternatively, the eight can be paired in

$$\frac{\binom{8}{2,2,2,2}}{4!} = \frac{\frac{8!}{2\cdot 2\cdot 2\cdot 2}}{4!} = 105 \text{ ways.}$$

The final answer is then $990 \cdot 105 = 103950$.

Problem 11. Six regular octagons each with sides of length 2 are placed in a two-by-three array and inscribed in a square as shown. The area of the square can be written in the form $m + n\sqrt{2}$, where m and n are positive integers. Find $m + n$.

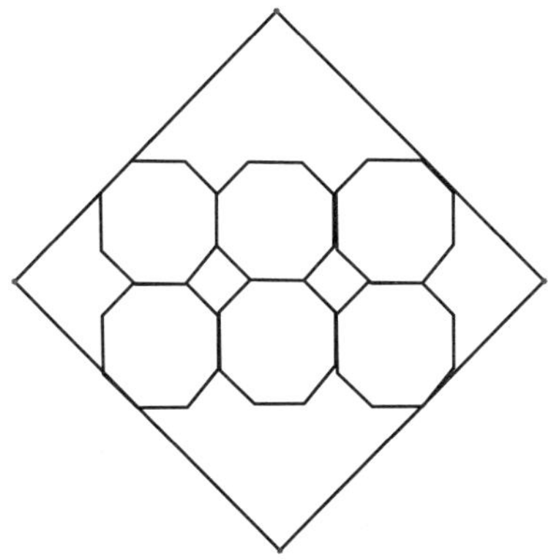

Answer: 194

Draw lines parallel to one side of the square that pass through the vertices of the octagons as shown. Label four adjacent lines passing through adjacent vertices of the octagons A, B, C, and D. The distance from line B to line C is the length of the side of one of the octagons, so it is 2. The distance from C to D is the length of one leg of an isosceles right triangle with hypotenuse of length 2. This length is $\dfrac{2}{\sqrt{2}} = \sqrt{2}$.

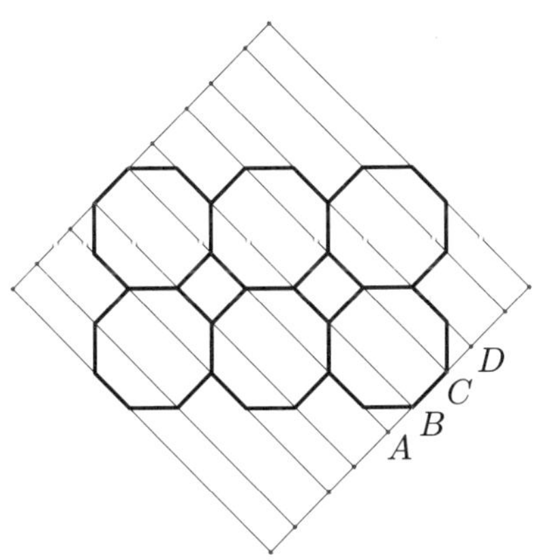

It follows that the length of the side of the square is

$$4 \cdot 2 + 5 \cdot \sqrt{2} = 8 + 5\sqrt{2}.$$

Therefore, the area of the square is

$$\begin{aligned}(8+5\sqrt{2})^2 &= 8^2 + 2 \cdot 8 \cdot 5\sqrt{2} + (5\sqrt{2})^2 \\ &= 64 + 80\sqrt{2} + 50 = 114 + 80\sqrt{2}.\end{aligned}$$

The requested sum is $114 + 80 = 194$.

Problem 12. We record one number every workday: Monday, Tuesday, Wednesday, Thursday, and Friday. On the first Monday, the number we record is ten and a half. On every Tuesday and every Thursday, the number we record is one third of what it was on the previous workday. On every Monday, Wednesday, and Friday, the number we record is double what it was on the previous workday. This goes on forever. What is the sum of all of the numbers we will record?

Answer: 252

If the number we record on a Monday is x, then on Tuesday it is $\frac{x}{3}$, on Wednesday it is $\frac{2x}{3}$, on Thursday it is $\frac{2x}{9}$, on Friday it is $\frac{4x}{9}$, and on the following Monday it is $\frac{8x}{9}$. Thus, the weekly total is

$$x + \frac{x}{3} + \frac{2x}{3} + \frac{2x}{9} + \frac{4x}{9} = \frac{8}{3}x,$$

and the weekly total for the following week will be $\frac{8}{9}$ as large. Then the sum of the weekly totals is a geometric series with first term $\frac{10.5 \cdot 8}{3}$ and common ratio $\frac{8}{9}$. The sum of all the terms in this geometric series is given by

$$\sum_{n=0}^{\infty} a \cdot r^n = \frac{a}{1-r} = \frac{10.5 \cdot 8}{3} \cdot \frac{1}{1 - \frac{8}{9}} = 10.5 \cdot 24 = 252.$$

Problem 13. If you roll six fair dice, let p be the probability that exactly five different numbers appear on the upper faces of the six dice. If $p = \frac{m}{n}$, where m and n are relatively prime positive integers, find $m+n$.

Answer: 133

The number of ways of rolling a sequence of six numbers is 6^6, and each of these ways is equally likely to occur. To choose a way of rolling a sequence of six numbers where exactly five different numbers appear you can

1) choose the one number from 1 to 6 that will not appear (6 ways)

2) choose the number that will appear twice (5 ways)

3) choose a permutation of the six numbers with one repeated number ($\frac{6!}{2!}$ ways)

The number of sequences of six numbers where exactly five different numbers appear is then the product $6 \cdot 5 \cdot \frac{6!}{2!}$. The desired probability is

$$\frac{6 \cdot 5 \cdot 6!}{6^6 \cdot 2!} = \frac{6 \cdot 5 \cdot 6 \cdot 5 \cdot 4 \cdot 3 \cdot 2 \cdot 1}{6 \cdot 6 \cdot 6 \cdot 6 \cdot 6 \cdot 6 \cdot 2 \cdot 1} = \frac{25}{108}$$

giving a final answer of $25 + 108 = 133$.

Problem 14. A circular track with diameter 500 is externally tangent at a point A to a second circular track with diameter 1700. Two runners start at point A at the same time and run at the same speed. The first runner runs clockwise along the smaller track, while the second runner runs clockwise along the larger track. There is a first time after they begin running when the positions of the two runners together with the point A are three distinct collinear points. At that time each runner will have run a distance of $\frac{m\pi}{n}$, where m and n are relatively prime positive integers. Find $m + n$.

Answer: 2128

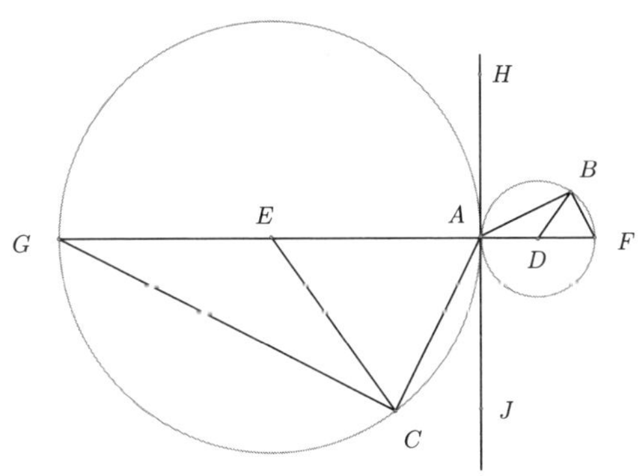

At a given time let the first runner be at position B along the smaller track, and the second runner be at position C along the larger track. Let D be the center of the smaller track, and E be the center of the larger track. Let F and G be the end points diametrically opposite of the point A on the smaller and larger tracks, respectively. Let $\overline{HJ}$ be the line tangent to two tracks at A as shown. Let t be the central angle ADB. Since the length of arc along the smaller track between A and B is equal to the length of arc along the larger track between A and C, it follows that central angle AEC is $\frac{5}{17}t$. Angle AFB is equal to angle

HAB, and both are equal to $\frac{5}{17}t$. Similarly, the angles AGC and JAC are both equal to $\frac{5}{34}t$. Thus, the angle measured clockwise from $\overline{AC}$ to $\overline{AB}$ is $\left[\pi - \frac{5}{34}t\right] + \frac{t}{2} = \frac{6}{17}t + \pi$. The points A, B, and C are collinear exactly when $\frac{6}{17}t + \pi$ is a multiple of π. This first occurs when $t = \frac{17\pi}{6}$, or when the first runner has run one complete lap around the first track plus an additional angle of $\frac{5\pi}{6}$ into the second lap. At that point, the second runner has run through an angle of $\frac{5\pi}{6}$ into the first lap on the larger track which is a distance of $\frac{5\pi}{6} \cdot \frac{1700}{2} = \frac{2125\pi}{3}$. So the requested answer is $2125 + 3 = 2128$.

Problem 15. For each positive integer n let $f(n)$ be the sum of the digits of n. For example, $f(16) = 7$, $f(f(78)) = 6$, and $f(f(f(5978))) = 2$. Find the least positive integer n such that $f(f(f(n)))$ is not a one-digit number.

Answer: 19999999999999999999999

First note that the smallest number whose digits add to a number k is always smaller than the smallest number whose digits add to the number $k+1$. The smallest n such that $f(n)$ is not a one-digit number is 19, a number whose digits add to one more than 9. The smallest number n such that $f(n)$ is 19 is 199, a number whose digits add to one more than $18 = 2\cdot 9$. The smallest n such that $f(n)$ is 199 is a number whose digits add to one more than $198 = 22 \cdot 9$. This would be the number which begins with the digit 1 and ends with 22 digits of 9, and it follows from the initial observation that it must be the smallest n such that $f(f(f(n)))$ is not a one-digit number.

Problem 16. Square $ABCD$ has side length 7. Let A_1, B_1, C_1, D_1 be points on rays $\overrightarrow{AB}$, $\overrightarrow{BC}$, $\overrightarrow{CD}$, $\overrightarrow{DA}$, respectively, where each point is 3 units from the end of the ray so that $A_1B_1C_1D_1$ forms a second square, as shown. Similarly, let A_2, B_2, C_2, D_2 be points on segments $\overline{A_1B_1}$, $\overline{B_1C_1}$, $\overline{C_1D_1}$, $\overline{D_1A_1}$, respectively, forming another square, where A_2 divides segment $\overline{A_1B_1}$ into two pieces whose lengths are in the same ratio as AA_1 is to A_1B. Continue this process to construct square $A_nB_nC_nD_n$ for each positive integer n. Find the total of all the perimeters of all of the squares.

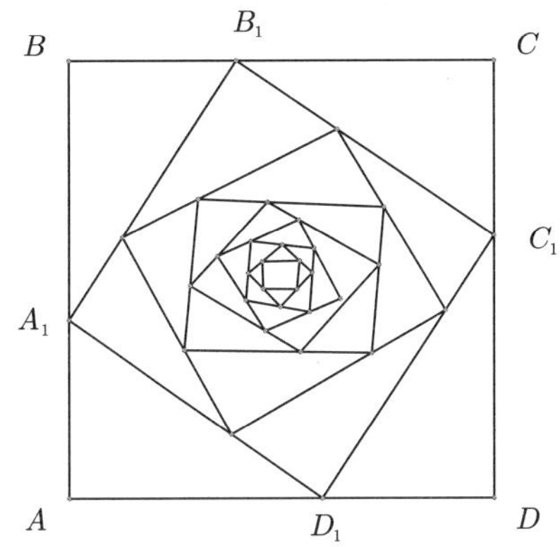

Answer: 98

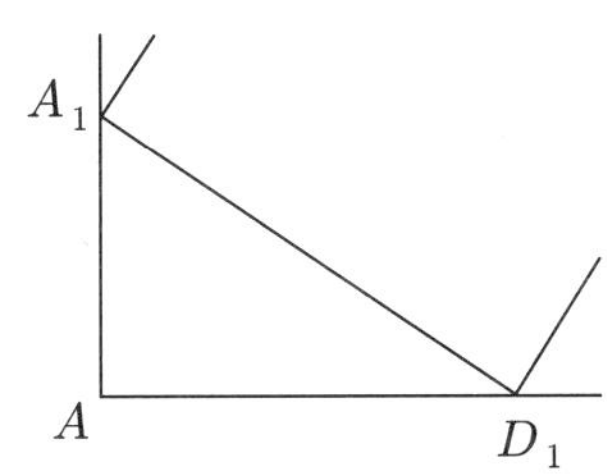

Since the length of $\overline{AD_1}$ is length 4 and $\overline{AA_1}$ is length 3, the Pythagorean Theorem gives that the length of $\overline{A_1D_1}$ is $\sqrt{3^2+4^2}=5$. It follows that the sides of square $A_1B_1C_1D_1$ are $\frac{5}{7}$ the length of the sides of square $ABCD$. Similarly, the sides of each square are $\frac{5}{7}$ the length of the sides of the previous square. Since the perimeter of square $ABCD$ is $4\cdot 7=28$, the sum of the lengths of the perimeters is given by the infinite sum

$$28\left[1+\frac{5}{7}+\left(\frac{5}{7}\right)^2+\left(\frac{5}{7}\right)^3+\left(\frac{5}{7}\right)^4+\dots\right]$$

which is a geometric series that converges to

$$\frac{28}{1-\frac{5}{7}}=\frac{7\cdot 28}{7-5}=98.$$

Problem 17. The number $24! = 620,448,401,733,239,439,360,000$ ends in four zeros, and $25! = 15,511,210,043,330,985,984,000,000$ ends in six zeros. Thus there is no integer n such that $n!$ ends in exactly five zeros. Let S be the set of all k such that for no integer n does $n!$ end in exactly k zeros. If the numbers in S are listed in increasing order, 5 will be the first number. Find the 100th number in that list.

Answer: 504

The least element of S results from the fact that 25! has two more factors of 5 than 24! because 25 contains more than one factor of 5. Similarly, there will be one or more entry in S for each $n!$ where n is a multiple of 25, one extra for each multiple of 125, and a another extra for each multiple of 625. We can estimate the 100th smallest number of S to be the one resulting from the factorial of the 80th multiple of 25 or 2000. Up to 2000 there are 80 multiples of 25, 16 multiples of 125, and 3 multiples of 625, so there are $80 + 16 + 3 = 99$ entries in S resulting from factorials up to 2000. Thus, the 100th smallest number of S results from the 81st multiple of 25 or 2025.
2025! has

$$\left\lfloor \frac{2025}{5} \right\rfloor + \left\lfloor \frac{2025}{25} \right\rfloor + \left\lfloor \frac{2025}{125} \right\rfloor + \left\lfloor \frac{2025}{625} \right\rfloor$$
$$= 405 + 81 + 16 + 3 = 505 \text{ zeros.}$$

The 100th number of zeros that is not obtained is, thus, 504.

Problem 18. The diagram below contains eight line segments, all the same length. Each of the angles formed by the intersections of two segments is either a right angle or a 45 degree angle. If the outside square has area 1000, find the largest integer less than or equal to the area of the inside square.

Answer: 171

Let the intersection points be labeled as in the following diagram.

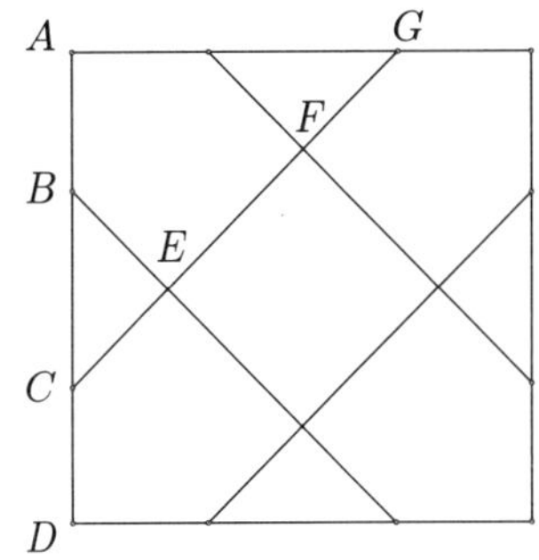

Let the side of the large square, $\overline{AD}$, have length s. Since triangle ACG is an isosceles right triangle with hypotenuse length s, the length of $\overline{AC}$ is $\frac{s}{\sqrt{2}}$. Similarly, $\overline{BD}$ has length $\frac{s}{\sqrt{2}}$. Therefore, $\overline{BC}$ is $\overline{AC}$ plus $\overline{BD}$ minus $\overline{AD}$ or $\frac{2s}{\sqrt{2}} - s = [\sqrt{2} - 1]s$. Triangle EBC is also an isosceles right triangle, so the length of side $\overline{CE}$ is

$$\frac{BC}{\sqrt{2}} = \frac{[\sqrt{2} - 1]s}{\sqrt{2}}.$$

This must also be the length of $\overline{FG}$, and since $\overline{CG}$ has length s, the length of $\overline{EF}$ is

$$s - 2\frac{[\sqrt{2} - 1]s}{\sqrt{2}} = s - [2 - \sqrt{2}]s = [\sqrt{2} - 1]s.$$

Thus, the area of the small inside square which equals the square of the length of $\overline{EF}$ is $([\sqrt{2} - 1]s)^2 = [3 - 2\sqrt{2}]s^2$. The value of $\sqrt{2}$ is approximately 1.4142, so $2\sqrt{2}$ is approximately 2.8284, and $1 - 2\sqrt{2}$ is close to 0.1716. It is given that s^2 is 1000, so it follows that the area of the smaller square is about 171.6, and the requested answer is 171.

Problem 19. One side of a triangle has length 75. Of the other two sides, the length of one is double the length of the other. What is the maximum possible area for this triangle?

Answer: 1875

Let y be half of 75 so that the side of the triangle with known length is $2y$. Let the other two sides of the triangle be $2x$ and $4x$. For $2y$, $2x$, and $4x$ to be the sides of a triangle, x must be between $\frac{y}{3}$ and y. Then the perimeter of the triangle is

$2x + 4x + 2y = 6x + 2y$, and the semiperimeter is $s = 3x + y$. By Heron's Formula, the square of the area of the triangle is

$$s(s-2y)(s-2x)(s-4x) = (3x+y)(3x-y)(y+x)(y-x)$$
$$= (9x^2 - y^2)(y^2 - x^2) = -9x^4 + 10y^2x^2 - y^4.$$

This is a quadratic function of x^2. Since the quadratic function $ax^2 + bx + c$ takes on its extreme value at $-\dfrac{b}{2a}$, the area has a maximum when

$$x^2 = -\frac{10y^2}{2(-9)} = \frac{5y^2}{9}.$$

This gives the maximum value for the square of the area of

$$-9\left(\frac{5y^2}{9}\right)^2 + 10y^2\left(\frac{5y^2}{9}\right) - y^4 = \frac{16y^4}{9},$$

and the maximum area is

$$\frac{4y^2}{3} = 3\left(\frac{2y}{3}\right)^2 = 3(25)^2 = 1875.$$

Alternatively, let the side of the triangle with length 75 be placed in the coordinate plane with end points at (0,0) and (75, 0). If the third vertex of the triangle is at point (x, y), then one has

$$2\sqrt{x^2+y^2} = \sqrt{(75-x)^2+y^2}$$

which simplifies to $(x+25)^2 + y^2 = 50^2$. This is the equation for a circle centered at $(-25, 0)$ with radius 50. It follows that the maximum height of the described triangle is 50 giving the maximum area of $\dfrac{75 \cdot 50}{2} = 1875$.

Problem 20. Find the least positive integer n such that the decimal representation of the binomial coefficient $\dbinom{2n}{n}$ ends in four zeros.

Answer: 313

Since $\dbinom{2n}{n} = \dfrac{(2n)!}{n! \cdot n!}$, the number of zeros at the end of the number will be equal to the minimum of the number of factors of 2 in the number and the number of factors of 5 in the number. The number of factors of 5 in $\dbinom{2n}{n}$ is $\displaystyle\sum_{n=1}^{\infty}\left(\left\lfloor\frac{2n}{5^j}\right\rfloor - 2\left\lfloor\frac{n}{5^j}\right\rfloor\right)$

where the term $\left\lfloor \frac{2n}{5^j} \right\rfloor - 2\left\lfloor \frac{n}{5^j} \right\rfloor$ is 0 for each j except in the cases when $n \pmod{5^j} \geq \frac{5^j}{2}$, and then the term is equal to 1. For $\binom{2n}{n}$ to end in four zeros, it must have at least four factors of 5. This $2n$ happens when $n \pmod{5^j} \geq \frac{5^j}{2}$ is true for at least four values of j. The first time this could occur is when $n \geq \frac{5^4}{2} = 312\frac{1}{2}$ or when $n = 313$. But $313 \pmod{5^j} \geq \frac{5^j}{2}$ is satisfied when $j = 1, 2$, and 3 as well as when $j = 4$. It remains to check that the number of factors of 2 in $\binom{626}{313}$ is also at least four, but this follows from the fact that $313 \pmod{2^j} \geq \frac{2^j}{2}$ is true when $j = 1, 4, 5$, and 6.

Problem 21. The area of the quadrilateral with vertices at the four points in three dimensional space (0,0,0), (2,6,1), $(-3, 0, 3)$, $(-4, 2, 5)$ is the number $\frac{m}{n}$, where m and n are relatively prime positive integers. Find $m + n$.

Answer: 53

Note that the four given points (x, y, z) are on the same plane since their coordinates all satisfy the linear equation

$$2x - y + 2z = 0.$$

Any figure in this plane can be projected to the $X - Y$ plane by sending points (x, y, z) to (x, y), and this projection reduces the area of the figure by a factor of two-thirds. Indeed, the triangle with vertices at (0, 0, 0), (1, 2, 0), and (0, 2, 1) has sides with lengths $\sqrt{2}$, $\sqrt{5}$, and $\sqrt{5}$, and so, by Heron's formula, the triangle has area

$$\begin{aligned}\sqrt{\left(\sqrt{5} + \frac{\sqrt{2}}{2}\right)\left(\sqrt{5} - \frac{\sqrt{2}}{2}\right)\left(\frac{\sqrt{2}}{2}\right)\left(\frac{\sqrt{2}}{2}\right)} &= \sqrt{\left(5 - \frac{1}{2}\right)\frac{1}{2}} \\ &= \sqrt{\frac{9}{4}} = \frac{3}{2}\end{aligned}$$

while its projection to the triangle with vertices at $(0, 0)$, $(1, 2)$, and $(0, 2)$ clearly has area 1. So, the given quadrilateral has an area which is $\frac{3}{2}$ the area of the triangle in the $X-Y$ plane with vertices at $(0, 0)$, $(2, 6)$, $(-3, 0)$, and $(-4, 2)$. There are many ways to calculate the area of this quadrilateral. One is to form the matrix $\begin{bmatrix} 0 & 0 \\ 2 & 6 \\ -4 & 2 \\ -3 & 0 \\ 0 & 0 \end{bmatrix}$ and find one half of its determinant given by the sum

$$(0\cdot 0+2\cdot 2+-4\cdot 0+-3\cdot 0)-(0\cdot 0+-3\cdot 2+-4\cdot 6+2\cdot 0)=34$$

giving an area of 17. A second way is to use Pick's Theorem which says to take the number of lattice points on the inside of the polygon (14) plus half the number of lattice points on the boundary of the polygon $(8/2 = 4)$ minus 1 to give the same result of 17. It follows that the original quadrilateral has area $17\cdot\frac{3}{2}=\frac{51}{2}$ and the required sum is 53.

Problem 22. Let $*$ be a binary operation on the positive integers satisfying

$$(a+b)*c=(a*c)+(b*c) \quad \text{and} \quad a*(b+c)=(a*b)*c$$

for all a, b, c. Given that $5*5=160$, find the value of $7*7$.

Answer: 896

First we claim that for all natural numbers a and b it holds that $a*b=a\cdot(1*b)$. This follows by induction on a since it is clearly true when a is 1, and if it is true for some a, one gets that

$$(a+1)*b=a*b+1*b=a\cdot(1*b)+1*b=(a+1)\cdot(1*b).$$

Secondly, we claim that for all natural numbers b it holds that $1*b=(1*1)^b$. This follows by induction on b since it is clearly true when b is 1, and if it is true for some b, one gets that

$$1*(b+1)=(1*b)*1=(1*1)^b*1.$$

Now from the first claim, this equals $(1*1)^b\cdot(1*1)$ which equals $(1*1)^{b+1}$. Combine these two claims to conclude that

$$a*b=a\cdot(1*1)^b.$$

So $5*5=5\cdot(1*1)^5=160$, and $(1*1)^5=32$ which implies that $1*1=2$. From this, conclude that $a*b=a\cdot 2^b$ and, thus, $7*7=7\cdot 2^7=896$.

Problem 23. Circle B has radius 2008 and is tangent to horizontal line A at point P. Circle C_1 has radius 1 and is tangent both to circle B and to line A at a point to the right of point P. Circle C_2 has radius larger than 1 and is tangent to line A and both circles B and C_1. For $n > 1$, circle C_n is tangent to line A and both circles B and C_{n-1}. Find the greatest n such that this sequence of circles can be constructed through circle C_n where the n circles are all tangent to line A at points to the right of P.

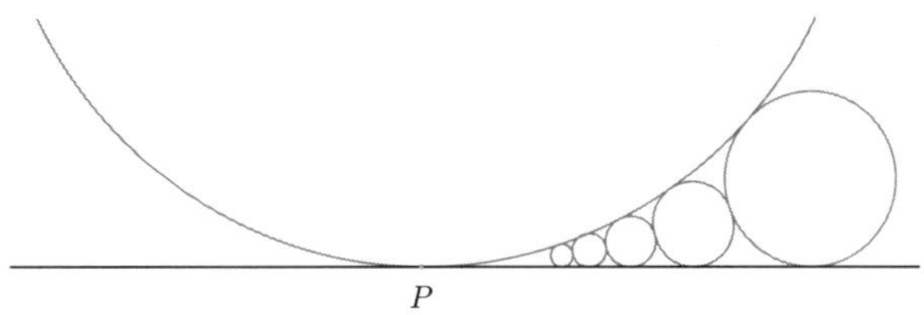

Answer: 45

Let circle C_1 be tangent to line A at point R. Let D and E be the centers of circles B and C_1, respectively. Let S be the point on line segment $\overline{DP}$ one unit from P. Segment $\overline{DE}$ has length $2008 + 1 = 2009$. Segment $\overline{DS}$ has length $2008 - 1 = 2007$. Then segment $\overline{PR}$ has the same length as segment $\overline{SE}$ which has length

$$\sqrt{2009^2 - 2007^2} = \sqrt{2(4016)} = 2\sqrt{2008}.$$

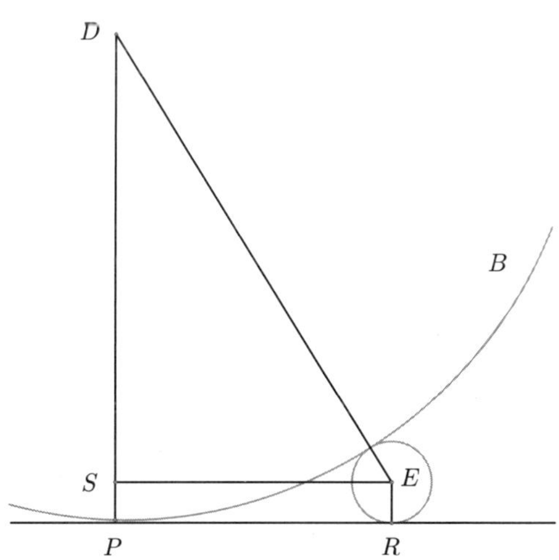

Perform an inversion of the plane through a circle centered at P with radius $2008\sqrt{2}$. This sends each point X to a second point,

X', a distance from P so that the ray from P through X goes through X'.

$$\frac{(2008\sqrt{2})^2}{PX} = \frac{2 \cdot 2008^2}{PX}$$

Thus, R goes to a point R' on line A at a distance

$$\frac{2 \cdot 2008^2}{2\sqrt{2008}} = 2008\sqrt{2008}$$

from P.

Inversions map circles and lines into circles and lines. Thus, line A gets transformed into line A, and circle B gets transformed into a line parallel to line A. The point of B diametrically opposed to point P ends up a distance $\dfrac{2 \cdot 2008^2}{2(2008)} = 2008$ from P, so circle B is transformed into a line parallel to line A and a distance 2008 from A. Circle C_1 gets transformed into circle C_1' and is tangent to the two parallel lines, so C_1' has diameter 2008. Similarly, each circle C_n gets transformed into a circle C_n' tangent to the two parallel lines. The transformed diagram contains a sequence of congruent circles, each tangent to the two parallel lines and to the previous circle.

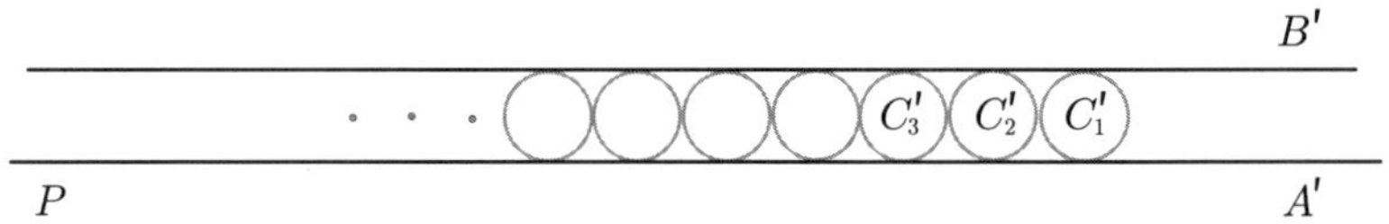

The points of tangency of the circles must fit between R' and P which is a distance of $2008\sqrt{2008}$. The distance between the points of tangency is the diameter of the circles, 2008. So the number of circles which fit is the smallest integer larger than $\frac{2008\sqrt{2008}}{2008} = \sqrt{2008}$ which is between 44 and 45, so the answer is 45.

Problem 24. Each of the distinct letters in the following addition problem represents a different digit. Find the number represented by the word $MEET$.

$$\begin{array}{cccccc} P & U & R & P & L & E \\ & & C & O & M & E & T \\ & & + & M & E & E & T \\ \hline Z & Z & Z & Z & Z & Z \end{array}$$

Answer: 9221

The solution to this problem involves some organized searching. First note that the value of Z must be one more than the value of P and, therefore, must be at least 2. Clearly, P, C, M, and Z cannot be zero. Also, note that the value of T cannot be zero or else E and Z would have the same value. So, the value of Z determines P. Then the value of T determines E, which in turn determines L, which determines M. Usually, this is as far as we need to go to find a contradiction. So, consider the different possible values for Z.

CASE $Z = 2$. Then $P = 1$. $T = 3 \to E = 5 \to L = 1 = P$ no. $T = 4 \to E = 4 = T$ no. $T = 5 \to E = 2 = Z$ no. $T = 6 \to E = 0 \to L = 1 = P$ no. $T = 7 \to E = 8 \to L = 4 \to M = 1 = P$ no. $T = 8 \to E = 6 \to L = 8 = T$ no. $T = 9 \to E = 4 \to L = 2 = Z$ no.

CASE $Z = 3$. Then $P = 2$. $T = 1 \to E = 1 = T$ no. $T = 4 \to E = 5 \to L = 2 = P$ no. $T = 5 \to E = 3 = Z$ no. $T = 6 \to E = 1 \to L = 0 \to M = 0 = L$ no. $T = 7 \to E = 9 \to L = 3 \to M = 0$ no. $T = 8 \to E = 7 \to L = 7 = E$ no. $T = 9 \to E = 5 \to L = 1 \to M = 3 = Z$ no.

CASE $Z = 4$. Then $P = 3$. $T = 1 \to E = 2 \to L = 0 \to M = 9 \to R + O$ ends in a 4 and $C + U$ ends in a 2. This can be done with the remaining numbers in more than one way including $R = 6$, $O = 8$, $C = 5$, and $U = 7$ to give a solution

$$\begin{array}{cccccc} 3 & 7 & 6 & 3 & 0 & 2 \\ & 5 & 8 & 9 & 2 & 1 \\ & & 9 & 2 & 2 & 1 \\ \hline 4 & 4 & 4 & 4 & 4 & 4 \end{array}$$

An exhaustive search (very exhausting search) shows that there are no other solutions for $MEET$, but it is sufficient to find this one.

Problem 25. Let S be a point chosen at random in the interior of the square $ABCD$. Let P be the probability that segments $\overline{AS}$, $\overline{SB}$, $\overline{AC}$ are congruent to the sides of a triangle. Then P can be written as $\dfrac{a - \pi\sqrt{b} - \sqrt{c}}{d}$, where a, b, c, d are positive integers and d is as small as possible. Find $ab + cd$.

Answer: 160

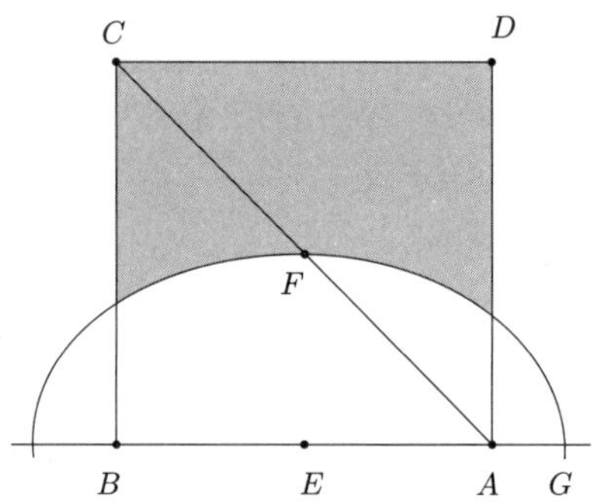

Let the midpoint of the segment $\overline{AB}$ be the point E, and the point at the center of the square be F. For the lengths of $\overline{AS}$, $\overline{SB}$, and $\overline{AC}$ to be the lengths of the sides of a triangle, the sum of the lengths of $\overline{AS}$ and $\overline{SB}$ must exceed the length of $\overline{AC}$. Thus, the lengths of $\overline{AS}$ and $\overline{SB}$ must exceed the lengths of $\overline{AF}$ plus $\overline{FB}$. It follows that S must lie outside of an ellipse passing through the point F where the ellipse has foci at A and B. Let G be the point where the ellipse intersects ray $\overrightarrow{EA}$ as shown.

Without loss of generality let the length of $\overline{AE}$ be 1. Then $\overline{EF}$ has length 1, and $\overline{AF}$ and $\overline{BF}$ both have length $\sqrt{2}$. By the properties of an ellipse, the lengths of $\overline{BG}$ and $\overline{AG}$ add to the same as the lengths of $\overline{AF}$ and $\overline{BF}$. Thus, $2\sqrt{2} = BG + AG = (EG + 1) + (EG - 1) = 2EG$, so $\overline{EG}$ has length $\sqrt{2}$. Stretch the above figure vertically upward by a factor of $\sqrt{2}$ so that the ellipse becomes a circle with radius $\sqrt{2}$ as shown.

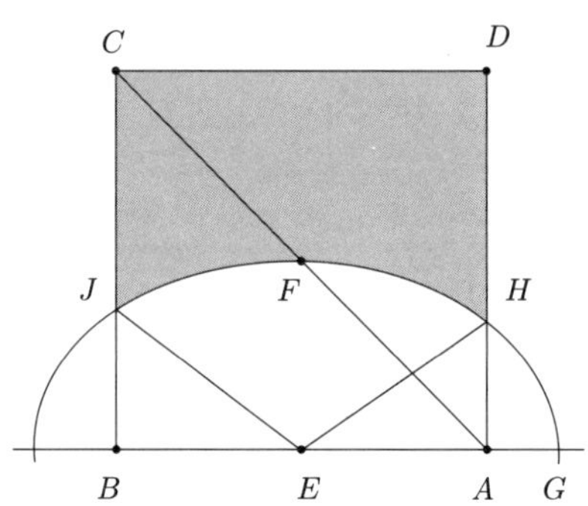

Let H be the intersection of $\overline{AD}$ and the circle, and J be the intersection of $\overline{BC}$ and the circle. Since the circle has radius $\sqrt{2}$ and segments $\overline{EA}$ and $\overline{BE}$ have length 1, it follows that

$\overline{AH}$ and $\overline{BJ}$ also have length 1. Thus, angles BEJ and AEH are both 45 degrees, and JEH is a right angle. It follows that the non-shaded region of the rectangle $ABCD$ is made up of a quarter of a circle and two congruent right isosceles triangles. The non-shaded region, therefore, has area

$$\frac{\pi\sqrt{2}^2}{4} + 2\frac{1\cdot 1}{2} = \frac{\pi+2}{2}.$$

The corresponding region in the original non-stretched version of the diagram must be smaller by a factor of $\sqrt{2}$, or be

$$\frac{\pi+2}{2\sqrt{2}} = \frac{\pi\sqrt{2}+2\sqrt{2}}{4}.$$

Since the original square has area 4, the probability that S lies in the shaded region of the original diagram is

$$1 - \frac{\frac{\pi\sqrt{2}+2\sqrt{2}}{4}}{4} = \frac{16-\pi\sqrt{2}-2\sqrt{2}}{16} = \frac{16-\pi\sqrt{2}-\sqrt{8}}{16}.$$

Thus, the requested sum is

$$ab + cd = (16)(2) + (8)(16) = 160.$$

2.11 2008 Middle School

Problem 1. Find the greatest prime factor in the sum of the two largest two-digit prime numbers.

Answer: 31

The two largest two-digit prime numbers are 89 and 97 whose sum is $186 = 2 \times 3 \times 31$.

Problem 2. Starting on April 15, 2008, you can go one day backward and one day forwards to get the calendar dates 14 and 16. If you go 15 days backward and 15 days forward, you get the calendar dates 31 (from March) and 30 (from April). Find the least positive integer k so that if you go k days backward and k days forward you get two calendar dates that are the same.

[This problem was originally given in 2008, a leap year.]

Answer: 76

For k less than 15, the backward date will be smaller than the forward date. For k from 15 to 45, the backward day is in March and cannot be equal to the forward day since one of the two dates will be an even number and the other an odd number. When $k = 47$, the days are 29 (from February) and 31 (from May). For k between 48 and 74, one date will be even while the other is odd. When k is 75, the backward date is 31 (from January) and the forward date is 29 (from June). Thus, when k is 76, both the backward and forward dates are 30.

Problem 3. There were 891 people voting at precinct 91. There were 20 percent more female voters than male voters. How many female voters were there?

Answer: 486

If there were M male voters and F female voters, then the problem states that

$$M + F = 891 \text{ and } M + 0.20M = F, \text{ or } F = 1.2M.$$

Combine these to get $1.2M + M = 891$ and $2.2M = 891$. Conclude that

$$M = \frac{891}{2.2} = \frac{8910}{22} = 405 \quad \text{and} \quad F = 891 - 405 = 486.$$

Problem 4. While driving his car, Ken pulled off the road to get gasoline when he was $\frac{7}{12}$ of the way through his trip. After driving another eleven miles, he noticed that he was $\frac{13}{20}$ of the way through his trip. How many miles long was his entire trip?

Answer: 165

If the entire trip was length x, the distances given in the problem imply

$$\frac{7}{12}x + 11 = \frac{13}{20}x.$$

Thus,

$$11 = \frac{13}{20}x - \frac{7}{12}x = \frac{3(13) - 5(7)}{60}x = \frac{39 - 35}{60}x = \frac{1}{15}x.$$

We conclude that the entire trip was length $15(11) = 165$.

Problem 5. What is the measurement in degrees of the angle formed by the minute and hour hands when a clock reads 12:12?

Answer: 66

Since the minute hand makes one revolution every hour, and the hour hand makes a revolution every 12 hours, the minute hand turns 12 times faster than the hour hand. At 12:00 the minute hand and the hour hand point in the same direction. In 12 minutes the minute hand covers $\frac{12}{60} = \frac{1}{5}$ of a revolution covering an arc of $\frac{360}{5} = 72$ degrees. The hour hand covers an arc only $\frac{1}{12}$ as large or 6 degrees. Thus, the angle between the two hands is 72 - 6 = 66 degrees.

Problem 6. The fraction

$$\frac{2}{\dfrac{2}{\dfrac{2}{\dfrac{2}{2+1}+1}+1} + 1 + \dfrac{2}{1 + \dfrac{2}{1 + \dfrac{2}{2+1}}}}$$

can be written in the form $\frac{m}{n}$, where m and n are relatively prime positive integers. Find $2m + n$.

Answer: 75

$$\frac{2}{\dfrac{2}{\dfrac{2}{\dfrac{2}{2+1}+1}+1}+1+\dfrac{2}{1+\dfrac{2}{1+\dfrac{2}{2+1}}}}-\frac{2}{\dfrac{2}{\dfrac{2}{\dfrac{2}{3}+1}+1}+1+\dfrac{2}{1+\dfrac{2}{1+\dfrac{2}{3}}}}$$

$$=\frac{2}{\dfrac{2}{\dfrac{2}{\frac{5}{3}}+1}+1+\dfrac{2}{1+\dfrac{2}{\frac{5}{3}}}}=\frac{2}{\dfrac{2}{\dfrac{6}{5}+1}+1+\dfrac{2}{1+\dfrac{6}{5}}}=\frac{2}{\dfrac{2}{\frac{11}{5}}+1+\dfrac{2}{\frac{11}{5}}}$$

$$=\frac{2}{\dfrac{10}{11}+1+\dfrac{10}{11}}=\frac{22}{10+11+10}=\frac{22}{31}.$$

So, the requested value is $2 \cdot 22 + 31 = 75$.

Problem 7. The diagram below shows an isosceles triangle with base 21 and height 28. Inscribed in the triangle is a square. Find the area of the shaded region inside the triangle and outside of the square.

Answer: 150

Let the vertices of the triangle be labeled A, B, C as shown. Let D and E be the points where the corners of the square meets sides $\overline{AB}$ and $\overline{BC}$, respectively. Suppose that the square has side length x.

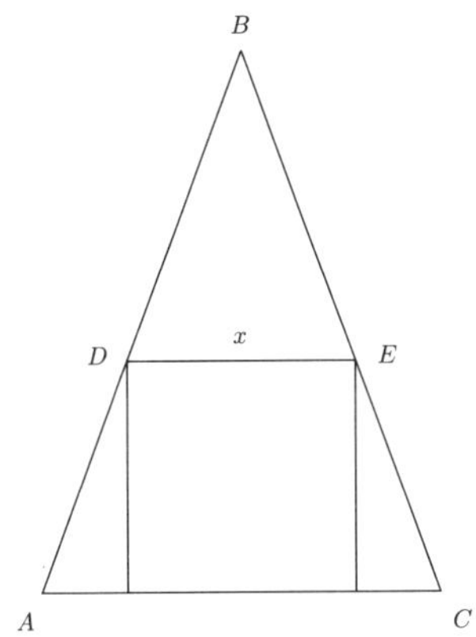

Since triangle DBE is similar to triangle ABC, the height of triangle DBE is $\frac{28}{21}x = \frac{4}{3}x$. It follows that the height of triangle ABC is equal to the length of the side of the square plus the height of triangle DBE, or $x + \frac{4}{3}x = \frac{7}{3}x = 28$. Therefore, x is 12. The area of triangle ABC minus the area of the square is

$$\frac{21 \cdot 28}{2} - 12^2 = 294 - 144 = 150.$$

Problem 8. At Mallard High School there are three intermural sports leagues: football, basketball, and baseball. There are 427 students participating in these sports: 128 play on football teams, 291 play on basketball teams, and 318 play on baseball teams. If exactly 36 students participate in all three of the sports, how many students participate in exactly two of the sports?

Answer: 238

The inclusion-exclusion principal states that the number of elements in the union of three sets A, B, and C is given by

$$\#(A \cup B \cup C) = \#(A) + \#(B) + \#(C) - \#(A \cap B) - \#(A \cap C) - \#(B \cap C) + \#(A \cap B \cap C).$$

If we let A be the set of students playing football, B be the set of students playing basketball, and C be the set of students playing baseball, we get

$$427 = 128 + 291 + 318 - \#(A \cap B) - \#(A \cap C) - \#(B \cap C) + 36.$$

So

$$\#(A \cap B) + \#(A \cap C) + \#(B \cap C) = 128 + 291 + 318 - 427 + 36 = 346.$$

Since we are interested in the number of students playing exactly two sports, we must subtract the 36 students from each intersection to give

$$\#(A\cap B)-36+\#(A\cap C)-36+\#(B\cap C)-36 = 346-3(36) = 238.$$

Alternatively, one could complete a Venn diagram showing the number of students playing in the three sports.

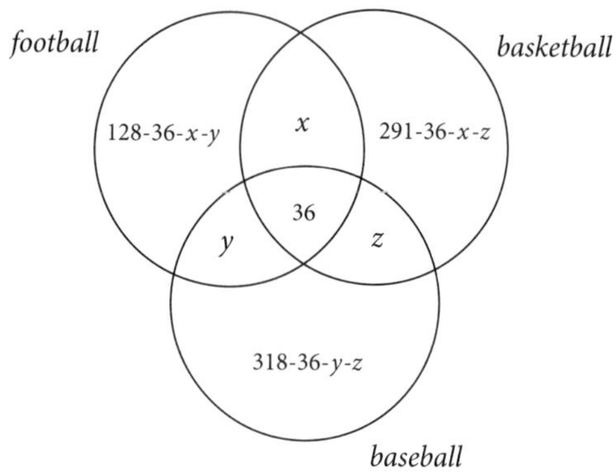

Then the total number of students participating in the three sport is

$$\begin{aligned}427 &= (128-36-x-y)+(291-36-x-z)\\ &\quad+(318-36-y-z)+x+y+z+36\\ &= 665-(x+y+z).\end{aligned}$$

So, the number of students playing exactly two sports is

$$x+y+z = 665-427 = 238.$$

Problem 9. One container of paint is exactly enough to cover the inside of an old rectangle which is three times as long as it is wide. If we make a new rectangle by shortening the old rectangle by 18 feet and widening it by 8 feet as shown below, one container of paint is also exactly enough to cover the inside of the new rectangle. Find the length in feet of the perimeter of the new rectangle.

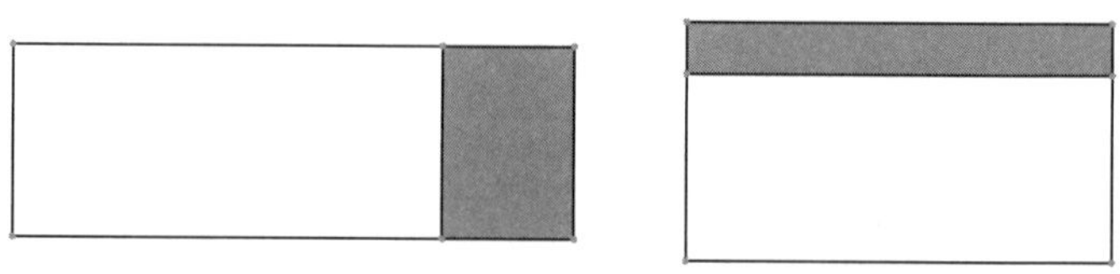

Answer: 172

Let the width of the old rectangle be x so that its length is $3x$. The new rectangle has width $x+8$ and length $3x-18$. Since the same amount of paint covers both rectangles, the two rectangles have the same area, and it follows that $3x \cdot x = (3x-18)(x+8)$ and $3x^2 = 3x^2 + 6x - 144$. This simplifies to $x = 24$. Thus, the old rectangle is 24 by 72, and the new rectangle is 32 by 54. The new rectangle has perimeter of $2(32+54) = 172$.

Problem 10. A 16×16 square sheet of paper is folded once in half horizontally and once in half vertically to make an 8×8 square. This square is again folded in half twice to make a 4×4 square. This square is folded in half twice to make a 2×2 square. This square is folded in half twice to make a 1×1 square. Finally, a scissor is used to make cuts through both diagonals of all the layers of the 1×1 square. How many pieces of paper result?

Answer: 544

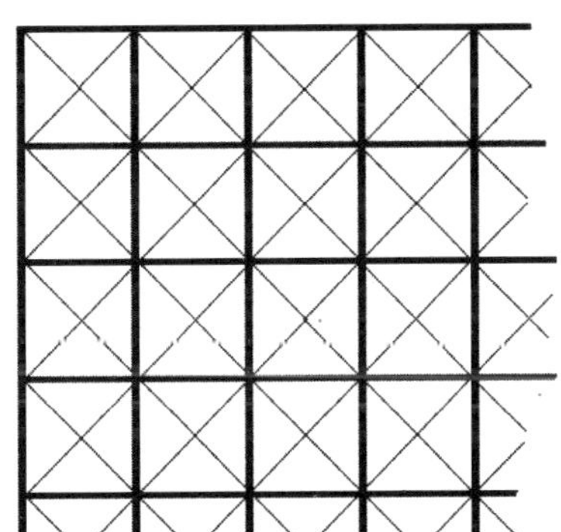

Instead of cutting along the diagonals of the 1×1 square, draw the diagonals of each of the 1×1 squares in every layer of the folded sheet. Then unfold the sheet of paper to reveal a 16×16 square with diagonals drawn on each of the 1×1 squares that make it up. It is then easy to see that if the diagonals had been cut, there would be exactly one piece of paper for each edge of a 1×1 square in the 16×16 grid including the edges on the side of the paper. To count the edges, notice that each vertical line of this grid has 16 edges of a 1×1 square along it. There are 17 such vertical lines. Similarly, each horizontal line of this grid has 16 edges of a 1×1 square along it, and there are 17 such horizontal lines. Thus, the total number of pieces of paper is $2 \times 16 \times 17 = 544$.

Problem 11. When Tim was Jim's age, Kim was twice as old as Jim. When Kim was Tim's age, Jim was 30. When Jim becomes Kim's age, Tim will be 88. When Jim becomes Tim's age, what will be the sum of the ages of Tim, Jim, and Kim?

Answer: 221

Let the current ages of Tim, Jim, and Kim be represented by T, J, and K, respectively. Then Tim was Jim's age $T-J$ years ago, so $K-(T-J) = 2[J-(T-J)]$ which simplifies to $T-3J+K = 0$. Kim was Tim's age $K-T$ years ago, so $J-(K-T) = 30$ which simplifies to $T+J-K = 30$. Jim will be Kim's age in $K-J$ years, so $T+(K-J) = 88$ which simplifies to $T-J+K = 88$. Adding the last two of these equations shows $2T = 118$, so Tim is currently 59. It follows that $K-J = 29$ and $K-3J = -59$, so $2J = 29+59 = 88$, and $J = 44$. Now it follows that $K = 73$. So, when Jim is Tim's age in $T-J = 15$ years, the sum of their ages will be $(59+15)+(44+15)+(73+15) = 221$.

Problem 12. A city is laid out with a rectangular grid of roads with 10 streets numbered from 1 to 10 running east-west and 16 avenues numbered from 1 to 16 running north-south. All streets end at First and Sixteenth Avenues, and all avenues end at First and Tenth Streets. A rectangular city park is bounded on the north and south by Sixth Street and Eighth Street, and bounded on the east and west by Fourth Avenue and Twelfth Avenue. Although there are no breaks in the roads that bound the park, no road goes through the park. The city paints a crosswalk from every street corner across any adjacent road. Thus, where two roads cross such as at Second Street and Second Avenue, there are four crosswalks painted, while at corners such as First Street and First Avenue, there are only two crosswalks painted. How many crosswalks are there painted on the roads of this city?

Answer: 544

At the four corners of the grid (where First Street and Tenth Street meet First Avenue and Sixteenth Avenue) there are two crosswalks. At all other intersections on these four roads there are three crosswalks. There are also three crosswalks at the two sides of the park on Seventh Street and the two sides of the park along Fifth through Eleventh Avenues. There are no crosswalks in the park along Seventh Street on Fifth through Eleventh Avenues. So of the $10 \times 16 = 160$ possible intersections in the city accounting for a possible $4 \times 160 = 640$ crosswalks, we need to subtract $4 \times 2 + 44 \times 1 + 16 \times 1 + 7 \times 4 = 96$. This leaves $640 - 96 = 544$ crosswalks.

Problem 13. Let $A_1A_2A_3 \ldots A_{12}$ be a regular dodecagon. Find the number of right triangles whose vertices are in the set

$$\{A_1, A_2, A_3, \ldots, A_{12}\}.$$

Answer: 60

The regular dodecagon can be inscribed in a circle, and, thus, for any right triangle whose vertices are in $\{A_1, A_2, A_3, \ldots, A_{12}\}$ the right angle will subtend an arc of a semicircle. The dodecagon has six diagonals which are diameters of the circle, so there are six pairs of vertices that can serve as end points for the hypotenuse of a right triangle. For each of these six diameters, there are ten vertices off the diameter which can be chosen as the other vertex of the triangle giving $6 \cdot 10 = 60$ triangles.

Problem 14. Ralph is standing along a road which heads straight east. If you go nine miles east, make a left turn, and travel seven miles north, you will find Pamela with her mountain bike. At exactly the same time that Ralph begins running eastward along the road at 6 miles per hour, Pamela begins biking in a straight line at 10 miles per hour. Pamela's direction is chosen so that she will reach a point on the road where Ralph is running at exactly the same time Ralph reaches that same point. Let m and n be relatively prime positive integers such that $\frac{m}{n}$ is the number of hours that it takes Pamela and Ralph to meet. Find $m + n$.

Answer: 29

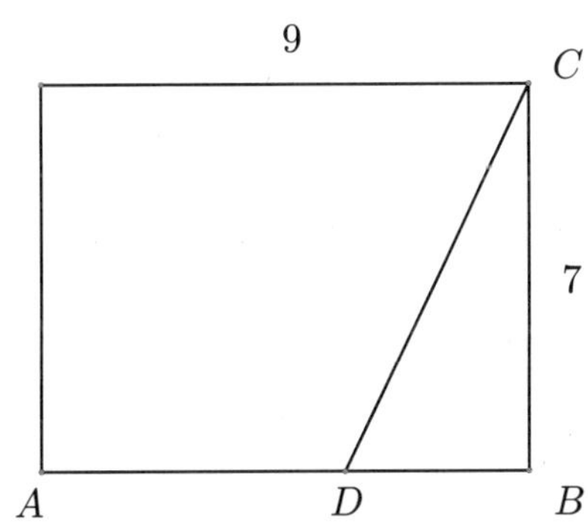

The positions are shown in this diagram where Ralph begins running at point A toward point B, and Pamela begins biking at point C toward point D between A and B.

If they meet at point D after a time t, then Ralph will have run a distance $AD = 6t$, and Pamela will have biked a distance $CD = 10t$. By the Pythagorean Theorem $DB^2 + BC^2 = CD^2$ or $(9-6t)^2 + 7^2 = (10t)^2$. This reduces to $64t^2 + 108t - 130 = 0$ which has the positive solution $t = \frac{13}{16}$. Thus, the requested sum is $13 + 16 = 29$.

Problem 15. Each of the distinct letters in the following subtraction represents a different digit. Find the number represented by the word $TEAM$.

$$\begin{array}{rrrrrr} P & U & R & P & L & E \\ - & C & O & M & E & T \\ \hline & & T & E & A & M \end{array}$$

Answer: 6852

Considering that $TEAM$ is a four digit number which is the difference between a six digit number and a five digit number, P must represent 1, U must represent 0, and C must represent 9. The equivalent addition problem is

$$\begin{array}{rrrrrr} & C & O & M & E & T \\ & & T & E & A & M \\ \hline P & U & R & P & L & E \end{array} \qquad \begin{array}{rrrrrr} & 9 & O & M & E & T \\ & & T & E & A & M \\ \hline 1 & 0 & R & 1 & L & E \end{array}$$

Observe that there must be a carry from the thousands column to the ten-thousands column and a carry from the hundreds column to the thousands column. There must also be at least one other carry, for it not, then $T + M = E$ and $E + A = L$ so $T + M + A = L$. But considering the available digits $\{2, 3, 4, 5, 6, 7, 8\}$, $T+M+A$ must be at least $2+3+4 = 9$ which is too big to be L. If both the ones and tens columns have carries, then $T + M = 10 + E$, $E + A + 1 = 10 + L$, and $M + E + 1 = 11$ from which one gets $T + M + E + A + M + E + 2 = E + L + 31$ which reduces to $T+2M+A+E = L+29$, and since $M+E = 10$, one gets $T + A = E + L + 11$. But $T + A$ is at most $7 + 8 = 15$ while $E + L$ is at least $2 + 3 = 5$ implying $15 \geq 5 + 11$ which is false. Therefore, there is exactly one carry from either the ones and tens column.

If there is a carry from the ones column, then $T+M=E+10$, $E+A+1=L$, and $M+E=11$. Then $T+(11-E)=E+10$ or $T+1=2E$, and since T cannot exceed 8, E cannot exceed 4. Also, since $M+E=11$ and M cannot exceed 8, E must be at least 3. It is not hard to eliminate the possibility that $E=3$ or $E=4$ since the value of E determines the value of M and T. It follows that there is no carry from the ones column, and there is a carry from the tens column.

Thus, $T+M=E$, $E+A=L+10$, $M+E+1=11$, and $O+T+1=R+10$. Then $T+(11-E)=E+10$ or $T+1=2E$, Thus, E must exceed 5, and T must be even. If $E=6$, it follows that $M=4$, and $T=2$. Then the only choices for the tens column are $A=7$ and $L=3$, and there are no possible values for O and R which allow a carry in the thousands column. If $E=7$, it follows that $M=3$ and $T=4$. Again, there is no way to apportion the remaining four digits to the remaining four positions. Finally, if $E=8$, it follows that $M=2$ and $T=6$. It is then possible to apportion the remaining four digits $\{3,4,5,7\}$ to the remaining letters $\{A,L,O,R\}$ as shown here:

$$\begin{array}{r} 9\,7\,2\,8\,6 \\ 6\,8\,5\,2 \\ \hline 1\,0\,4\,1\,3\,8 \end{array}$$

2.12 2009 High School

Problem 1. In five years, Tom will be twice as old as Cindy. Thirteen years ago, Tom was three times as old as Cindy. How many years ago was Tom four times as old as Cindy?

Answer: 19

Let Tom be age t, and Cindy be age c. Then the first two statements of the problem say

$$t + 5 = 2(c + 5) \text{ and } t - 13 = 3(c - 13).$$

The first of these two equations simplifies to $t = 2c + 5$. Substituting this into the second equation gives

$$(2c + 5) - 13 = 3(c - 13)$$

which simplifies to $2c - 8 = 3c - 39$ and $c = 31$. So, $t = 2c + 5 = 2(31) + 5 = 67$. Cindy is now 31, and Tom is 67. The answer to the problem is n where $67 - n = 4(31 - n)$. This simplifies to $67 - n = 124 - 4n$ and $3n = 57$. Thus, $n = 19$, and Tom was four times as old as Cindy 19 years ago.

Problem 2. Find the least positive integer n such that for every prime number p, $p^2 + n$ is never a prime.

Answer: 5

$2^2 + 1 = 5$, $3^2 + 2 = 11$, $2^2 + 3 = 7$, and $3^2 + 4 = 13$ are all prime numbers showing that n cannot be 1, 2, 3, or 4. When p is 2, $p^2 + 5 = 9$ which is not prime. When p is a prime greater than 2, $p^2 + 5$ is an even number greater than 2, so it is not prime. Thus, 5 is the least integer satisfying the given condition.

Problem 3. In the diagram $ABCDEFG$, is a regular heptagon. Shown is the star $AEBFCGD$. The degree measure of the obtuse angle formed by $\overline{AE}$ and $\overline{CG}$ is $\frac{m}{n}$, where m and n are relatively prime positive integers. Find $m + n$.

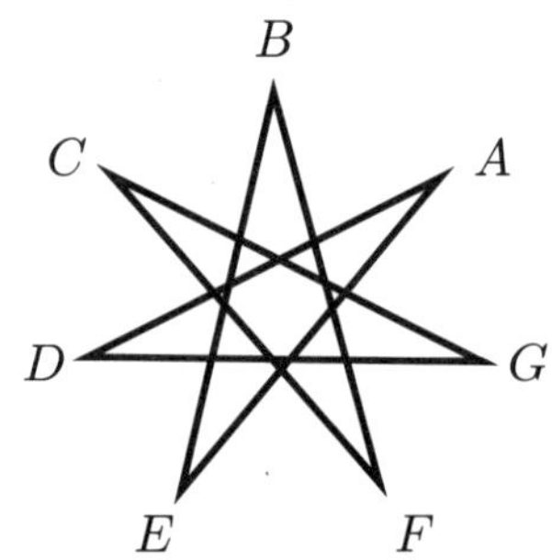

Answer: 727

Let H be the point at the intersection of segments $\overline{AE}$ and $\overline{CG}$. The desired angle AHC is an exterior angle of triangle AHG, so the measure of the desired angle is equal to the sum of the measures of angle HGA and angle HAG. These two angles are congruent. Angle GAB is an angle of a regular heptagon, so its measure is $\frac{180\cdot(7-2)}{7}$. The vertices of the regular heptagon lie on a circle, and the arc of that circle from point D to point E measures $\frac{360}{7}$, and, thus, angle DAE measures $\frac{180}{7}$. Therefore, the desired angle AHC is twice the angle HAG, and this is the difference between the measures of angle BAG and angle DAE. The desired measure is $\frac{180\cdot 5}{7} - \frac{180}{7} = \frac{720}{7}$. The requested sum is $720 + 7 = 727$.

Alternatively, one can rotate the line $\overline{AE}$ to line $\overline{BE}$, then rotate line $\overline{BE}$ to line $\overline{BF}$, then rotate line $\overline{BF}$ to line $\overline{CF}$, and finally rotate line $\overline{CF}$ to line $\overline{CG}$. Each of these four rotations is, as seen above, a rotation through an angle of $\frac{180}{7}$. The four rotations end up rotating line $\overline{AE}$ through an angle of $4 \cdot \frac{180}{7} = \frac{720}{7}$.

Problem 4. There are three bags of marbles. Bag two has twice as many marbles as bag one. Bag three has three times as many marbles as bag one. Half the marbles in bag one, one third the marbles in bag two, and one fourth the marbles in bag three are green. If all three bags of marbles are dumped into a single pile, $\frac{m}{n}$ of the marbles in the pile would be green, where m and n are relatively prime positive integers. Find $m + n$.

Answer: 95

Assume that the first bag contains K marbles. Then the second bag contains $2K$ marbles, and the third bag contains $3K$ marbles. The total number of green marbles is $\frac{1}{2}\cdot K+\frac{1}{3}\cdot 2K+\frac{1}{4}\cdot 3K$, and the total number of marbles is $K + 2K + 3K = 6K$. The desired fraction is therefore

$$\frac{\frac{1}{2}\cdot K+\frac{1}{3}\cdot 2K+\frac{1}{4}\cdot 3K}{6K} = \frac{\frac{1}{2}+\frac{2}{3}+\frac{3}{4}}{6} = \frac{23}{72}.$$

The requested sum is $23 + 72 = 95$.

Problem 5. Find n so that $\left(4^{n+7}\right)^3 = \left(2^{n+23}\right)^4$.

Answer: 25

Simplify the equation $\left(4^{n+7}\right)^3 = \left(2^{n+23}\right)^4$ to get $\left(2^2\right)^{3(n+7)} = 2^{4(n+23)}$ and $2^{6(n+7)} = 2^{4(n+23)}$. It follows that $6n+42 = 4n+92$ and $2n = 50$. Thus, $n = 25$.

Problem 6. Wiles county contains eight townships as shown on the map. If there are four colors available, in how many ways can the the map be colored so that each township is colored with one color and no two townships that share a border are colored with the same color?

Answer: 576

Label the townships as shown. There are four choices for colors for township A, and three choices for colors for township B. If townships C and D have the same color, then there are two choices for the color of C and D, and there are two choices for colors for each of the remaining townships E, F, G, and H. If townships C and D have different colors, then there are two choices for the color of C and one choice for a color for D. Since B, C, and D will all have different colors, there is only one choice of color for township E. Then there are two choices for colors for each of the remaining townships F, G, and H. The total number of ways to color the eight townships is then $4 \cdot 3 \cdot 2 \cdot 2 \cdot 2 \cdot 2 \cdot 2 + 4 \cdot 3 \cdot 2 \cdot 1 \cdot 2 \cdot 2 \cdot 2 = 4 \cdot 3 \cdot 2 \cdot 3 \cdot 2 \cdot 2 \cdot 2 = 576$.

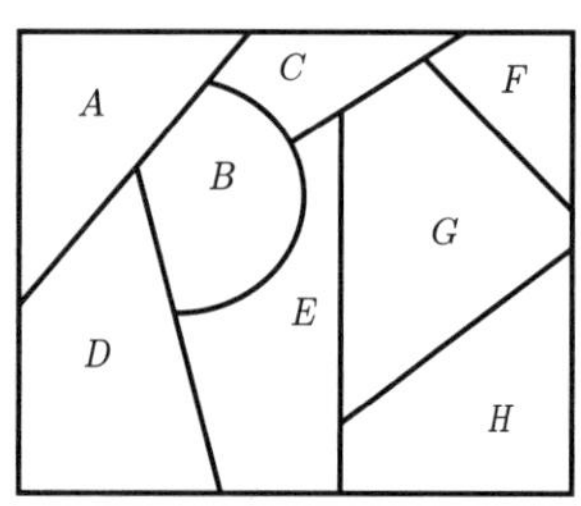

Problem 7. Figure $ABCD$ is bounded by a semicircle CDA and a quarter circle ABC. Given that the distance from A to C is 18, find the area of the figure.

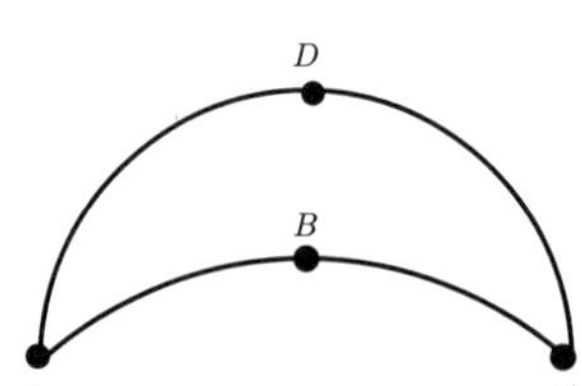

Answer: 81

Let F be the center of the semicircle, and E be the center of the quarter circle as shown. The requested area is the area of semicircle $AFCD$ minus the area of quarter circle $AECB$ plus the area of the triangle AEC. Since FC is 9 and EC is $9\sqrt{2}$, the requested area is

$$\frac{9^2\pi}{2} - \frac{(9\sqrt{2})^2\pi}{4} + \frac{18 \cdot 9}{2} = 81.$$

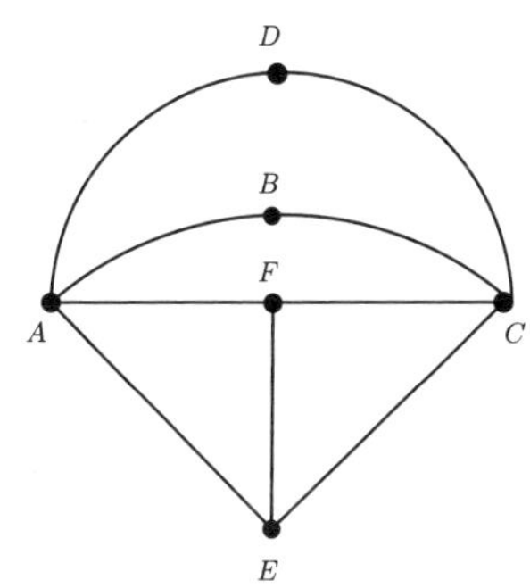

Problem 8. Find the least positive integer that has exactly 20 positive integer divisors.

Answer: 240

If p is a prime number, then the prime power p^a has $a+1$ divisors. A product of prime powers $p_1^{a_1} p_2^{a_2} \cdots p_k^{a_k}$ has

$$(a_1 + 1)(a_2 + 1) \cdots (a_k + 1)\text{divisors.}$$

For a number to have 20 positive integer divisors,

$$(a_1 + 1)(a_2 + 1) \cdots (a_k + 1)$$

must be $20 = 2 \cdot 10 = 4 \cdot 5 = 2 \cdot 2 \cdot 5$. The minimal positive integers corresponding to these four possibilities are

$$2^{19} = 524,288, \ 2^9 \cdot 3 = 1536, \ 2^4 \cdot 3^3 = 432, \text{ and } 2^4 \cdot 3 \cdot 5 = 240.$$

The least of these is 240.

Problem 9. Bill bought 13 notebooks, 26 pens, and 19 markers for 25 dollars. Paula bought 27 notebooks, 18 pens, and 31 markers for 31 dollars. How many dollars would it cost Greg to buy 24 notebooks, 120 pens, and 52 markers?

Answer: 88

If notebooks cost x dollars, pens cost y dollars, and markers cost z dollars, then the problem states that

$$13x + 26y + 19z = 25 \text{ and } 27x + 18y + 31z = 31.$$

It follows that the cost in dollars to Greg would be

$$\begin{aligned} 24x + 120y + 52z &= 6(13x + 26y + 19z) - 2(27x + 18y + 31z) \\ &= 6(25) - 2(31) = 88. \end{aligned}$$

Problem 10. Towers grow at points along a line. All towers start with height 0 and grow at the rate of one meter per second. As soon as any two adjacent towers are each at least 1 meter tall, a new tower begins to grow at a point along the line exactly half way between those two adjacent towers. Before time 0 there are no towers, but at time 0 the first two towers begin to grow at two points along the line. Find the total of all the heights of all towers at time 10 seconds.

Answer: 1033

Two towers begin to grow at time 0. At time 1, the first two towers reach a height of 1 meter, so one new tower begins to grow. At time 2, the new tower reaches a height of 1, so two new towers begin to grow, one between the new tower and the first old tower, and one between the new tower and the second old tower. Each second thereafter, twice as many new towers begin to grow, so at time 3, 2^2 towers begin, at time 4, 2^3 towers begin, and so forth until at time 9, 2^8 towers begin to grow. A tower that begins to grow at time k reaches a height of $10 - k$ by time 10. Thus, the total height of all towers at time 10 is

$$2 \cdot 10 + \sum_{k=1}^{9} 2^{k-1}(10-k) = 20 + \sum_{k=1}^{9} \sum_{j=1}^{k} 2^{j-1}$$

$$= 20 + \sum_{k=1}^{9} (2^k - 1) = 20 + 2^{10} - 11 = 9 + 1024 = 1033.$$

Problem 11. The four points $A(-1,2)$, $B(3,-4)$, $C(5,-6)$, $D(-2,8)$ lie in the coordinate plane. Compute the minimum possible value of $PA + PB + PC + PD$ over all points P.

Answer: 23

Let P_0 be the intersection of $\overline{AC}$ and $\overline{BD}$. Note that P_0 is between A and C, and also between B and D. Therefore, if P is not P_0, then by the triangle inequality either $PA + PC > P_0A + P_0C$ or $PB + PD > P_0B + P_0D$, or both. Thus, the quantity has its minimum value at P_0, and this quantity is equal to $AC + BD$.

$$AC = \sqrt{[5-(-1)]^2 + [-6-2]^2} = 10$$
$$BD = \sqrt{[-2-3]^2 + [8-(-4)]^2} = 13.$$

So the minimum value of the desired quantity is $10 + 13 = 23$.

Problem 12. What is the least possible sum of two positive integers a and b where $a \cdot b = 10!$?

Answer: 3810

$10! = 1 \cdot 2 \cdot 3 \cdot 4 \cdot 5 \cdot 6 \cdot 7 \cdot 8 \cdot 9 \cdot 10 = 2^8 \cdot 3^4 \cdot 5^2 \cdot 7$. To make the sum $a+b$ as small as possible, a and b should be as close to each other as possible where $a \cdot b$ contains 8 factors of 2, 4 factors of 3, 2 factors of 5, and a factor of 7. This can be found by finding the best way to approximate 7 as even powers of 2, 3, and 5. For example, 7 is close to 3^2, closer to $\frac{5^2}{2^2}$, but very close to $\frac{2^6}{3^2}$. Thus, the optimal factoring of 10! is found by taking $2 \cdot 3 \cdot 5$ and multiplying it by either 2^6 or $3^2 \cdot 7$ to get the factoring $(2^7 \cdot 3 \cdot 5)(2 \cdot 3^3 \cdot 5 \cdot 7) = 1920 \cdot 1890$. Thus, the least sum is $1920 + 1890 = 3810$.

Problem 13. Greta is completing an art project. She has twelve sheets of paper: four red, four white, and four blue. She also has twelve paper stars: four red, four white, and four blue. She randomly places one star on each sheet of paper. The probability that no star will be placed on a sheet of paper that is the same color as the star is $\frac{m}{n}$, where m and n are relatively prime positive integers. Find $n - 100m$.

Answer: 25

$\frac{m}{n}$ is the probability that if we separate the twelve stars into a red set, a white set, and a blue set, all size four, no star gets placed into the set of its color. The number of equally likely ways of placing 12 stars into 3 distinguishable sets of 4 is given by the multinomial coefficient

$$\binom{12}{4,4,4} = \frac{12!}{4! \cdot 4! \cdot 4!}.$$

Now count the number of ways of placing the stars so that no star ends up in a set with the same color as the star. In such a placement, let r be the number of red stars in the white set. Then there are $4-r$ blue stars in the white set, so there will be r blue stars in the red set. Then there are $4-r$ white stars in the red set, so there will be r white stars in the blue set. It follows that the number of placements with r red stars in the white set is $\binom{4}{r}^3$. Thus, the number of placements of stars so that no star ends up in a set with the same color as the star is

$$\sum_{r=0}^{4} \binom{4}{r}^3 = 1^3 + 4^3 + 6^3 + 4^3 + 1^3 = 173.$$

Because

$$\binom{12}{4,4,4} = \frac{12 \cdot 11 \cdot 10 \cdot 9 \cdot 8 \cdot 7 \cdot 6 \cdot 5 \cdot 4 \cdot 3 \cdot 2 \cdot 1}{4 \cdot 3 \cdot 2 \cdot 1 \cdot 4 \cdot 3 \cdot 2 \cdot 1 \cdot 4 \cdot 3 \cdot 2 \cdot 1} = 17,325,$$

the requested difference is $17325 - 100 \cdot 173 = 25$.

Problem 14. Let $ABCD$ be a trapezoid with $\overline{AB}$ parallel to $\overline{CD}$, base $\overline{AB}$ has length 1, and base $\overline{CD}$ has length 41. Let points X and Y lie on sides $\overline{AD}$ and $\overline{BC}$, respectively, such that $\overline{XY}$ is parallel to $\overline{AB}$ and $\overline{CD}$, and $\overline{XY}$ has length 31. Let m and n be relatively prime positive integers such that the ratio of the area of $ABYX$ to the area of $CDXY$ is $\frac{m}{n}$. Find $m + 2n$.

Answer: 10

Let E be the intersection of line $\overline{AD}$ and line $\overline{BC}$. Then triangles EAB, EXY, and EDC are similar. If triangle EAB has area α, then EXY has area $31^2\alpha$, and EDC has area $41^2\alpha$. It follows that the desired ratio is

$$\frac{31^2\alpha - \alpha}{41^2\alpha - 31^2\alpha} = \frac{961 - 1}{1681 - 961} = \frac{960}{720} = \frac{4}{3},$$

and the requested sum is $4 + 2 \cdot 3 = 10$.

Problem 15. What is the remainder when 7^{8^9} is divided by 1000?

Answer: 801

The Euler ϕ function tells how many positive integers less than a given integer are relatively prime to that integer. Thus, $\phi(1000) = \phi(8) \cdot \phi(125) = 4 \cdot 100 = 400$. Since for positive integers a and n the power $a^{\phi(n)}$ is congruent to 1 mod n, it follows that $7^{400} \equiv 1 \pmod{1000}$. So, $7^{8^9} \pmod{1000} \equiv 7^{8^9 \pmod{400}}$. Here

$$8^9 \equiv 2^{27} \equiv 2^{11}2^{11}2^5 \equiv 2048 \cdot 2048 \cdot 32 \equiv 48^2 \cdot 32 \equiv 96^2 \cdot 8$$
$$\equiv (100-4)^2 \cdot 8 \equiv 16 \cdot 8 \equiv 128 \pmod{400}.$$

Then

$$7^{8^9} \equiv 7^{128} \pmod{1000}.$$

Note that $7^2 = 49$ so

$$\begin{aligned} 7^4 &= (50-1)^2 = 2500 - 2 \cdot 50 + 1 = 2401 \equiv 401 \pmod{1000} \\ 7^8 &= (400+1)^2 = 160,000 + 800 + 1 \equiv 801 \pmod{1000} \\ 7^{16} &= (800+1)^2 = 640,000 + 1600 + 1 \equiv 601 \pmod{1000} \\ 7^{32} &= (600+1)^2 = 360,000 + 1200 + 1 \equiv 201 \pmod{1000} \\ 7^{64} &= (200+1)^2 = 40,000 + 400 + 1 \equiv 401 \pmod{1000}. \end{aligned}$$

It follows that $7^{8^9} \equiv 801 \pmod{1000}$.

Problem 16. Let $z = \cos \frac{1}{1000} + i \sin \frac{1}{1000}$. Find the least positive integer n so that z^n has an imaginary part that exceeds $\frac{1}{2}$.

Answer: 524

By DeMoivre's Theorem, $z^n = \cos \frac{n}{1000} + i \sin \frac{n}{1000}$. So n must satisfy $\sin \frac{n}{1000} > \frac{1}{2}$, and $\frac{n}{1000} > \frac{\pi}{6}$. Thus,

$$n > \frac{1000\pi}{6} > \frac{3141.59}{6} > 523.$$

Therefore, the smallest n satisfying the condition is $n = 524$.

Problem 17. How many ordered triples (a, b, c) of odd positive integers satisfy $a + b + c = 25$?

Answer: 78

Let (a, b, c) be an ordered triple of odd positive integers satisfying $a + b + c = 25$. Then $(a-1) + (b-1) + (c-1) = 22$ and $(x, y, z) = \left(\frac{a-1}{2}, \frac{b-1}{2}, \frac{c-1}{2}\right)$ is an ordered triple of nonnegative integers satisfying $x + y + z = 11$. Since there is a one-to-one correspondence between the ordered triples (a, b, c) in the problem and the ordered triples of nonnegative integers (x, y, z) satisfying $x + y + z = 11$, the answer is the number of ways to place 11 indistinguishable items into three distinguishable containers. This problem can be solved using the well-known *stars and bars* technique of counting the arrangements of 11 stars and two separating bars or $\binom{11+2}{2} = \binom{13}{2} = \frac{13 \cdot 12}{2} = 78$.

Alternatively, a solution to $a + b + c = 25$ in non-negative integers corresponds to any linear arrangement of 25 stars and 2 bars. One can obtain such an arrangement by starting with a list of 27 starts and changing two of the stars into bars. For the arrangement to correspond to a solution involving only odd integers, the two stars changed to bars must be stars in even numbered positions. There are 13 even numbered positions. The number of ways of choosing two such positions is, as above, $\binom{13}{2} = \frac{13 \cdot 12}{2} = 78$.

Problem 18. In triangle ABC let D be the point on $\overline{AB}$ so that $\overline{CD}$ is an altitude of the triangle, and let E be the point on $\overline{BC}$ so that $\overline{AE}$ bisects angle BAC. Let G be the intersection of $\overline{AE}$ and $\overline{CD}$, and let point F be the intersection of side $\overline{AC}$ and the ray $\overrightarrow{BG}$. If $AB = 28$, $AC = 14$, and $CD = 10$, then the length of $\overline{CF}$ can be written as $\frac{k - m\sqrt{p}}{n}$, where k, m, n, p are positive integers, k and n are relatively prime, and p is not divisible by the square of any prime. Find $k - m + n + p$.

Answer: 623

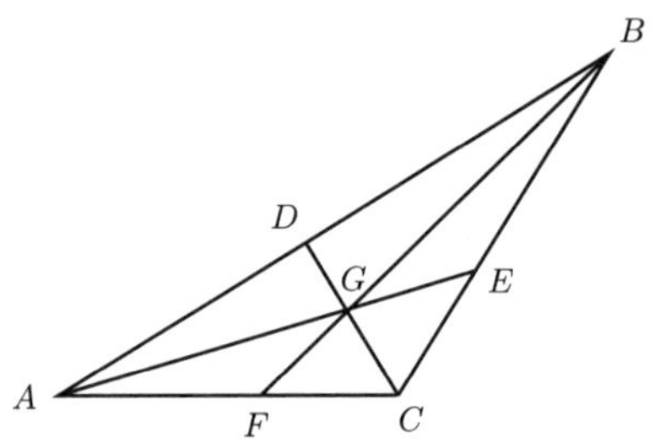

By the Pythagorean theorem $\overline{AD}$ has length $\sqrt{14^2 - 10^2} = 4\sqrt{6}$. Then $\overline{BD}$ has length $28 - 4\sqrt{6}$. By the angle bisector theorem, $\frac{BE}{CE} = \frac{AB}{AC} = 2$. It then follows from Ceva's Theorem that

$$1 = \frac{AD}{BD} \cdot \frac{BE}{CE} \cdot \frac{CF}{AF} = \frac{4\sqrt{6}}{28 - 4\sqrt{6}} \cdot 2 \cdot \frac{CF}{14 - CF}.$$

Thus, $14 - CF = \dfrac{2\sqrt{6}}{7 - \sqrt{6}} \cdot CF$ or

$$CF = 14 \cdot \frac{7 - \sqrt{6}}{7 + \sqrt{6}} = 14 \cdot \frac{7^2 - 14\sqrt{6} + 6}{7^2 - 6} = \frac{770 - 196\sqrt{6}}{43}.$$

The requested sum is then $770 - 196 + 43 + 6 = 623$.

Problem 19. If a and b are complex numbers such that $a^2 + b^2 = 5$ and $a^3 + b^3 = 7$, then their sum, $a + b$, is real. The greatest possible value for the sum $a + b$ is $\frac{m+\sqrt{n}}{2}$ where m and n are integers. Find n.

Answer: 57

Let $s = a + b$ and $r = a \cdot b$. Then

$$s^2 - 2r = (a + b)^2 - 2a \cdot b = a^2 + b^2 = 5$$

and

$$s^3 - 3r \cdot s = (a + b)^3 - 3a \cdot b \cdot (a + b) = a^3 + b^3 = 7.$$

It follows that

$$14 = 2s^3 - 3(2r)s = 2s^3 - 3(s^2 - 5)s = 2s^3 - 3s^3 + 15s = 15s - s^3.$$

Then

$$s^3 - 15s + 14 = (s - 1)(s^2 + s - 14) = 0.$$

Therefore, s can be 1 or $\frac{-1\pm\sqrt{1+4\cdot 14}}{2} = \frac{-1\pm\sqrt{57}}{2}$. The greatest possible value for $a + b = s$ is $\frac{-1+\sqrt{57}}{2}$, and the requested answer is 57.

Problem 20. Five men and seven women stand in a line in random order. Let m and n be relatively prime positive integers so that $\frac{m}{n}$ is the probability that each man stands next to at least one woman. Find $m + n$.

Answer: 287

Twelve people stand in a line. There are $\binom{12}{5} = 792$ equally likely ways to select the five positions where men will stand. To find the requested probability, count the number of these arrangements where each man stands next to at least one woman. Clearly, in such an arrangement no three men can stand together in the line, and no two men can stand together at either end of the line. The five men can all stand separately forming groups arranged as 1-1-1-1-1, or two can stand together forming groups arranged as 2-1-1-1, 1-2-1-1, 1-1-2-1, or 1-1-1-2, or two groups of men can stand together forming groups arranged as 2-2-1, 2-1-2, or 1-2-2.

For men to stand in groups 1-1-1-1-1, four women must stand between these men, and the other three women can stand in any of the six positions to the left of, to the right of, or between the men. The number of such arrangements can be counted using the well known stars and bars technique with three stars (women) and five bars (men) to give $\binom{8}{3} = 56$ arrangements where no two men stand together.

For men to stand in groups 2-1-1-1, three women must stand between the groups of men and one woman must stand to the left of the group of two men. The remaining three women can stand in any of five positions. Thus, there are $\binom{7}{3} = 35$ such arrangements. There are also 35 arrangements with the groupings 1-1-1-2. For men to stand in groups 1-2-1-1, three women must stand between the groups of men, and the remaining four women can stand in any of five positions. Thus, there are $\binom{8}{4} = 70$ such arrangements. There are also 70 arrangements with the groupings 1-1-2-1. This accounts for $35 + 35 + 70 + 70 = 210$ arrangements with one group of two men standing together.

For men to stand in groups 2-2-1, two women must stand between the groups of men and one woman must stand to the left of all the men. The remaining four women can stand in any of four positions. Thus, thee are $\binom{7}{3} = 35$ such arrangements. There are also 35 arrangements the groupings 1-2-2. For men to stand in groups 2-1-2, two women are needed to stand between the groups of men, and two women are needed to stand at the far right and far left of the men. The remaining three women can stand in any of four positions. Thus, there are $\binom{6}{3} = 20$ arrangements with the men grouped 2-1-2. This accounts for 35 $+ 35 + 20 = 90$ arrangements with two groups of men standing together.

The total number of arrangements of positions where men can stand so that every man stands next to at least one woman is $56 + 210 + 90 = 356$, and the required probability is $\frac{356}{792} = \frac{89}{198}$. The requested sum is $89 + 198 = 287$.

Alternatively, one can count the number of ways the men can stand so that at least one man does not stand next to a woman. If two men stand on the left end of the line, the other three men can stand in any of 10 positions in $\binom{10}{3} = 120$ ways. Similarly, there are 120 ways for two men to stand on the right end of the line, but 8 of these have two men at each end of the line. The total number of ways for two men to stand at the end of the line is $120 + 120 - 8 = 232$. For exactly three men to stand together, not at the end of the line, the three men can be placed in any of the six spaces between the women, and the other two men can stand individually in any of seven other positions or stand together in any of five positions. So the number of ways exactly three men can stand together is $6\left(\binom{7}{2}+5\right) = 6(21+5) = 156$. Exactly four men can stand together in one of the six spaces between the women, and the fifth man can stand in one of seven other spaces for a total of $6 \cdot 7 = 42$ ways. Finally, there are six ways for all five men to stand together. This gives a total of $232 + 156 + 42 + 6 = 436$. This leaves $792 - 436 = 356$ ways for the men to stand so that each man stands next to at least one woman. As above, this gives the answer 287.

Problem 21. A cylinder radius 12 and a cylinder radius 36 are held tangent to each other with a tight band. The length of the band is $m\sqrt{k} + n\pi$ where m, k, and n are positive integers and k is not divisible by the square of any prime. Find $m + k + n$.

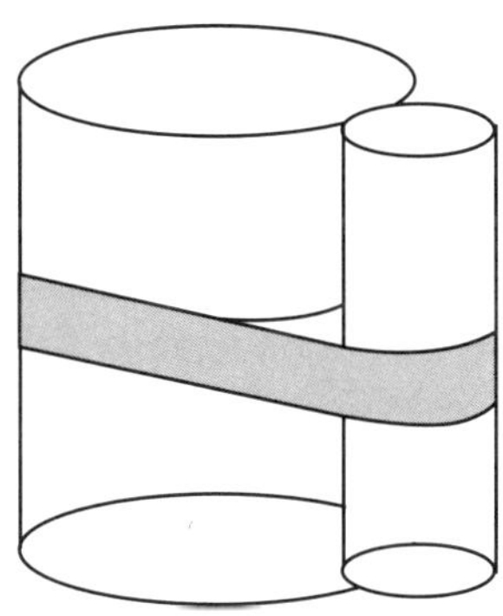

Answer: 107

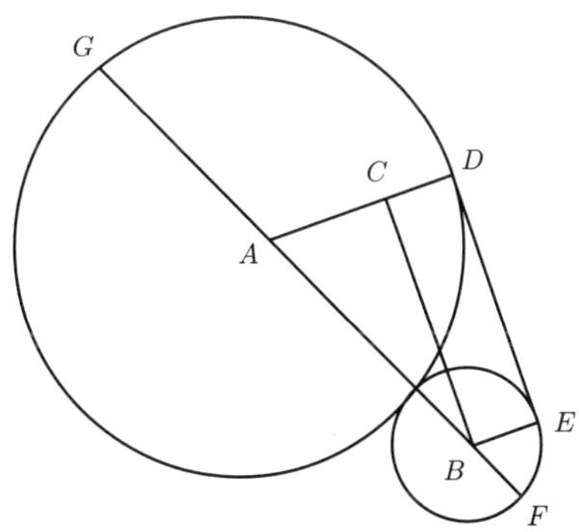

The diagram shows a cross section of the two cylinders and the band. Here the line segment $\overline{FG}$ passes through A and B, the centers of the large cylinder and small cylinder, respectively. Points D and E are where the band is tangent to the large cylinder and small cylinder, respectively. Then $\overline{DE}$ is perpendicular to radii $\overline{AD}$ and $\overline{BE}$. Let C be the point on $\overline{AD}$ so that $\overline{CB}$ is parallel to $\overline{DE}$. Thus, $BCDE$ is a rectangle.

It follows that $\overline{BE}$ and $\overline{CD}$ are both length 12, and $\overline{AC}$ is length $36 - 12 = 24$. $\overline{AB}$ is length $36 + 12 = 48$. Thus, triangle ABC is a 30-60-90 right triangle with $\overline{BC}$ having length $24\sqrt{3}$ which is also the length of $\overline{DE}$. The length of the band is twice the distance that follows the arc from F to E, the line segment from E to D, and the arc from D to G. Because angle FBE is 60 degrees and angle DAG is 120 degrees, the band has length

$$2\left(\frac{2 \cdot 12\pi}{6} + 24\sqrt{3} + \frac{2 \cdot 36\pi}{3}\right) = 48\sqrt{3} + 56\pi.$$

The requested sum is $48 + 3 + 56 = 107$.

Problem 22. A parabola, a line perpendicular to the parabola's axis of symmetry, and three similar isosceles triangles each with a base on the line and vertex on the parabola are shown in the following diagram. The two smaller triangles are congruent and each have one base vertex on the parabola and one base vertex shared with the larger triangle. The ratio of the height of the larger triangle to the height of the smaller triangles is $\frac{a+\sqrt{b}}{c}$ where a, b, c are positive integers, and a and c are relatively prime. Find $a+b+c$.

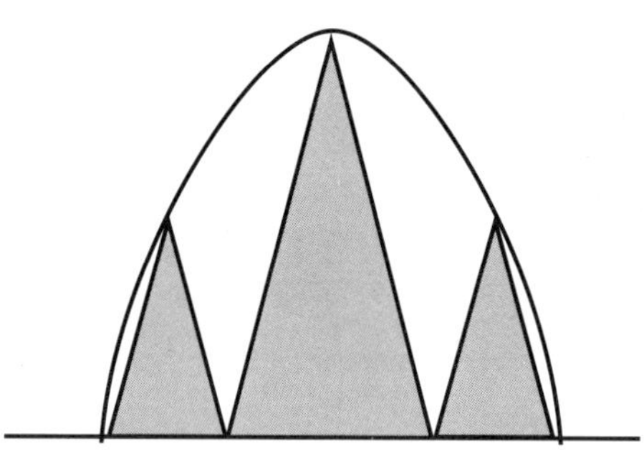

Answer: 20

Place the parabola on a coordinate plane so it is symmetric around the y axis, and the bases of the triangles are along the x axis. Since all parabolas are similar, one can assume that the points on the parabola satisfy $1 - x^2 = y$. Let the larger triangle have base length $2r$ and each smaller triangle have base length $2s$. Then the large triangle has vertices at $(\pm r, 0)$ and $(0, 1)$ while the smaller triangle to the right has vertices at $(r, 0)$, $(1, 0)$, and $(r + s, \frac{s}{r})$. Then the distance from the origin to the right most vertex of the smaller triangle is $1 = r+2s$. The top vertex of the smaller triangle is on the parabola, so it satisfies $1-(r+s)^2 = \frac{s}{r}$. Substituting $r = 1 - 2s$ into the latter equation and simplifying yields $2s^2 - 5s + 1 = 0$ which has solution $s = \frac{5-\sqrt{17}}{4}$ implying $r = \frac{\sqrt{17}-3}{2}$. Since the triangles are similar, the ratio of the height of the large triangle to the height of the small triangles is

$$\frac{r}{s} = \frac{2(\sqrt{17}-3)}{5-\sqrt{17}} = \frac{2(\sqrt{17}-3)}{5-\sqrt{17}} \cdot \frac{5+\sqrt{17}}{5+\sqrt{17}} = \frac{1+\sqrt{17}}{2}.$$

The requested sum is $1 + 17 + 2 = 20$.

Problem 23. Square $ABCD$ has side length 4. Points E and F are the midpoints of sides $\overline{AB}$ and $\overline{CD}$, respectively. Eight 1 by 2 rectangles are placed inside the square so that no two of the eight rectangles overlap (see diagram). If the arrangement of eight rectangles is chosen randomly, then there are relatively prime positive integers m and n so that $\frac{m}{n}$ is the probability that none of the rectangles crosses the line segment $\overline{EF}$ (as in the arrangement on the right). Find $m + n$.

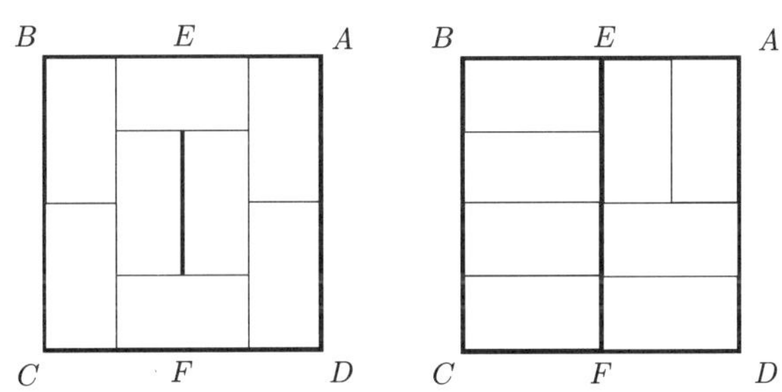

Answer: 61

First, count the number of ways to place the rectangles so that none of them cross the segment $\overline{EF}$. In these arrangements four 1 by 2 rectangles cover the 2 by 4 rectangle $AEFD$, and four 1 by 2 rectangles cover the 2 by 4 rectangle $EBCF$. Four 1 by 2 rectangles can cover a 2 by 4 rectangle in five ways:

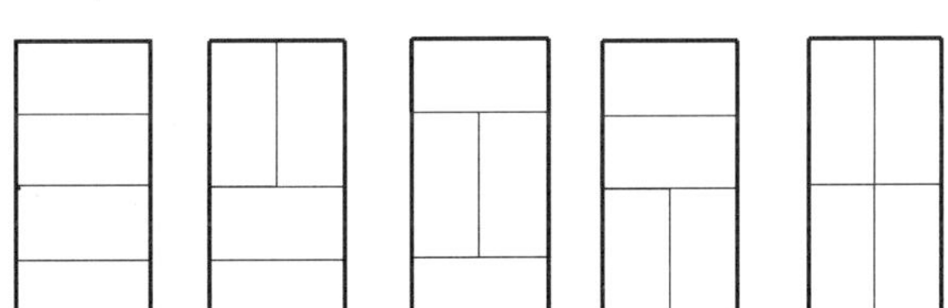

Thus, there are $5 \cdot 5 = 25$ ways to cover the 4 by 4 rectangle without crossing the segment $\overline{EF}$.

Now count the number of ways to cover the 4 by 4 rectangle while crossing the segment $\overline{EF}$. The segment $\overline{EF}$ can be crossed by either two 1 by 2 rectangles or by four 1 by 2 rectangles. The possible arrangements are:

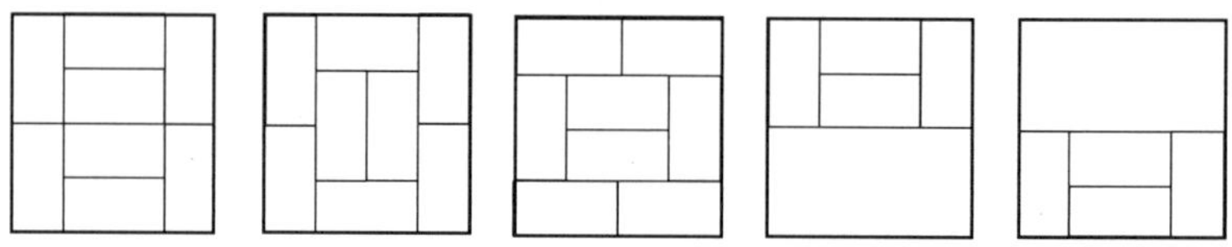

There is 1 way to cross with 4 rectangles. | There is 1 way to cross with rectangles at the top and bottom. | There is 1 way to cross with 2 rectangles in the middle. | There are 4 ways to cross with 2 rectangles at the top. | There are 4 ways to cross with 2 rectangles at the bottom.

Thus, there are 11 ways to cover the 4 by 4 rectangle while crossing the segment $\overline{EF}$. It follows that the probability that a randomly selected arrangement does not cross the segment $\overline{EF}$ is

$$\frac{25}{25+11} = \frac{25}{36}.$$

The requested sum is $25 + 36 = 61$.

Problem 24. A right circular cone pointing downward forms an angle of $60°$ at its vertex. Sphere S with radius 1 is set into the cone so that it is tangent to the side of the cone. Three congruent spheres are placed in the cone on top of S so that they are all tangent to each other, to sphere S, and to the side of the cone. The radius of these congruent spheres can be written as $\frac{a+\sqrt{b}}{c}$ where a, b, c are positive integers such that a and c are relatively prime. Find $a + b + c$.

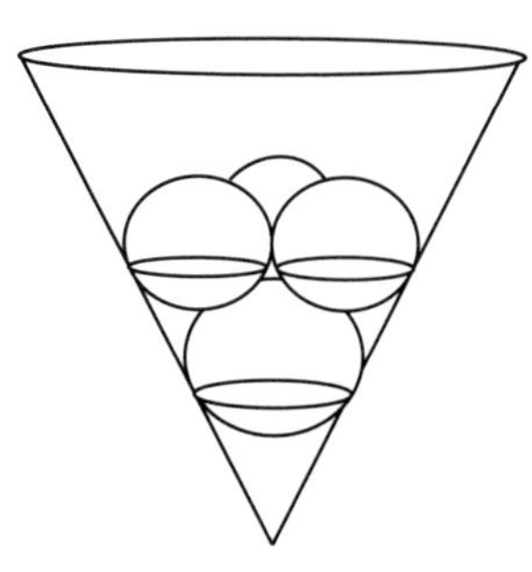

Answer: 364

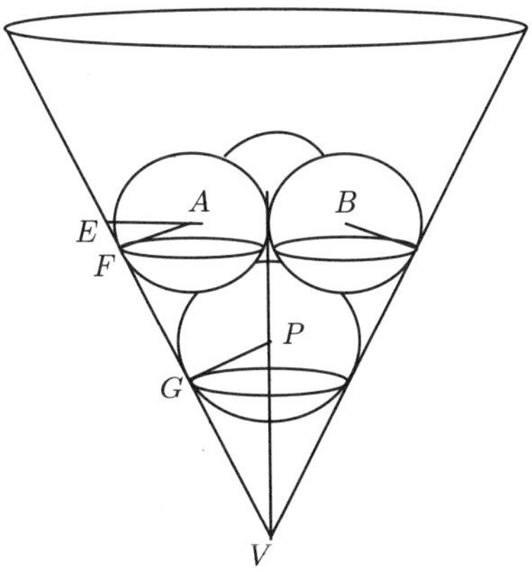

Let P be the center of the first sphere, and A, B, and C be the centers of the three congruent spheres. Let D be the circumcenter of the triangle ABC, H be the foot of the altitude of ABC on side $\overline{AB}$, and E be the intersection of ray $\overrightarrow{DA}$ and the cone. Let F be the point where the sphere centered at A touches the cone, V be the vertex of the cone, and G be the point on $\overline{FV}$ where S is tangent to the cone. Let r be the radius of the three congruent spheres. Triangle VPG is a 30-60-90 triangle, so segment $\overline{PV}$ has length 2. Triangle AEF is also a 30-60-90 triangle, so segment $\overline{AE}$ has length $\frac{2r}{\sqrt{3}}$.

Triangle ABC is equilateral with side length $2r$, so its altitude is $r\sqrt{3}$, and since the circumcenter D is one-third the distance from H to B, the circumradius is

$$\sqrt{r^2 + \frac{r^2}{3}} = \frac{2r}{\sqrt{3}}.$$

It follows that the length of $\overline{DE}$ is $\frac{4r}{\sqrt{3}}$, and since triangle VED is another 30-60-90 triangle, the length of $\overline{DV}$ is $4r$. The length of $\overline{DP}$ is then $4r-2$, the length of $\overline{AP}$ is $r+1$, and the Pythagorean Theorem gives that

$$\left(\frac{2r}{\sqrt{3}}\right)^2 + (4r-2)^2 = (r+1)^2 .$$

This reduces to $49r^2 - 54r + 9 = 0$ giving solutions $r = \frac{27\pm\sqrt{288}}{49}$. The requested sum is $27+288+49 = 364$. Note that the second answer gives the radii of three small congruent spheres which fit underneath the sphere of radius 1 rather than on top of it.

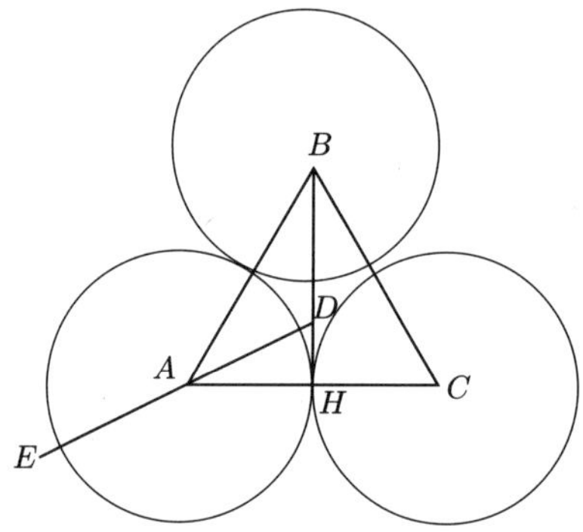

Problem 25. The polynomial $P(x) = a_0 + a_1 x + a_2 x^2 + \cdots + a_8 x^8 + 2009x^9$ has the property that $P\left(\frac{1}{k}\right) = \frac{1}{k}$ for $k = 1, 2, 3, 4, 5, 6, 7, 8, 9$. There are relatively prime positive integers m and n such that $P\left(\frac{1}{10}\right) = \frac{m}{n}$. Find $n - 10m$.

Answer: 20090

Let $Q(x) = x^9\left(P\left(\frac{1}{x}\right) - \frac{1}{x}\right)$. This defines $Q(x)$ to be a polynomial for all $x \neq 0$, and so it can be considered to be a polynomial for all values of x. The polynomial Q is zeros at $x = 1$, 2, 3, 4, 5, 6, 7, 8, 9, and $Q(0) = 2009$. Thus,

$$Q(x) = -2009\frac{(x-1)(x-2)(x-3)\cdots(x-9)}{9!}.$$

Since

$$P\left(\frac{1}{x}\right) = \frac{Q(x)}{x^9} + \frac{1}{x},$$

the value of $P(\frac{1}{10})$ is

$$\frac{Q(10)}{10^9} + \frac{1}{10} = \frac{-2009}{10^9} + \frac{1}{10} = \frac{10^8 - 2009}{10^9}.$$

The requested value is $n - 10m = 10^9 - 10(10^8 - 2009) = 20090$.

2.13 2009 Middle School

Problem 1. The pentagon below has three right angles. Find its area.

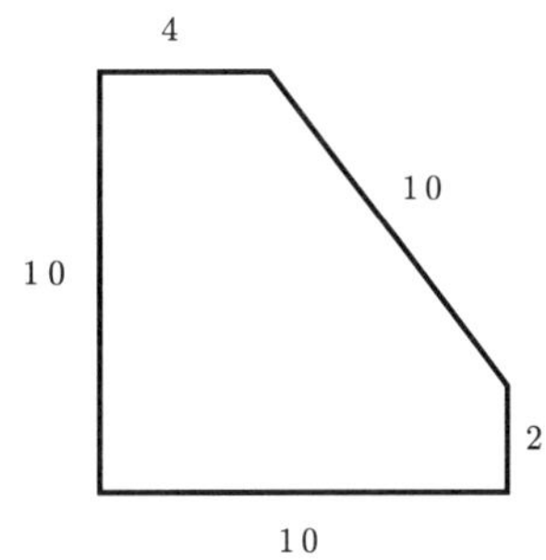

Answer: 76

By adding a 6 : 8 : 10 right triangle, the pentagon becomes a square as shown. Thus, the requested area is the area of the square minus the area of the triangle which equals

$$10^2 - \frac{6 \cdot 8}{2} = 100 - 24 = 76.$$

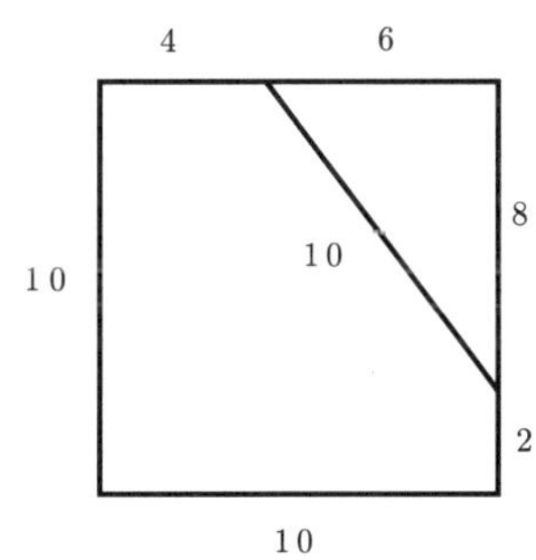

Problem 2. Let $p_1 = 2, p_2 = 3, p_3 = 5, \dots$ be the sequence of prime numbers. Find the least positive even integer n so that $p_1+p_2+p_3+\cdots+p_n$ is not a prime.

Answer: 8

When n is 2, the sum is $2 + 3 = 5$ which is prime. When n is 4, the sum is $2 + 3 + 5 + 7 = 17$ which is prime. When n is 6, the sum is $2 + 3 + 5 + 7 + 11 + 13 = 41$ which is prime. But when n is 8, the sum is $2 + 3 + 5 + 7 + 11 + 13 + 17 + 19 = 77$ which is 7 times 11, so it is not prime. Thus, the answer is 8.

Problem 3. The Purple Comet*!* Math Meet runs from April 27 through May 3, so the sum of the calendar dates for these seven days is $27 + 28 + 29 + 30 + 1 + 2 + 3 = 120$. What is the largest sum of the calendar dates for seven consecutive Fridays occurring at any time in any year?

Answer: 142

Seven consecutive Fridays either include four Fridays from a single month or five Fridays from a single month. If five Fridays occur in a month, the largest sum of the those five dates would be

$$3 + 10 + 17 + 24 + 31 = 85.$$

If two of the seven Fridays are in the following month, they would have a date sum of $7 + 14 = 21$. No choice of Fridays in the first month would allow the sum of the two dates from the second month to be greater, so the largest total possible when there are five Fridays one month followed by two in the next month is $85 + 21 = 106$. If five Fridays occur in one month with two Fridays in the previous month, the two Fridays in the previous month could have a total of at most $20 + 27 = 47$ giving a total of $85 + 47 = 132$. If five Fridays occur in a month with one Friday in the previous month and one in the following month, the largest total possible is

$$27 + 85 + 7 = 119.$$

If four Fridays occur in a month, the largest sum of those four dates would be

$$10 + 17 + 24 + 31 = 82.$$

The largest sum for three Fridays in the next month is

$$7 + 14 + 21 = 42.$$

No choice of dates in the first month can result in a higher sum of three dates in the second month, so the largest total possible when there are four Fridays in the first month is $82 + 42 = 124$. If there are four Fridays in a month along with three Fridays in the previous month, the largest sum of those four dates from the second month would be

$$7 + 14 + 21 + 28 = 70.$$

This occurs exactly when one gets the largest possible sum of three dates from the first month, $17+24+31 = 72$. The total for the seven Fridays is then $72+70 = 142$. If one or two Fridays are moved from the first month to the month following the month with four Fridays, either the 17 is replace by a 7 or the 17 and 24 are replaced by 7 and 14. In either case, the sum is smaller. Thus, the largest sum occurs when the first Friday of the seven consecutive Fridays falls on the 17-th day of a 31 day month giving the total 142.

Problem 4. John, Paul, George, and Ringo baked a circular pie. Each cut a piece that was a sector of the circle. John took one-third of the whole pie. Paul took one-fourth of the whole pie. George took one-fifth of the whole pie. Ringo took one-sixth of the whole pie. At the end the pie had one sector remaining. Find the measure in degrees of the angle formed by this remaining sector.

Answer: 18

The total number of degrees removed by the four out of the 360 degree circle was

$$360\left(\frac{1}{3}+\frac{1}{4}+\frac{1}{5}+\frac{1}{6}\right) = 120+90+72+60 = 342.$$

Thus, the number of degrees remaining is $360 - 342 = 18$.

Problem 5. A train car held 6000 pounds of mud which was 88 percent water. Then the train car sat in the sun, and some of the water evaporated so that now the mud is only 82 percent water. How many pounds does the mud weigh now?

Answer: 4000

The original mud was 88 percent water, so it was 12 percent dirt, or $6000(.12) = 720$ pounds dirt. The mud now weighs m and is 82 percent water, so it is 18 percent dirt. Thus, $m(.18) = 720$ and $m = \frac{720}{.18} = \frac{72000}{18} = 4000$ pounds.

Problem 6. Find n so that $20^{2009} = 10^{2000} \cdot 40^9 \cdot 2^n$.

Answer: 1991

The given equation is equivalent to

$$2^{4018} \cdot 5^{2009} = 2^{2000} \cdot 5^{2000} \cdot 2^{27} \cdot 5^9 \cdot 2^n.$$

The same number of factors of 5 appear on each side of this equation. For the number of factors of 2 to be the same, one needs $4018 = 2000+27+n$. Thus, $n = 4018-2000-27 = 1991$.

Problem 7. How many distinct four letter arrangements can be formed by rearranging the letters found in the word **FLUFFY**? For example, FLYF and ULFY are two possible arrangements.

Answer: 72

Consider the arrangements that contain four distinct letters (that is, with no repeated F). There are $4! = 24$ such arrangements. Then consider the arrangements that contain two F's. To select such an arrangement, one must choose the two letters to go along with the two F's in one of $\binom{3}{2} = 3$ ways and then choose a rearrangement of the two chosen letters with the two repeated F's in one of $\frac{4!}{2!} = 12$ ways. Thus, there are $3 \cdot 12 = 36$ of these arrangements. Finally, consider the arrangements that contain all three F's. To select such an arrangement, one must choose a fourth letter in one of 3 ways and then choose a rearrangement of the chosen letter with the three repeated F's in one of $\frac{4!}{3!} = 4$ ways. Thus, there are $3 \cdot 4 = 12$ of these arrangements. The total number of arrangement is, therefore, $24 + 36 + 12 = 72$.

Problem 8. Find the number of non-congruent scalene triangles whose sides all have integral length, and the longest side has length 11.

Answer: 20

Let the sides of such a triangle be $a < b < c = 11$. Then, by the triangle inequality, $7 \leq b \leq 10$ and $c - b < a < b$. Thus, as b decreases by 1, the range of a decreases by 2. For b = 10 we have $2 \leq a \leq 9$, hence the number of triangles is $8 + 6 + 4 + 2 = 20$.

Problem 9. One plant is now 44 centimeters tall and will grow at a rate of 3 centimeters every two years. A second plant is now 80 centimeters tall and will grow at a rate of 5 centimeters every six years. In how many years will the plants be the same height?

Answer: 54

The first plant grows at a rate of $\frac{3}{2}$ centimeters per year, so after t years it has reached a height of $44 + \frac{3}{2}t$. The second plant grows at a rate of $\frac{5}{6}$ centimeters per year, so after t years it has reached a height of $80 + \frac{5}{6}t$. These two heights are equal when $44 + \frac{3}{2}t = 80 + \frac{5}{6}t$. This simplifies to $\left(\frac{3}{2} - \frac{5}{6}\right) t = 80 - 44$ or $\frac{2}{3}t = 36$ and $t = \frac{3}{2} \cdot 36 = 54$.

Alternatively, the first plant grows 9 centimeters every 6 years, so it gains on the second plant at a rate of 4 centimeters every

6 years. It needs to grow $80 - 44 = 36 = 9 \cdot 4$ more centimeters than the second plant to catch up to the second plant. Thus, it must grow for $9 \cdot 6 = 54$ years to catch up.

Problem 10. The diagram shows a 20 by 20 square $ABCD$. Points E, F, G are equally spaced on side $\overline{BC}$, while points H, I, J, K on side $\overline{DA}$ are placed so that the triangles BKE, EJF, FIG, and GHC are isosceles. Points L and M are midpoints of the sides $\overline{AB}$ and $\overline{CD}$, respectively. Find the total area of the shaded regions.

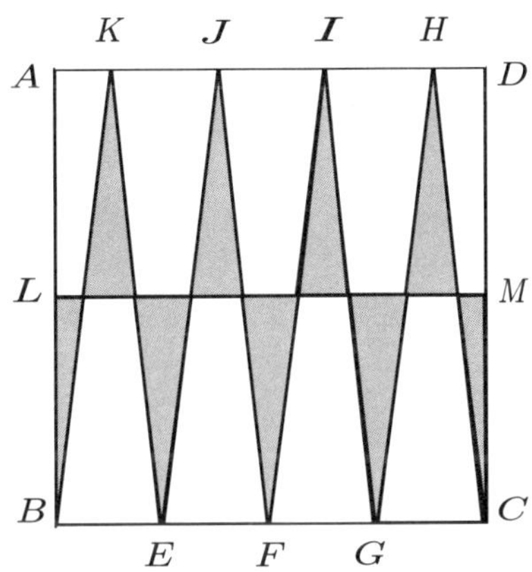

Answer: 100

Each of the shaded triangles has area equal to half of the length of its base times its height. Each of these triangles has height 10, and the sum of the bases of the triangles is 20. Thus, the combined areas of the shaded triangles is $\frac{10 \cdot 20}{2} = 100$. Note that the answer remains the same even if the points are not equally spaced, and the triangles are not isosceles.

Problem 11. Aisha went shopping. At the first store she spent 40 percent of her money plus four dollars. At the second store she spent 50 percent of her remaining money plus 5 dollars. At the third store she spent 60 percent of her remaining money plus six dollars. When Aisha was done shopping at the three stores, she had two dollars left. How many dollars did she have with her when she started shopping?

Answer: 90

Working backwards, at the third store Aisha spent 60 percent of her money plus six dollars and ended up with two dollars.

Thus, she entered the third store with $\frac{6+2}{1-.6} = 20$ dollars. At the second store Aisha spent 50 percent of her money plus 5 dollars and ended up with 20 dollars. Thus, she entered the store with $\frac{20+5}{1-.5} = 50$ dollars. At the first store Aisha spent 40 percent of her money plus 4 dollars and ended up with 50 dollars. Thus, she entered the store with $\frac{50+4}{1-.4} = 90$ dollars.

Problem 12. In isosceles triangle ABC sides $\overline{AB}$ and $\overline{BC}$ have length 125, while side $\overline{AC}$ has length 150. Point D is the midpoint of side $\overline{AC}$. Point E is on side $\overline{BC}$ so that $\overline{BC}$ and $\overline{DE}$ are perpendicular. Similarly, F is on side $\overline{AB}$ so that $\overline{AB}$ and $\overline{DF}$ are perpendicular. Find the area of triangle DEF.

Answer: 1728

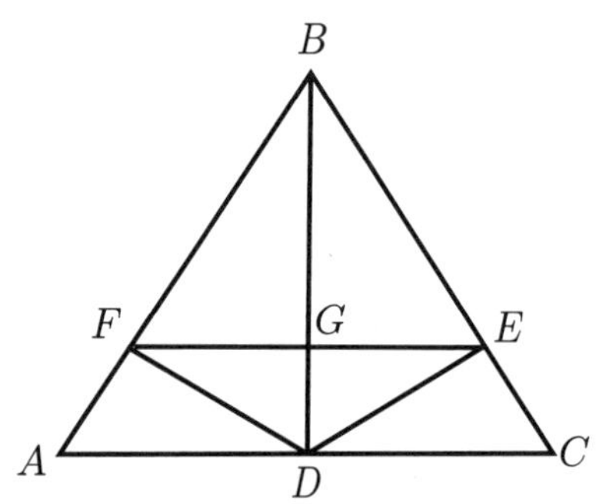

Let G be the intersection of $\overline{EF}$ and $\overline{BD}$ as in the diagram. Note that both triangles ADB and CDB are $3:4:5$ right triangles. Each of the triangles CED, DEB, EGB, AFD, DFB, and FGB are right triangles that share an acute angle with either ADB or CDB, and, thus, they are all similar $3:4:5$ right triangles. From this similarity, side $\overline{BD}$ in triangle CDB has length $125 \cdot \frac{4}{5} = 100$. Then side $\overline{BE}$ in triangle DEB has length $100 \cdot \frac{4}{5} = 80$. Similarly, $\overline{GB}$ in triangle EGB has length $80 \cdot \frac{4}{5} = 64$, $\overline{GE}$ has length $80 \cdot \frac{3}{5} = 48$, and $\overline{GD}$ has length $BD - BG = 100 - 64 = 36$. The area of triangle DEF is the product of the lengths of $\overline{GE}$ and $\overline{GD}$ or $48 \cdot 36 = 1728$.

Problem 13. How many subsets of the set $\{1, 2, 3, \ldots, 12\}$ contain exactly one or two prime numbers?

Answer: 1920

The prime numbers from set $A = \{1, 2, 3, ..., 12\}$ form the subset $B = \{2, 3, 5, 7, 11\}$ while $C = A \setminus B = \{1, 4, 6, 8, 9, 10, 12\}$. Set

B has 5 elements while set C contains 7 elements. A subset of A that contains exactly one or two prime numbers is the union of a subset of B of size one or two with any subset of set C. Set B has 5 subsets of size one and $\binom{5}{2} = 10$ subsets of size two. Set C has $2^7 = 128$ subsets. It follows that there are $(5+10) \cdot 128 = 1920$ acceptable sets.

Problem 14. Rectangle $ABCD$ measures 70 by 40. Eighteen points (including A and C) are marked on the diagonal $\overline{AC}$, dividing the diagonal into 17 congruent pieces. Twenty-two points (including A and B) are marked on the side $\overline{AB}$, dividing the side into 21 congruent pieces. Seventeen non-overlapping triangles are constructed as shown. Each triangle has two vertices that are two of these adjacent marked points on the side of the rectangle, and one vertex that is one of the marked points along the diagonal of the rectangle. Only the left 17 of the 21 congruent pieces along the side of the rectangle are used as bases of these triangles. Find the sum of the areas of these 17 triangles.

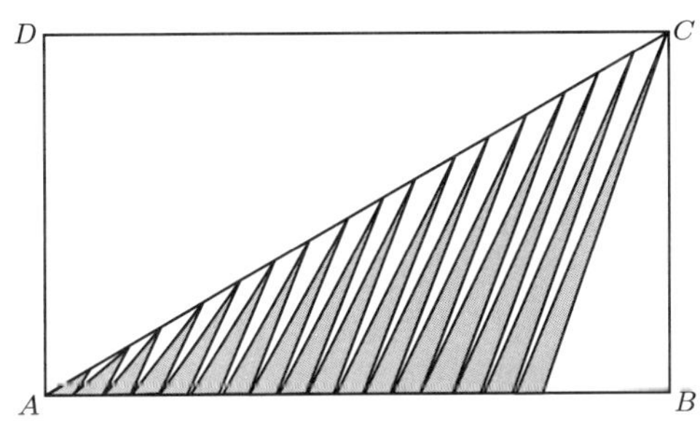

Answer: 600

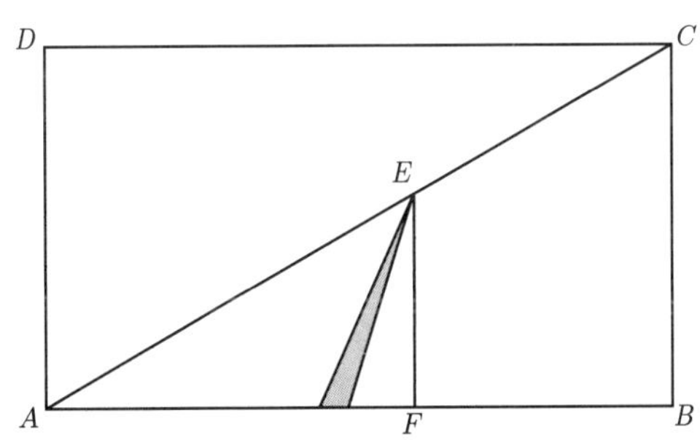

Let the diagonal of the rectangle have length d. Consider the k-th triangle from the left as shown with vertex E along the diagonal. Let F be the point on $\overline{AB}$ so that $\overline{EF}$ is perpendicular to $\overline{AB}$. Then the height of the triangle is the length of $\overline{EF}$ which is, by similar triangles,

$$AE \cdot \frac{BC}{AC} = \frac{kd}{17}\frac{40}{d} = \frac{40k}{17}.$$

The area of the triangle is then given by

$$\frac{1}{2}\frac{70}{21}\frac{40k}{17} = \frac{40 \cdot 70}{2 \cdot 21 \cdot 17}k.$$

The sum of the areas of the 17 triangles is

$$\sum_{k=1}^{17} \frac{40 \cdot 70}{2 \cdot 21 \cdot 17}k = \frac{40 \cdot 70}{2 \cdot 21 \cdot 17}\frac{17 \cdot 18}{2} = 600.$$

Note that we assumed that side $\overline{AB}$ was the side that measures 70 rather than 40, but the answer is the same if $\overline{AB}$ is the side that measures 40.

Problem 15. We have twenty-seven 1 by 1 cubes. Each face of every cube is marked with a positive integer so that two opposite faces (top and bottom, front and back, left and right) are always marked with an even number and an odd number where the even number is twice that of the odd number. The twenty-seven cubes are put together to form one 3 by 3 cube as shown. When two cubes are placed face-to-face, adjoining faces are always marked with an odd number and an even number where the even number is one greater than the odd number. Find the sum of all of the numbers on all of the faces of all the 1 by 1 cubes.

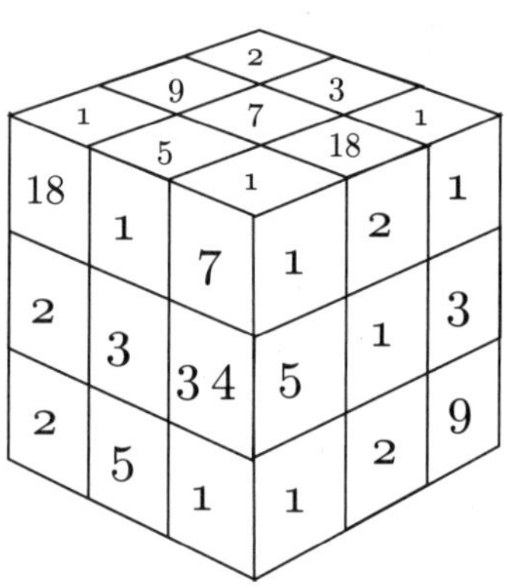

Answer: 1377

Suppose one face of a 1 by 1 cube is on the outside of the 3 by 3 cube, and that face is marked with the odd number n. Then the face opposite that face on the 1 by 1 cube is marked with $2n$. The face on the next cube adjacent to the face with $2n$ must be marked with $2n-1$. Its opposite face must be marked $4n-2$. The face adjacent to it must be marked $4n-3$, and the face opposite this face on its 1 by 1 cube must be marked $8n-6$. For example, if a 1 appears on the side of the big cube, then $8 \cdot 1 - 6 = 2$ will appear on the opposite side of the big cube. The sum of the six numbers starting with n and ending on the opposite side of the big cube with $8n-6$ is $n + (2n) + (2n - 1) + (4n-2) + (4n-3) + (8n-6) = 21n - 12$. The diagram shows fourteen faces marked with either a 1 or 2 = 8(1) - 6, five faces marked with either a 3 or an 18 = 8(3) - 6, four faces marked with either a 5 or a 34 = 8(5) - 6, two faces marked with a 7, and two faces marked with a 9. Thus, the 27 odd numbers appearing on the outside of the 3 by 3 cube add up to $14 \cdot 1 + 5 \cdot 3 + 4 \cdot 5 + 2 \cdot 7 + 2 \cdot 9 = 81$. For each of these odd numbers n on the outside of the 3 by 3 cube, there are six other sides of 1 by 1 cubes adding up to $21n - 12$. It follows that the sum of all the numbers marked on all the sides of 1 by 1 cubes is $21 \cdot 81 - 27 \cdot 12 = 81(21-4) = 81 \cdot 17 = 1377$.

2.14 2010 High School

Problem 1. Let x satisfy $(6x+7)+(8x+9) = (10+11x)+(12+13x)$. There are relatively prime positive integers m and n so that $x = -\frac{m}{n}$. Find $m+n$.

Answer: 8

The equation simplifies to $(6+8)x+(7+9) = (10+12)+(11+13)x$ or $14x + 16 = 22 + 24x$. Thus, $-10x = 6$ and $x = -\frac{3}{5}$. The requested sum is $3 + 5 = 8$.

Problem 2. The prime factorization of $12 = 2 \cdot 2 \cdot 3$ has three prime factors. Find the number of prime factors in the factorization of

$$12! = 12 \cdot 11 \cdot 10 \cdot 9 \cdot 8 \cdot 7 \cdot 6 \cdot 5 \cdot 4 \cdot 3 \cdot 2 \cdot 1.$$

Answer: 19

$$12! = 12 \cdot 11 \cdot 10 \cdot 9 \cdot 8 \cdot 7 \cdot 6 \cdot 5 \cdot 4 \cdot 3 \cdot 2 \cdot 1$$
$$= (2 \cdot 2 \cdot 3) \cdot 11 \cdot (2 \cdot 5) \cdot (3 \cdot 3) \cdot (2 \cdot 2 \cdot 2) \cdot 7 \cdot (2 \cdot 3) \cdot 5 \cdot (2 \cdot 2) \cdot 3 \cdot 2 \cdot 1$$

which is a total for 19 prime factors.

Problem 3. The grid below contains six rows with six points in each row. Points that are adjacent either horizontally or vertically are a distance two apart. Find the area of the irregularly shaped ten sided figure shown.

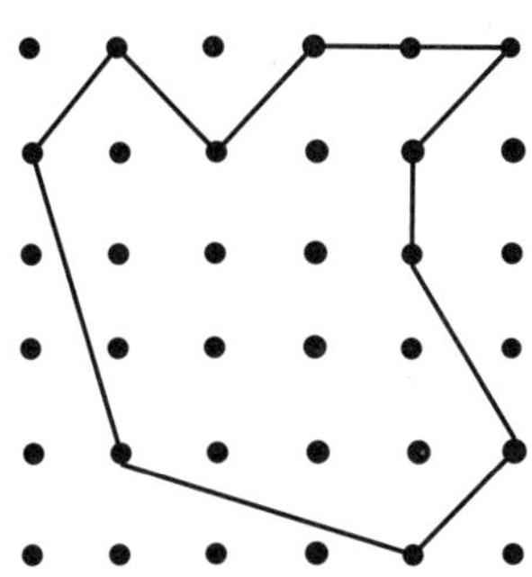

Answer: 66

The entire grid forms a square with side length 10, so the grid has area 100. The figure shown, thus, has area 100 minus the areas of two trapezoids and four right triangles shaded in the diagram shown below. Both trapezoids have bases of length 2 and 8 with height 2, so each has area $\frac{2+8}{2}2 = 10$. The largest right triangle has legs with lengths 2 and 6, so it has area $\frac{2\cdot 6}{2} = 6$. Two small right triangles each have two legs of length 2, so each has area $\frac{2\cdot 2}{2} = 2$. The last right triangle has a hypotenuse of length 4 and an altitude to the hypotenuse of length 2, so its area is $\frac{2\cdot 4}{2} = 4$. Therefore, the area of the ten sided figure is $100 - (10 + 10 + 6 + 2 + 2 + 4) = 100 - 34 = 66$.

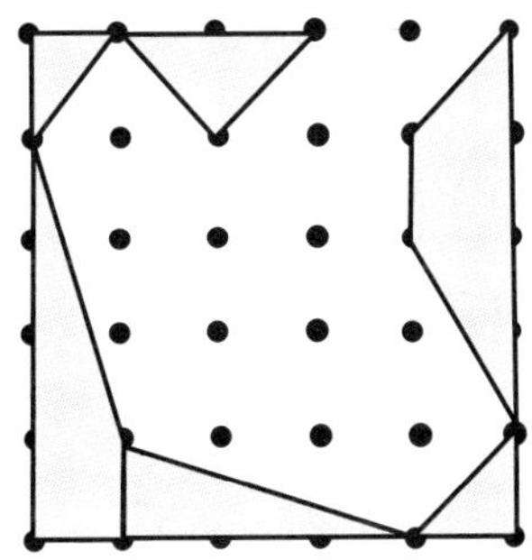

Alternatively, assume instead that the points on the grid are only one unit apart from each other rather than two units. Pick's Theorem says that because there are 12 grid points inside the polygon and 11 grid points on the boundary of the polygon, the area of the polygon is $12+\frac{11}{2}-1 = \frac{33}{2}$. But the distance between points on the grid is 2, so the area is $2^2 = 4$ times as large or $\frac{33}{2}\cdot 4 = 66$.

Problem 4. Sally's salary in 2006 was \$37,500. For 2007 she got a salary increase of x percent. For 2008 she got another salary increase of x percent. For 2009 she got a salary decrease of $2x$ percent. Her 2009 salary is \$34,825. Suppose instead, Sally had gotten a $2x$ percent salary decrease for 2007, an x percent salary increase for 2008, and an x percent salary increase for 2009. What would her 2009 salary be then?

Answer: 34825

Let $y = \frac{x}{100}$. Sally's 2009 salary is given by

$$37000(1 + y)(1 + y)(1 - 2y) = 34825.$$

If the salary increases and decreases are reversed, her 2009 salary would be $37000(1-2y)(1+y)(1+y)$ which is still equal to 34825 since multiplication is a commutative operation. This happens if x is approximately 13.4144 percent.

Problem 5. If a and b are positive integers such that $a \cdot b = 2400$, find the least possible value of $a + b$.

Answer: 98

The number 2400 has 36 positive integer divisors. One can check all 18 pairs of these divisors, although the minimum sum will occur when the pair of divisors are closest to $\sqrt{2400}$ which is between 48 and 49. The pairs of divisors closest to this is 48 and 50 whose sum is 98 which is the least possible.

Problem 6. Evaluate the sum $1+2-3+4+5-6+7+8-9+\cdots+208+209-210$.

Answer: 7245

Grouping the numbers in the sum in groups of three gives

$$1 + 2 - 3 + 4 + 5 - 6 + 7 + 8 - 9 + \cdots + 208 + 209 - 210$$

$$= (1+2-3) + (4+5-6) + (7+8-9) + \cdots + (208+209-210)$$

$$= 0 + 3 + 6 + \cdots + 207 = 3(0 + 1 + 2 + \cdots + 69).$$

Since the sum of the first n positive integers is given by $\frac{n(n+1)}{2}$, the desired sum is $3 \cdot \frac{69 \cdot 70}{2} = 7245$.

Problem 7. Find the sum of the digits in the decimal representation of the number $5^{2010} \cdot 16^{502}$.

Answer: 7

$5^{2010} \cdot 16^{502} = 5^{2010} \cdot \left(2^4\right)^{502} = 5^{2010} \cdot 2^{2008} = 5^2 \cdot 10^{2008}$ which is the number 25 followed by 2008 zeros. Thus, the sum of its digits is $2 + 5 = 7$.

Problem 8. The diagram below shows some small squares each with area 3 enclosed inside a larger square. Squares that touch each other do so with the corner of one square coinciding with the midpoint of a side of the other square. Find the integer n such that the area of the shaded region inside the larger square but outside the smaller squares is $\sqrt{n}$.

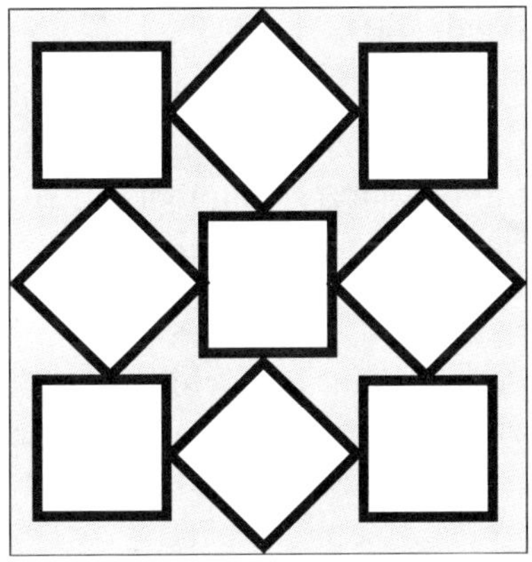

Answer: 288

Each of the small squares has side length $\sqrt{3}$ and diagonal length $\sqrt{3}\sqrt{2} = \sqrt{6}$. Since the large square has sides equal to the sum of the lengths of two of the diagonals and one of the sides of the small squares, the area of the large square is

$$\left(\sqrt{3} + 2\sqrt{6}\right)^2 = 3 + 4 \cdot \sqrt{3}\sqrt{6} + 4 \cdot 6 = 27 + 12\sqrt{2}.$$

Each small square has area 3, and there are nine small squares with a total area of 27. Thus, the desired area is $12\sqrt{2} = \sqrt{288}$. The requested answer is 288.

Problem 9. Find the positive integer n such that $\frac{80-6\sqrt{n}}{n}$ is the reciprocal of $\frac{80+6\sqrt{n}}{n}$.

Answer: 64

If the two numbers are reciprocals, their product is

$$1 = \frac{80 - 6\sqrt{n}}{n} \cdot \frac{80 + 6\sqrt{n}}{n} = \frac{80^2 - 6^2 n}{n^2} \text{ or } n^2 + 36n - 80^2 = 0$$

which has solutions

$$n = \frac{-36 \pm \sqrt{36^2 + 4 \cdot 80^2}}{2} = \frac{-36 \pm 4\sqrt{9^2 + 40^2}}{2} = -18 \pm 2 \cdot 41.$$

Thus, the positive solution is $2 \cdot 41 - 18 = 64$.

Problem 10. The set S contains nine numbers. The mean of the numbers in S is 202. The mean of the five smallest of the numbers in S is 100. The mean of the five largest numbers in S is 300. What is the median of the numbers in S?

Answer: 182

Let the median be m. The sum of all the numbers in S is $9 \cdot 202$. The sum of the five smallest numbers in S is $5 \cdot 100$. The sum of the five largest numbers in S is $5 \cdot 300$. Thus,

$$m = 5 \cdot 100 + 5 \cdot 300 - 9 \cdot 202 = 500 + 1500 - 1818 = 182.$$

Problem 11. A jar contains one white marble, two blue marbles, three red marbles, and four green marbles. If you select two of these marbles without replacement, the probability that both marbles will be the same color is $\frac{m}{n}$, where m and n are relatively prime positive integers. Find $m+n$.

Answer: 11

There are ten marbles in all, so there are $\binom{10}{2} = 45$ equally likely ways to select two marbles. There is no way to select two white marbles, $\binom{2}{2} = 1$ way to select two blue marbles, $\binom{3}{2} = 3$ ways to select two red marbles, and $\binom{4}{2} = 6$ ways to select two green marbles. Thus, the probability of selecting two marbles of the same color is $\frac{0+1+3+6}{45} = \frac{10}{45} = \frac{2}{9}$. The requested sum is $2 + 9 = 11$.

Problem 12. A good approximation of π is 3.14. Find the least positive integer d such that if the area of a circle with diameter d is calculated using the approximation 3.14, the error will exceed 1.

Answer: 51

The answer is the least positive integer d such that

$$\left(\frac{d}{2}\right)^2 (\pi - 3.14) \geq 1.$$

That is,

$$d^2 \geq \frac{4}{\pi - 3.14} \approx \frac{4}{0.00159} \approx 2511.$$

The value of d is thus 51.

Problem 13. Let S be the set of all 10-term arithmetic progressions that include the numbers 4 and 10. For example,

$$(-2, 1, 4, 7, 10, 13, 16, 19, 22, 25)$$

and

$$\left(10, 8\frac{1}{2}, 7, 5\frac{1}{2}, 4, 2\frac{1}{2}, 1, -\frac{1}{2}, -2, -3\frac{1}{2}\right)$$

are both members of S. Find the sum of all values of a_{10} for each

$$(a_1, a_2, a_3, a_4, a_5, a_6, a_7, a_8, a_9, a_{10}) \in S,$$

that is,

$$\sum_{(a_1,a_2,a_3,\ldots,a_{10})\in S} a_{10}.$$

Answer: 630

An arithmetic progression containing the terms 4 and 10 is completely determined by the positions in the progression of the 4 and 10. For each $(a_1, a_2, a_3, a_4, a_5, a_6, a_7, a_8, a_9, a_{10}) \in S$ there is exactly one $(b_1, b_2, b_3, b_4, b_5, b_6, b_7, b_8, b_9, b_{10}) \in S$ where the positions of the 4 and the 10 are reversed. For example, the sequence $(-2, 1, 4, 7, 10, 13, 16, 19, 22, 25)$ and the sequence $(16, 13, 10, 7, 4, 1, -2, -5, -8, -11)$ have their 4 and 10 terms in opposite positions. Then $(a_1 + b_1, a_2 + b_2, a_3 + b_3, a_4 + b_4, a_5 + b_5, a_6 + b_6, a_7 + b_7, a_8 + b_8, a_9 + b_9, a_{10} + b_{10})$ is also an arithmetic progression, but since two of its terms are $4 + 10 = 14$, it is a constant progression with all of its terms equal to 14. In particular, $a_{10} + b_{10} = 14$. It follows that mean of all tenth terms of progressions in S is $\frac{14}{2} = 7$, and the requested sum of all tenth terms is 7 times the number of elements in S. Since an element of S is completely determined by the placement of the terms 4 and 10, and there are ten possible placements of the term 4, and nine possible placements of the term 10 once the 4 has been placed, the size of S is $10 \cdot 9 = 90$. Thus, the requested sum is $7 \cdot 90 = 630$.

Problem 14. There are positive integers b and c such that the polynomial $2x^2 + bx + c$ has two real roots which differ by 30. Find the least possible value of $b + c$.

Answer: 126

The quadratic formula says that the two roots of the polynomial are $\frac{-b+\sqrt{b^2-8c}}{4}$ and $\frac{-b-\sqrt{b^2-8c}}{4}$. These two roots differ by $\frac{\sqrt{b^2-8c}}{2} = 30$ or $b^2 - 8c - 3600$. Since c is positive, the smallest value of b and c occur when $3600 + 8c$ is the smallest perfect square multiple of 8 exceeding 60^2. This is $64^2 = 4096$ which occurs when $b = 64$ and $c = \frac{4096-3600}{8} = 62$. Thus, the least possible value of $b + c$ is $64 + 62 = 126$.

Problem 15. Find the least possible sum $a + b + c + d + e$, where a, b, c, d, e are positive integers satisfying the conditions

- each of the pairs of integers (a,b), (b,c), (c,d), (d,e) are **not** relatively prime
- all other pairs of the five integers **are** relatively prime.

Answer: 47

Since a and b are not relatively prime, there must be a prime number p_1 that divides both a and b. Similarly, there must be a prime number p_2 that divides both b and c, but since a and c are relatively prime, p_1 and p_2 are distinct. Similarly, there is a third prime p_3 that divides both c and d, and a fourth prime p_4 that divides both d and e, with all four primes being distinct. Thus, the smallest possible value for the sum $a+b+c+d+e$ is $p_1 + p_1 \cdot p_2 + p_2 \cdot p_3 + p_3 \cdot p_4 + p_4$. It follows that the four primes should be the smallest four prime numbers: 2, 3, 5, 7. It is also clear that p_1 and p_4 should be the largest of those four prime numbers. An assignment that minimizes the sum is $p_1 = 5, p_2 = 3, p_3 = 2, p_4 = 7$ resulting in the sum

$$5 + 5 \cdot 3 + 3 \cdot 2 + 2 \cdot 7 + 7 = 5 + 15 + 6 + 14 + 7 = 47.$$

Problem 16. Triangle ABC has sides lengths $AB = 39$, $BC = 57$, $CA = 70$, as shown. Median $\overline{AD}$ is divided into three congruent segments by points E and F. Lines $\overline{BE}$ and $\overline{BF}$ intersect side $\overline{AC}$ at G and H, respectively. Find the distance from G to H.

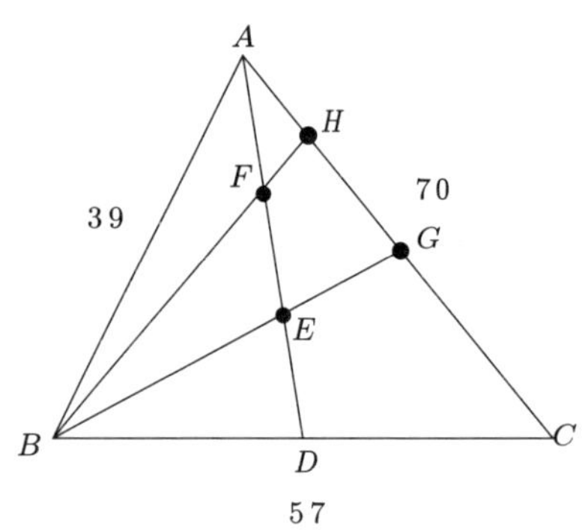

Answer: 21

More generally, let X be a point on median $\overline{AD}$, and let line $\overline{BX}$ intersect side $\overline{CA}$ at point Y as shown below. Let Z be a point on side $\overline{BC}$ so that $\overline{YZ}$ is parallel to $\overline{XD}$. Then triangles

BDX and BZY are similar, and triangles YZC and ADC are similar. This implies that

$$\frac{BZ}{YZ} = \frac{BD}{DX} \text{ and } \frac{CZ}{YZ} = \frac{CD}{AD},$$

so

$$\frac{BD}{DX} + \frac{CD}{AD} = \frac{BZ + CZ}{YZ} = \frac{BC}{YZ}.$$

Also,

$$\frac{YC}{YZ} = \frac{AC}{AD},$$

so

$$YC = YZ\frac{AC}{AD} = \frac{BC}{\frac{BD}{DX} + \frac{CD}{AD}} \cdot \frac{AC}{AD} = \frac{AC}{\frac{BD}{BC}\frac{AD}{DX} + \frac{CD}{BC}} = \frac{2}{\frac{AD}{DX} + 1} \cdot AC.$$

Applying this when $DX = \frac{1}{3} \cdot AD$ shows that

$$GC = \frac{2}{3+1} \cdot 70 = 35.$$

Applying it when $DX = \frac{2}{3} \cdot AD$ shows that

$$HC = \frac{2}{\frac{3}{2} + 1} \cdot 70 = 56.$$

Thus, $GH = 56 - 35 = 21$. Note that the solution does not depend on the lengths of $\overline{AB}$ and $\overline{BC}$.

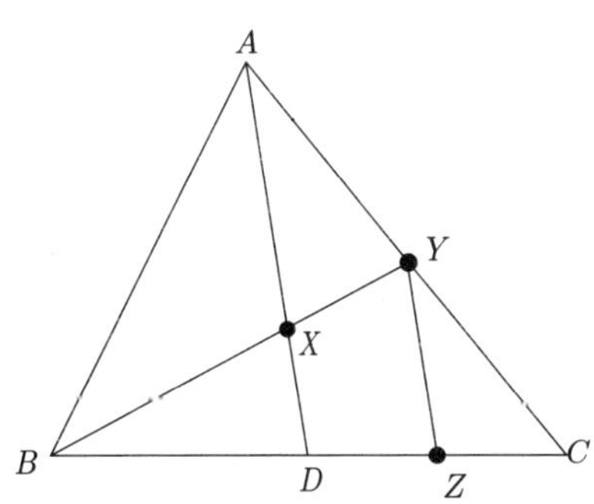

Alternatively, drop perpendiculars from A, G and E to $\overline{BH}$ and denote the feet of these perpendiculars by K, L and M, respectively. Triangles AFB and BFE have equal areas since $\overline{AF} \equiv \overline{FE}$, and the corresponding heights are the same. These triangles share side $\overline{BF}$ so the corresponding heights must be equal, $AK = EM$.

Triangles BEM and BGL are similar because they have two congruent angles, and thus $\frac{GL}{EM} = \frac{BG}{BE}$. Since E is one third the distance from D to A along a median, E is the centroid of triangle ABC, and $\frac{BG}{BE} = \frac{3}{2}$. Then the ratio of similarity must be the same for triangles LHG and KHA, thus $HG = \frac{3}{2}AH$. Since E is the centroid, $\overline{BG}$ is a median, and $AG = 35$, $HG = \frac{3}{5} \cdot 35 = 21$.

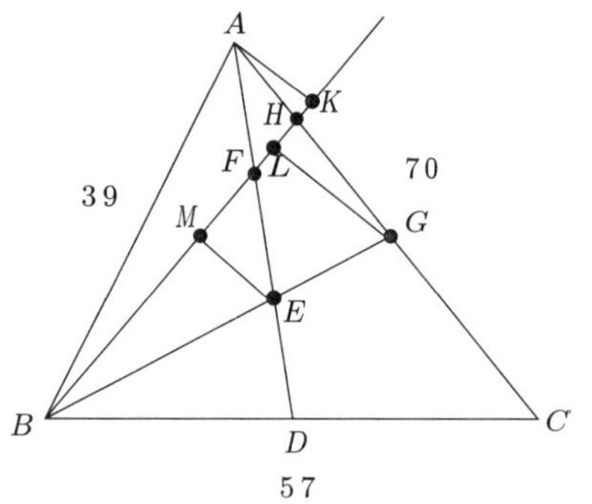

Problem 17. Alan, Barb, Cory, and Doug are on the golf team; Doug, Emma, Fran, and Greg are on the swim team; and Greg, Hope, Inga, and Alan are on the tennis team. These nine people sit in a circle in random order. The probability that no two people from the same team sit next to each other is $\frac{m}{n}$, where m and n are relatively prime positive integers. Find $m + n$.

Answer: 2521

Nine people can sit in a circle in 8! equally likely arrangements. For the arrangement to have members of the same team not sitting next to each other, Greg, who is on both the swim team and the tennis team, must sit next to both Barb and Cory. Similarly, Alan, who is on both the golf team and the tennis team, must sit next to both Emma and Fran, and Doug, who is on both the golf team and the swim team, must sit next to both Hope and Inga. For each of these groups of three people: {Barb,Greg,Cory}, {Emma,Alan,Fran}, {Hope,Doug,Inga}, there are two arrangements of the people in the group (for example, Barb,Greg,Cory and Cory,Greg,Barb). There are also $2! = 2$ arrangements of the three groups around the circle. It follows that there are $2 \cdot 2 \cdot 2 \cdot 2 = 16$ arrangements of the people so that no two members of the same team sit next to each other. The requested probability is $\frac{16}{8!} = \frac{16}{8\cdot7\cdot6\cdot5\cdot4\cdot3\cdot2\cdot1} = \frac{1}{7\cdot6\cdot5\cdot4\cdot3} = \frac{1}{2520}$. The requested sum is $1 + 2520 = 2521$.

Problem 18. When $4\cos\theta - 3\sin\theta = \frac{13}{3}$, it follows that

$$7\cos 2\theta - 24\sin 2\theta = \frac{m}{n},$$

where m and n are relatively prime positive integers. Find $m+n$.

Answer: 122

Because $3^2 + 4^2 = 5^2$, there is an angle ϕ so that $\cos\phi = \frac{4}{5}$ and $\sin\phi = \frac{3}{5}$. Recall the angle sum formula

$$\cos(\phi+\theta) = \cos\phi\cdot\cos\theta - \sin\phi\cdot\sin\theta$$

and the double angle formulas

$$\cos 2\theta = 2\cos^2\theta - 1 \text{ and } \sin 2\theta = 2\sin\theta\cos\theta.$$

Thus,

$$\cos 2\phi = 2\cos^2\phi - 1 = 2\left(\frac{4}{5}\right)^2 - 1 = \frac{7}{25}$$

and

$$\sin 2\phi = 2\sin\phi\cos\phi = 2\cdot\frac{4}{5}\cdot\frac{3}{5} = \frac{24}{25}.$$

From the given information one sees that

$$\cos(\phi+\theta) = \frac{4\cos\theta - 3\sin\theta}{5} = \frac{13}{15}.$$

Then,

$$\cos 2(\phi+\theta) = 2\cos^2(\phi+\theta) - 1 = 2\left(\frac{13}{15}\right)^2 - 1 = \frac{338}{225} - 1 = \frac{113}{225}.$$

Also,

$$\begin{aligned}\cos 2(\phi+\theta) &= \cos 2\phi\cdot\cos 2\theta - \sin 2\phi\cdot\sin 2\theta\\ &= \frac{7}{25}\cos 2\theta - \frac{24}{25}\sin 2\theta = \frac{1}{25}\frac{m}{n}.\end{aligned}$$

It follows that

$$\frac{m}{n} = 25\frac{113}{225} = \frac{113}{9}.$$

The requested sum is $113 + 9 = 122$.

Alternatively, set $x = \cos\theta$ and $y = \sin\theta$. Then it is known that

$$4x - 3y = \frac{13}{3} \text{ and } x^2 + y^2 = 1,$$

and, from the double angle formulas for $\sin\theta$ and $\cos\theta$, the requested value is

$$7\cos 2\theta - 24\sin 2\theta = 7(2x^2 - 1) - 24(2xy).$$

But then

$$\begin{aligned} 7(2x^2 - 1) - 24(2xy) &= 14x^2 - 7 - 48xy \\ &= -18(x^2 + y^2) + 2(4x - 3y)^2 - 7 \\ &= -18(1) + 2\left(\frac{13}{3}\right)^2 - 7 \\ &= \frac{113}{9}. \end{aligned}$$

Problem 19. The centers of the three circles A, B, and C are collinear with the center of circle B lying between the centers of circles A and C. Circles A and C are both externally tangent to circle B, and the three circles share a common tangent line. Given that circle A has radius 12 and circle B has radius 42, find the radius of circle C.

Answer: 147

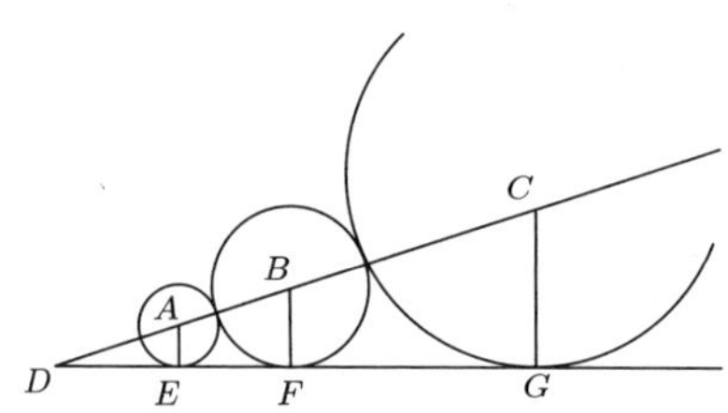

Let circles A, B, and C have centers A, B, C and radii a, b, and c, and be tangent to their common tangent at points E, F, and G, respectively. Let D be the intersection of the common tangent and the line containing the centers of the circles. Then triangles ADE, BDF, and CDG are similar. It follows that

$$\frac{a}{AD} = \frac{b}{BD} = \frac{b}{AD + a + b}.$$

This yields

$$\frac{AD}{a} = \frac{b+a}{b-a},$$

so by similar triangles

$$\frac{AD}{a} = \frac{CD}{c} = \frac{AD+a+2b+c}{c} = \frac{a\frac{b+a}{b-a}+a+2b+c}{c}.$$

Solving for c gives

$$c = \frac{a\frac{b+a}{b-a}+a+2b}{\frac{b+a}{b-a}-1} = \frac{a(b+a)+(a+2b)(b-a)}{(b+a)-(b-a)}$$

$$= \frac{b^2}{a} = \frac{42^2}{12} = 147.$$

Alternatively, note that there is a dilation centered at D that maps circle A to circle B. The constant of this dilation is $\frac{42}{12} = \frac{7}{2}$. The same dilation moves circle B to circle C, thus the radius of circle C is $\frac{7}{2} \cdot 42 = 147$.

Problem 20. How many of the rearrangements of the digits 123456 have the property that, for each digit, no more than two digits smaller than that digit appear to the right of that digit? For example, the rearrangement 315426 has this property because digits 1 and 2 are the only digits smaller than 3 which follow 3, digits 2 and 4 are the only digits smaller than 5 which follow 5, and digit 2 is the only digit smaller than 4 which follows 4.

Answer: 162

Begin with six open positions. To get an arrangement of digits having the desired property, choose a position to place the digit 6. The 6 can be placed in any of the 3 rightmost positions. Once the position of the 6 has been chosen, choose a position to place the digit 5. The 5 can be placed in any of the 3 rightmost positions not occupied by the 6. Continuing this way, the digits 6, 5, 4, and 3 can all be placed in one of 3 positions, the 3 rightmost positions which are left open. Finally, there will be 2 available positions to place the 2, and only one position left for the 1. Thus, the number of ways to choose an arrangement with the desired property is $3^4 \cdot 2 = 162$.

Problem 21. Let a be the sum of the numbers:

$$\begin{aligned}&99 \times 0.9\\&999 \times 0.9\\&9999 \times 0.9\\&\vdots\\&999\cdots 9 \times 0.9,\end{aligned}$$

where the final number in the list is 0.9 times a number written as a string of 101 digits all equal to 9. Find the sum of the digits in the number a.

Answer: 891

The list contains 100 numbers. The sum is

$$\begin{aligned}a &= 0.9 \times (99 + 999 + 9999 + \cdots + 999\cdots 9)\\&= 0.9 \times [(100 - 1) + (1000 - 1) + \cdots + (1000\cdots 0 - 1)]\\&= 0.9 \times \left[\sum_{n=0}^{101}(10^n - 1) - 9\right] = \frac{9}{10} \cdot \left[\sum_{n=0}^{101} 10^n - 111\right]\\&= \frac{9}{10} \cdot \left[\frac{10^{102} - 1}{10 - 1} - 111\right] = \frac{1}{10}\left[(10^{102} - 1) - 999\right]\\&= 10^{101} - 100.\end{aligned}$$

The decimal representation of this number is 101 digits long and consists of 99 nines followed by 2 zeros. The sum of the digits of this number is $99 \cdot 9 = 891$.

Problem 22. Ten distinct points are placed on a circle. All ten of the points are paired so that the line segments connecting the pairs do not intersect. In how many different ways can this pairing be done?

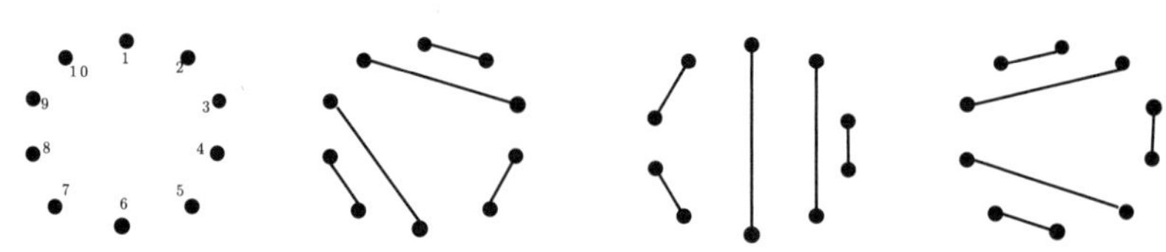

Answer: 42

More generally, let a_n be the number of ways one can draw n non-intersecting line segments between the pairings of $2n$ points on a circle. Clearly, $a_1 = 1$. If a_1 through a_n are known, a_{n+1} can be calculated as follows. Number the points in the circle in order from 1 to $2n+2$. Point 1 will need to be paired with one of the n points labeled with an even number so that there will be an even number of points lying on each side of the line segment drawn from point 1. If point 1 is paired with point $2k$, then there are $2k-2$ points left on one side of the segment, and $2n-2k$ points left on the other side. It follows that there are $a_{k-1} \cdot a_{n-k}$ ways to pair the remaining points. This gives the recursive formula

$$a_{n+1} = \sum_{k=1}^{n+1} a_{k-1} \cdot a_{n-k}$$

assuming that a_0 is defined to be 1. Thus,

$$\begin{aligned}
a_2 &= a_0 \cdot a_1 + a_1 \cdot a_0 = 2 \\
a_3 &= a_0 \cdot a_2 + a_1 \cdot a_1 + a_2 \cdot a_0 = 5 \\
a_4 &= a_0 \cdot a_3 + a_1 \cdot a_2 + a_2 \cdot a_1 + a_3 \cdot a_0 = 14 \\
a_5 &= a_0 \cdot a_4 + a_1 \cdot a_3 + a_2 \cdot a_2 + a_3 \cdot a_1 + a_4 \cdot a_0 = 42.
\end{aligned}$$

Thus, the answer is 42. These numbers are also the Catalan numbers

$$a_n = C(n) = \frac{\binom{2n}{n}}{n+1}.$$

Problem 23. A disk with radius 10 and a disk with radius 8 are drawn so that the distance between their centers is 3. Two congruent small circles lie in the intersection of the two disks so that they are tangent to each other and to each of the larger circles, as shown. The radii of the smaller circles are both $\frac{m}{n}$ where m and n are relatively prime positive integers. Find $m+n$.

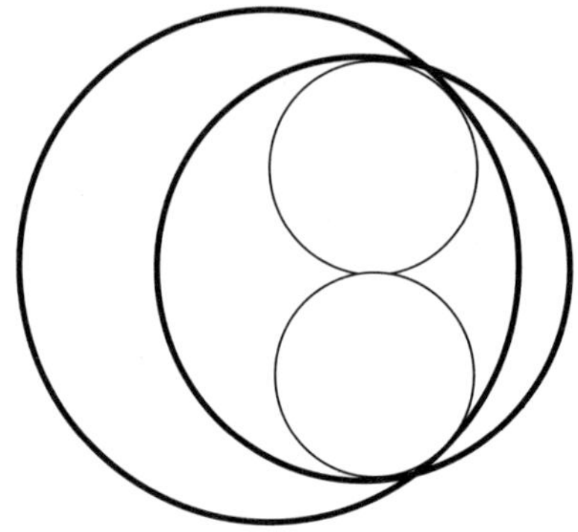

Answer: 19

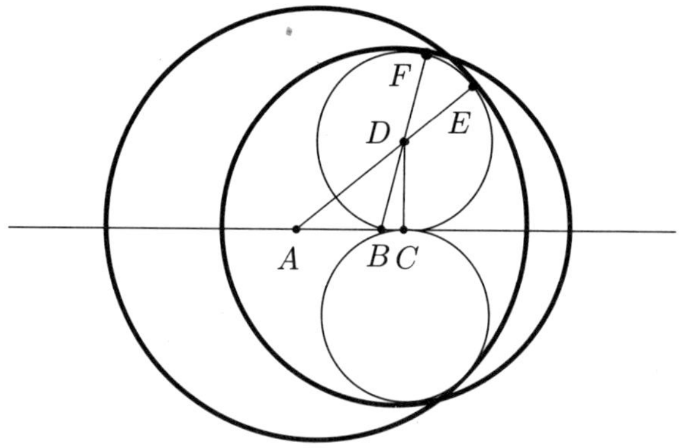

Let A be the center of the circle with radius 10, and B be the center of the circle with radius 8. The two small circles are tangent to each other and the line $\overline{AB}$ at point C. Let D be the center of one of the small circles, E be the point where that small circle is tangent to the circle centered at A, and F be the point where that small circle is tangent to the circle centered at B. The triangles ACD and BCD are right triangles with right angles at C. Let r be the radius of the small circles. Then $CD = r$. Since $AE = 10$, it follows from the Pythagorean Theorem that

$$AC = \sqrt{(10-r)^2 - r^2}.$$

Since $BF = 8$, it follows from the Pythagorean Theorem that

$$BC = \sqrt{(8-r)^2 - r^2}.$$

But $AC = BC + 3$, so

$$\sqrt{(10-r)^2 - r^2} = \sqrt{(8-r)^2 - r^2} + 3.$$

Square both sides to get

$$100 - 20r = 64 - 16r + 6\sqrt{(8-r)^2 - r^2} + 9.$$

Solving for the radical term and squaring again yields

$$6\sqrt{(8-r)^2 - r^2} = 27 - 4r$$

and

$$36\,(64 - 16r) = 27^2 - 216r + 16r^2.$$

This simplifies to $16r^2 + 360r - 1575 = 0$ which has one positive solution at

$$r = \frac{-360 + \sqrt{360^2 + 4 \cdot 16 \cdot 1575}}{2 \cdot 16}$$

$$= \frac{-360 + 24\sqrt{15^2 + 175}}{32} = \frac{-360 + 24 \cdot 20}{32} = \frac{15}{4}.$$

The requested sum is $15 + 4 = 19$.

Problem 24. Find the number of ordered pairs of integers (m, n) that satisfy

$$20m - 10n = mn.$$

Answer: 24

Rewrite the condition as $mn - 20m + 10n = 0$ and then

$$mn - 20m + 10n - 200 = -200$$

which factors as $(m + 10)(n - 20) = -200$.
Thus, (m, n) satisfies the desired condition if the product of $m + 10$ and $n - 20$ is -200. The 12 positive integer divisors of 200 are 1, 2, 4, 5, 8, 10, 20, 25, 40, 50, 100, and 200. If $m + 10$ is any one of these positive integer divisors of 200 or the negative of any one of them, then setting $n = \frac{-200}{m+10} + 20$ lets (m, n) satisfy the required condition. Thus, there are 24 ordered pairs satisfying the condition.

Problem 25. Let x_1, x_2, x_3 be the three roots of the polynomial $x^3 + 3x + 1$. There are relatively prime positive integers m and n so that

$$\frac{m}{n} = \frac{x_1^2}{(5x_2+1)(5x_3+1)} + \frac{x_2^2}{(5x_3+1)(5x_1+1)} + \frac{x_3^2}{(5x_1+1)(5x_2+1)}.$$

Find $m + n$.

Answer: 10

First note that since $x^3 + 3x + 1 = (x - x_1)(x - x_2)(x - x_3)$, we get $x_1 + x_2 + x_3 = 0$, $x_1x_2 + x_1x_3 + x_2x_3 = 3$, and $x_1x_2x_3 = -1$. It also follows that

$$0 = (x_1 + x_2 + x_3)^2 = x_1^2 + x_2^2 + x_3^2 + 2(x_1x_2 + x_1x_3 + x_2x_3)$$

and $x_1^2 + x_2^2 + x_3^2 = -6$.
Finally, since each x_j satisfies $x^3 + 3x + 1 = 0$, one gets that

$$x_1^3 + x_2^3 + x_3^3 = -3(x_1 + x_2 + x_3) - 3 = -3.$$

Now,

$$\begin{aligned}
&\frac{x_1^2}{(5x_2+1)(5x_3+1)} + \frac{x_2^2}{(5x_1+1)(5x_3+1)} + \frac{x_3^2}{(5x_1+1)(5x_2+1)} \\
&= \frac{x_1^2(5x_1+1) + x_2^2(5x_2+1) + x_3^2(5x_3+1)}{(5x_1+1)(5x_2+1)(5x_3+1)} \\
&= \frac{5(x_1^3+x_2^3+x_3^3) + (x_1^2+x_2^2+x_3^2)}{5^3(x_1x_2x_3) + 5^2(x_1x_2+x_1x_3+x_2x_3) + 5(x_1+x_2+x_3) + 1} \\
&= \frac{5(-3) - 6}{125(-1) + 25(3) + 5(0) + 1} \\
&= \frac{-21}{-125 + 75 + 1} \\
&= \frac{21}{49} = \frac{3}{7}.
\end{aligned}$$

The requested sum is $3 + 7 = 10$.

Problem 26. In the coordinate plane a parabola passes through the points (7,6), (7,12), (18,19), (18,48). The axis of symmetry of the parabola is a line with slope $\frac{r}{s}$ where r and s are relatively prime positive integers. Find $r + s$.

Answer: 71

First consider the parabola with equation $y = x^2$ and two parallel lines with equations $y = mx + b_1$ and $y = mx + b_2$ for constants m, $b_1 > 0$, and $b_2 > 0$. Note that the axis of symmetry of the parabola is the y axis. The first line intersects the parabola at two points with x coordinates $\frac{m \pm \sqrt{m^2+4b_1}}{2}$, and the second line at x coordinates $\frac{m \pm \sqrt{m^2+4b_2}}{2}$. It follows that the difference between the x coordinates of the intersection points to the left of the y axis is $\frac{\sqrt{m^2+4b_1}-\sqrt{m^2+4b_2}}{2}$ which is also the difference between the x coordinates of the intersection points to the right of the y axis. Since all parabolas are geometrically similar, this shows that when two parallel lines intersect a parabola, the difference in the distances from the axis of symmetry for two intersection points on one side of the axis is the same as the difference in distances from the axis of symmetry for the two intersection points on the other side of the axis.

Because the slope of the axis of symmetry will not change if the parabola is translated seven units in the negative x direction and 6 units in the negative y direction, it can be assumed that the given parabola passes through the four points (0,0), (0,6), (11,13), and (11,42). Assume that the axis of symmetry of the parabola has slope m. Then there are constants a and b so that the line parallel to the axis which passes through (11,13) has equation $y = mx + a$, and the line parallel to the axis which passes through (11,42) has equation $y = mx + b$. It follows that $13 = 11m + a$ and $42 = 11m + b$ which implies that $b - a = 42 - 13 = 29$. The result of the previous paragraph says that the lines $y = mx + b$ and $y = mx + 6$ must be the same distance apart as the lines $y = mx$ and $y = mx + a$. Thus, $b - 6 = 0 - a$ or $a + b = 6$. This lets us solve for b to get $b = \frac{35}{2}$. So, the line parallel to the axis of symmetry which passes through the point (11,42) also passes through the point $(0, \frac{35}{2})$. Conclude that the slope m of this line, and thus the slope of the axis of symmetry, is $\frac{42-\frac{35}{2}}{11} = \frac{49}{22}$. The requested sum is $49 + 22 = 71$.

Problem 27. Let a and b be real numbers satisfying

$$2(\sin a + \cos a)\sin b = 3 - \cos b.$$

Find $3\tan^2 a + 4\tan^2 b$.

Answer: 35

Apply the Cauchy-Schwarz inequality to vectors

$$\langle 2(\sin a + \cos a), 1\rangle \text{ and } \langle \sin b, \cos b\rangle$$

to get that

$$[4(\sin a + \cos a)^2 + 1](\sin^2 b + \cos^2 b)$$

is greater than

$$[2(\sin a + \cos a)\sin b + \cos b]^2 = 3^2.$$

This simplifies to

$$[4(1+2\sin a\cos a)+1]\left(\sin^2 b+\cos^2 b\right)=5+4\sin 2a\geq 9.$$

Thus, $\sin 2a \geq 1$, and it follows that $\sin 2a = 1$. Use the identity $\tan\frac{x}{2}=\frac{\sin x}{1+\cos x}$ to conclude that $\tan a=\frac{1}{1+0}=1$. Since equality occurs in the Cauchy-Schwartz inequality only when the vectors are parallel, conclude that $\frac{\sin b}{\cos b}=\frac{2(\sin a+\cos a)}{1}$ and $\tan^2 b = 4\left(\sin^2 a+2\sin a\cos a+\cos^2 a\right) = 4\left(1+\sin 2a\right) = 8$. Now $3\tan^2 a+4\tan^2 b=3(1)+4(8)=35$.

Problem 28. There are relatively prime positive integers m and n such that

$$\frac{m}{n}=\sum_{k=3}^{\infty}\frac{1}{k^5-5k^3+4k}.$$

Find $m+n$.

Answer: 97

The sum

$$\sum_{n=3}^{\infty}\frac{1}{k^5-5k^3+4k}=\sum_{k=3}^{\infty}\frac{1}{(k-2)(k-1)k(k+1)(k+2)}$$

$$=\frac{1}{4}\sum_{k=3}^{\infty}\frac{(k+2)-(k-2)}{(k-2)(k-1)k(k+1)(k+2)}$$

$$=\frac{1}{4}\sum_{k=3}^{\infty}\left[\frac{1}{(k-2)(k-1)k(k+1)}-\frac{1}{(k-1)k(k+1)(k+2)}\right].$$

This last summation telescopes leaving

$$\frac{1}{4}\frac{1}{(3-2)(3-1)3(3+1)}=\frac{1}{4\cdot 1\cdot 2\cdot 3\cdot 4}=\frac{1}{96}.$$

The requested sum is $1+96=97$.

Problem 29. Square $ABCD$ is shown in the diagram below. Points E, F, G are on sides $\overline{AB}$, $\overline{BC}$, $\overline{DA}$, respectively, so that lengths of $\overline{BE}$, $\overline{BF}$, $\overline{DG}$ are equal. Points H and I are the midpoints of segments $\overline{EF}$ and $\overline{CG}$, respectively. Segment $\overline{GJ}$ is the perpendicular bisector of segment $\overline{HI}$. The ratio of the areas of pentagon $AEHJG$ and quadrilateral $CIHF$ can be written as $\frac{m}{n}$, where m and n are relatively prime positive integers. Find $m+n$.

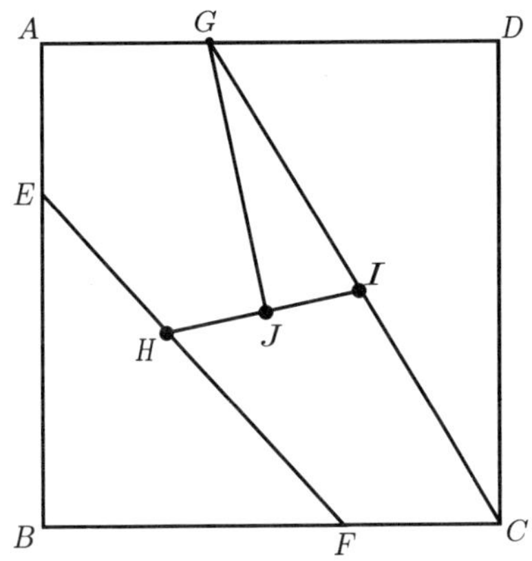

Answer: 175

Without loss of generality, let the square have side length 18. Set up coordinate axes so that $\overline{BC}$ is the x-axis and $\overline{AB}$ is the y-axis. Then the coordinates of point $A = (0, 18)$, $B = (0, 0)$, $C = (18, 0)$, and $D = (18, 18)$. Let the coordinates of point E be $(0, y)$. Then the coordinates of point $F = (y, 0)$, $G = (18-y, 18)$, $H = (\frac{y}{2}, \frac{y}{2})$, $I = (18 - \frac{y}{2}, 9)$, and $J = (9, \frac{18+y}{4})$. The line $\overline{GJ}$ is perpendicular to the line $\overline{HI}$, so the slope of $\overline{GJ} = \frac{\frac{54-y}{4}}{9-y}$ must be the negative reciprocal of the slope of $\overline{HI} = \frac{9-\frac{y}{2}}{18-y} = \frac{1}{2}$. It follows that $\frac{54-y}{4(9-y)} = -2$ so $y = 14$.

The areas of pentagon $AEHJG$ and quadrilateral $CIHF$ can be calculated from the coordinates of their vertices $A = (0, 18)$, $E = (0, 14)$, $H = (7, 7)$, $J = (9, 8)$, $G = (4, 18)$, $C = (18, 0)$, $I = (11, 9)$, and $F = (14, 0)$. This can be done in a number of ways including by calculating half the pseudo-determinants

$$\frac{1}{2}\begin{vmatrix} 0 & 18 \\ 0 & 14 \\ 7 & 7 \\ 9 & 8 \\ 4 & 18 \\ 0 & 18 \end{vmatrix} = \frac{1}{2}\left[\begin{matrix} (0\cdot 14 + 0\cdot 7 + 7\cdot 8 + 9\cdot 18 + 4\cdot 18) - \\ (0\cdot 18 + 4\cdot 8 + 9\cdot 7 + 7\cdot 14 + 0\cdot 18) \end{matrix}\right] = \frac{97}{2}$$

and

$$\frac{1}{2}\begin{vmatrix} 18 & 0 \\ 11 & 9 \\ 7 & 7 \\ 14 & 0 \\ 18 & 0 \end{vmatrix} = \frac{1}{2}\left[\begin{matrix} (18\cdot 9 + 11\cdot 7 + 7\cdot 0 + 14\cdot 0) - \\ (18\cdot 0 + 14\cdot 7 + 7\cdot 9 = 11\cdot 0) \end{matrix}\right] = 39.$$

The desired ratio of areas is $\frac{\frac{97}{2}}{39} = \frac{97}{78}$, and the requested sum is $97 + 78 = 175$.

Problem 30. Let x and y be real numbers satisfying

$$(x^2 + x - 1)(x^2 - x + 1) = 2(y^3 - 2\sqrt{5} - 1)$$

and

$$(y^2 + y - 1)(y^2 - y + 1) = 2(x^3 + 2\sqrt{5} - 1)$$

Find $8x^2 + 4y^3$.

Answer: 20

First note that the second equation is not the same as the first equation with the variables x and y interchanged. This shows that when the equations are satisfied, x and y are two distinct values. Rewrite the two equations as

$$x^4 - (x-1)^2 = 2y^3 - 4\sqrt{5} - 2$$

$$y^4 - (y-1)^2 = 2x^3 + 4\sqrt{5} - 2$$

and add the two equations to get $\left(x^4 - x^2 + 2x + 1 - 2x^3\right) + \left(y^4 - y^2 + 2y + 1 - 2y^3\right) = 0$ and $(x^2 - x - 1)^2 + (y^2 - y - 1)^2 = 0$. It follows that both $x^2 - x - 1 = 0$ and $y^2 - y - 1 = 0$ so x and y must be the two roots of $x^2 - x - 1$. The two values that satisfy the original equation are $x = \frac{1-\sqrt{5}}{2}$ and $y = \frac{1+\sqrt{5}}{2}$. Finally,

$$\begin{aligned} 8x^2 + 4y^3 &= 8\frac{1 - 2\sqrt{5} + (\sqrt{5})^2}{2^2} + 4\frac{1 + 3\sqrt{5} + 3(\sqrt{5})^2 + (\sqrt{5})^3}{2^3} \\ &= (12 - 4\sqrt{5}) + (8 + 4\sqrt{5}) = 20. \end{aligned}$$

2.15 2010 Middle School

Problem 1. If $125 + n + 135 + 2n + 145 = 900$, find n.

Answer: 165

The given equation simplifies to $405 + 3n = 900$ which implies $3n = 495$, so $n = 165$.

Problem 2. Three boxes each contain four bags. Each bag contains five marbles. How many marbles are there altogether in the three boxes?

Answer: 60

There are $3 \cdot 4 = 12$ bags each containing 5 marbles, so there are $5 \cdot 12 = 60$ marbles.

Problem 3. The sum

$$\frac{1}{1} + \frac{1}{2} + \frac{1}{3} + \frac{1}{4} + \frac{1}{5} + \frac{1}{6} = \frac{m}{n},$$

where m and n are relatively prime positive integers. Find $m+n$.

Answer: 69

Placing each fraction over a common denominator of 60, the sum becomes

$$\frac{60}{60} + \frac{30}{60} + \frac{20}{60} + \frac{15}{60} + \frac{12}{60} + \frac{10}{60} = \frac{147}{60} = \frac{49}{20}.$$

The requested sum is $49 + 20 = 69$.

Problem 4. The grid below contains five rows with six points in each row. Points that are adjacent either horizontally or vertically are a distance one apart. Find the area of the pentagon shown.

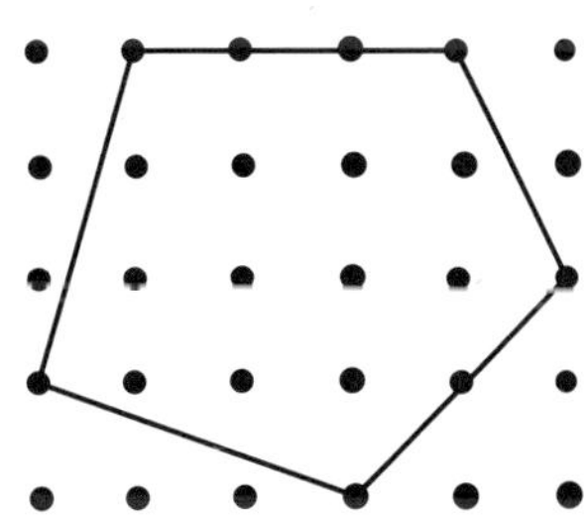

Answer: 14

The pentagon is a four by five rectangle with four right triangles removed as shown. Two of the right triangles have legs of length

1 and 3. They each have area $\frac{1\cdot 3}{2} = \frac{3}{2}$. The other two triangles have legs of length 1 and 2 and legs of length 2 and 2. These two triangles have areas $\frac{1\cdot 2}{2} = 1$ and $\frac{2\cdot 2}{2} = 2$, respectively. Thus, the pentagon has area $4 \cdot 5 - \frac{3}{2} - \frac{3}{2} - 1 - 2 = 14$.

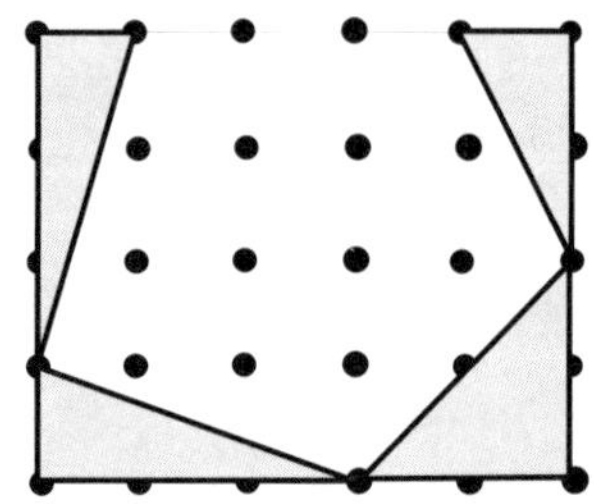

Alternatively, Pick's Theorem says that because there are 11 grid points inside the polygon and 8 grid points on the boundary of the polygon, the area of the polygon is $11 + \frac{8}{2} - 1 = 14$.

Problem 5. Find the least positive integer k so that $k+25973$ is a palindrome (a number which reads the same forward and backwards).

Answer: 89

The five digit palindrome beginning 259 is 25952 which does not exceed 25973, so the least palindrome exceeding 25973 must begin 260 and is 26062. The desired $k = 26062 - 25973 = 89$.

Problem 6. Find the sum of the prime factors of 777.

Answer: 47

The number 777 factors as $3 \cdot 7 \cdot 37$, so the requested sum is $3 + 7 + 37 = 47$.

Problem 7. If x and y are positive real numbers where x is p percent of y and y is $4p$ percent of x, find p.

Answer: 50

The problem states that $x = \frac{p}{100}y$ and $y = \frac{4p}{100}x$. Replacing x in the second equation with its value from the first equation yields $y = \frac{4p}{100} \cdot \frac{p}{100}y = \frac{4p^2}{100^2}y$. Since y is not zero, this reduces to $p^2 = \frac{100^2}{4}$, so $p = \frac{100}{2} = 50$.

Problem 8. There are exactly two four-digit numbers that are multiples of 3 and their first digit is double their second digit, their third digit is three more than their fourth digit, and their second digit is 2 less than their fourth digit. Find the difference of these two numbers.

Answer: 6333

If such a number has a second digit of m, then its first digit is $2m$, its fourth digit is $m+2$, and its third digit is $m+5$. The sum of the four digits is then $m+2m+(m+5)+(m+2)=5m+7$. To be a multiple of 3, the digits must add to a multiple of 3, so $5m+7$ must be a multiple of 3. This happens when m is 1, 4, or 7, but when m is 7, $m+5$ is not a digit. Thus, the two numbers are 2163 (when $m=1$) and 8496 (when $m=4$). The difference of these two numbers is 8496 - 2163 = 6333.

Problem 9. What percent of the numbers 1, 2, 3,. . . , 1000 are divisible by exactly one of the numbers 4 and 5?

Answer: 35

Exactly one quarter of the numbers 1, 2, 3, . . . 1000 are divisible by 4 which is $\frac{1000}{4}=250$ numbers. Exactly one fifth of the numbers are divisible by 5 which is $\frac{1000}{5}=200$ numbers. Some of the numbers are divisible by both 4 and 5. They are the ones divisible by 20, so there are exactly $\frac{1000}{20}=50$ numbers divisible by both 4 and 5. Thus, there are 250 - 50 numbers divisible by 4 and not divisible by 5, and there are 200 - 50 = 150 numbers divisible by 5 and not divisible by 4. This shows there are 200 + 150 = 350 numbers divisible by exactly one of 4 and 5 which is $\frac{350}{1000}\cdot 100=35$ percent of the numbers.

Problem 10. A baker uses $6\frac{2}{3}$ cups of flour when she prepares $\frac{5}{3}$ recipes of rolls. She will use $9\frac{3}{4}$ cups of flour when she prepares $\frac{m}{n}$ recipes of rolls, where m and n are relatively prime positive integers. Find $m+n$.

Answer: 55

Since the ratio of the number of recipes made to the cups of flour used must be constant, $\frac{\frac{5}{3}}{6\frac{2}{3}}=\frac{\frac{m}{n}}{9\frac{3}{4}}$. Thus, $\frac{\frac{5}{3}}{\frac{20}{3}}=\frac{\frac{m}{n}}{\frac{39}{4}}$ and $\frac{m}{n}=\frac{\frac{5}{3}}{\frac{20}{3}}\frac{39}{4}=\frac{39}{16}$. The requested sum is $39+16=55$.

Problem 11. There are two rows of seats with three side-by-side seats in each row. Two little boys, two little girls, and two adults sit in the six seats so that neither little boy sits to the side of either little girl. In how many different ways can these six people be seated?

Answer: 176

First observe that the two adults may not sit in the same row. Since there are three seats in each row, there are $3\cdot 3=9$ ways of selecting two seats for the adults to sit. If the two adults sit in the two center seats in each row, then the four children can sit in any of the other four seats. There are 2 ways for the adults to seat themselves in the two central seats, and there are $4! = 24$ ways for the children to sit in the four other seats. Thus, there are $2 \cdot 24 = 48$ ways for the people to sit with the adults in the two center seats.

If either adult sits in an end seat, then there are two seats next to that adult where either two little boys or two little girls will need to sit. Thus, the little boys will need to sit in one row, and the little girls will need to sit in the other row. There are $9 - 1 = 8$ ways to choose the pair of seats for the adults to sit so that at least one adult sits in an end seat. There are two ways for the adults to choose which of the two seats to sit in, so there are $2 \cdot 8 = 16$ ways for the adults to choose seats. There are two ways for the little boys to choose a row to sit in. Then there are $2 \cdot 2$ ways for the little boys and little girls to choose seats. This accounts for $16 \cdot 2 \cdot 2 \cdot 2 = 128$ ways for the six people to sit. Thus, the final count is $48 + 128 = 176$.

Problem 12. The diagram below shows twelve 30-60-90 triangles placed in a circle so that the hypotenuse of each triangle coincides with the longer leg of the next triangle. The fourth and last triangle in this diagram are shaded. The ratio of the perimeters of these two triangles can be written as $\frac{m}{n}$, where m and n are relatively prime positive integers. Find $m + n$.

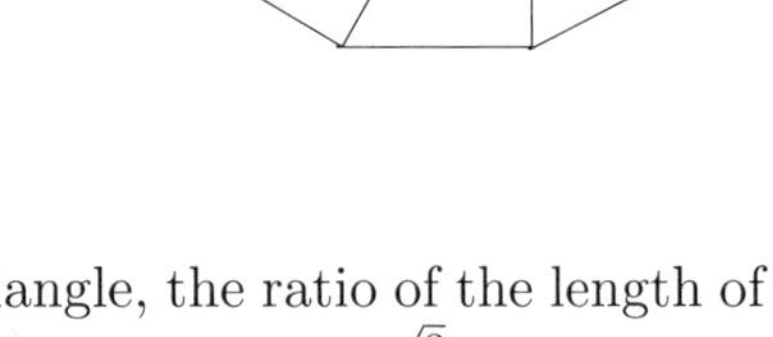

Answer: 337

In a 30-60-90 triangle, the ratio of the length of the longer leg to the length of the hypotenuse is $\frac{\sqrt{3}}{2}$. So, since all twelve triangles are similar, the ratio of the length of the hypotenuse of one of the triangles in the diagram to the length of the hypotenuse of the next larger triangle is the same $\frac{\sqrt{3}}{2}$. It follows that the perimeters of two adjacent triangles are in this same ratio. Thus, the ratio of the perimeter of the fourth triangle to the perimeter of the twelfth triangle is

$$\left(\frac{\sqrt{3}}{2}\right)^{12-4} = \frac{3^4}{2^8} = \frac{81}{256}.$$

The requested sum is $81 + 256 = 337$.

Problem 13. Find the number of sets A that satisfy the three conditions:

- A is a set of two positive integers
- each of the numbers in A is at least 22 percent the size of the other number
- A contains the number 30.

Answer: 129

The set A contains the number 30. The smallest integer which could be a member of A is $7 > 30 \cdot 0.22 = 6.6$. The greatest integer that could be in A is $136 < 30 \div 0.22 = 136.\overline{36}$. So the second number in A can be any integer between 7 and 136 except for the number 30 because the set $\{30, 30\}$ is a set of just one integer and not a set of two integers. Thus, the number of possible sets A is $136 - 6 - 1 = 129$.

Problem 14. Let $ABCD$ be a trapezoid where $\overline{AB}$ is parallel to $\overline{CD}$. Let P be the intersection of diagonal $\overline{AC}$ and diagonal $\overline{BD}$. If the area of triangle PAB is 16, and the area of triangle PCD is 25, find the area of the trapezoid.

Answer: 81

Let $CD = x$, and y be the length of the altitude of triangle PCD from P to $\overline{CD}$. Then $\frac{xy}{2} = 25$. Let α be the ratio of AB to CD. Note that triangles PAB and PCD are similar, so the area of triangle PAB must be $25\alpha^2 = 16$, and $\alpha = \frac{4}{5}$. The trapezoid has bases of lengths x and αx and a height of $y + \alpha y$. It follows that the trapezoid has area

$$\frac{x + \alpha x}{2}(y + \alpha y) = \frac{xy}{2}(1+\alpha)^2 = 25\left(1 + \frac{4}{5}\right)^2 = 25\left(\frac{9}{5}\right)^2 = 81.$$

Problem 15. In the arrangement

$$\begin{array}{lllll} 1 & & & & \\ 2 & 3 & & & \\ 4 & 5 & 6 & & \\ 7 & 8 & 9 & 10 & \\ 11 & 12 & 13 & 14 & 15 \\ \vdots & & & & \end{array}$$

what is the number that will appear directly below the number 2010?

Answer: 2073

The final number in row k is the kth triangular number $\frac{k(k+1)}{2}$. A little investigating shows that row 62 ends with the number $\frac{62 \cdot 63}{2} = 1953$, and row 63 ends with the number $\frac{63 \cdot 64}{2} = 2016$. It follows that 2010 is number $2010 - 1953 = 57$ in row 62. The number directly below 2010 is number 57 in row 63 which is $2016 + 57 = 2073$.

Problem 16. Half the volume of a 12 foot high cone-shaped pile is grade A ore while the other half is grade B ore. The pile is worth \$62. One-third of the volume of a similarly shaped 18 foot pile is grade A ore while the other two-thirds is grade B ore. The second pile is worth \$162. Two-thirds of the volume of a similarly shaped 24 foot pile is grade A ore while the other one-third is grade B ore. What is the value in dollars of the 24 foot pile?

Answer: 608

Let a 12 foot pile of grade A ore be worth x and a 12 foot pile of grade B ore be worth y. Then the problem states that $\frac{1}{2}x + \frac{1}{2}y = 62$ which simplifies to $x + y = 124$. Since the 18 foot pile has a volume $\left(\frac{18}{12}\right)^3 = \frac{27}{8}$ times that of the 12 foot pile, the problem states that $\frac{27}{8}\left(\frac{1}{3}x + \frac{2}{3}y\right) = 162$ which simplifies to $x + 2y = 144$. The 24 foot pile then has worth $8\left(\frac{2}{3}x + \frac{1}{3}y\right) = \frac{8}{3}(2x + y) = \frac{8}{3}(3(x + y) - (x + 2y)) = \frac{8}{3}(3 \cdot 124 - 144) = 608$. Note that $x = 104$ and $y = 20$, but these values are not needed to find the solution.

Problem 17. The diagram below shows a triangle divided into sections by three horizontal lines which divide the altitude of the triangle into four equal parts, and three lines connecting the top vertex with points that divide the opposite side into four equal parts. If the shaded region has area 100, find the area of the entire triangle.

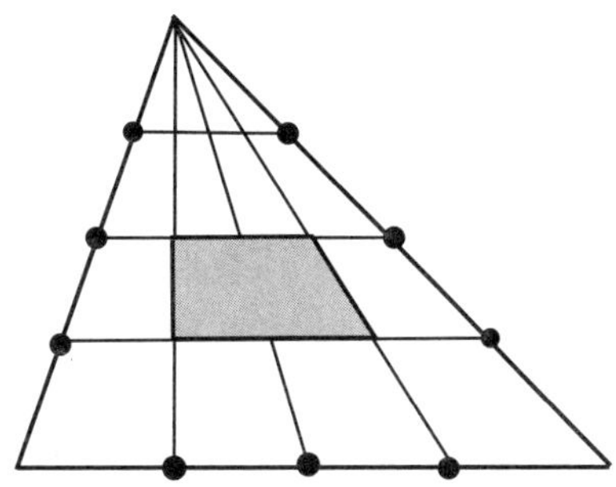

Answer: 640

Label the vertices of the triangle and intersection points as shown.

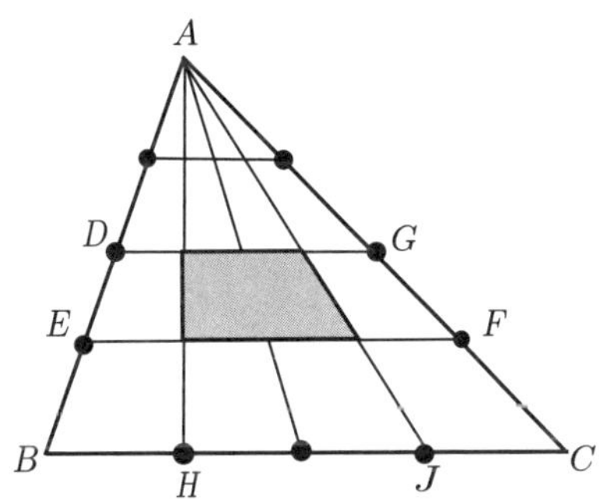

Let x be the area of triangle ABC. Triangle AEF is similar to triangle ABC with an altitude that is $\frac{3}{4}$ the length, so the area of triangle AEF is $x\left(\frac{3}{4}\right)^2$. Triangle ADG is similar to triangle ABC with an altitude that is $\frac{1}{2}$ the length, so the area of triangle ADG is $x\left(\frac{1}{2}\right)^2$. Finally, the shaded region has half the area of trapezoid $DEFG$ because both bases are half as long. Thus,

$$100 = \frac{1}{2}\left[x\left(\frac{3}{4}\right)^2 - x\left(\frac{1}{2}\right)^2\right]$$

and

$$100 = x \cdot \frac{1}{2} \cdot \left(\frac{9}{16} - \frac{1}{4}\right).$$

This gives

$$x = 100 \cdot 2 \cdot \frac{16}{5} = 640.$$

Problem 18. How many three-digit positive integers contain both even and odd digits?

Answer: 675

There are $999-99 = 900$ three digit positive integers. Since there are 5 odd digits, $5^3 = 125$ of the three-digit positive integers contain only odd digits. Since the first digit of a three-digit integer cannot be zero, there are $4 \cdot 5^2 = 100$ three-digit positive integers that contain only even digits. Thus, there are $900 - 125 - 100 = 675$ three-digit positive integers that contain both even and odd digits.

Problem 19. Square A is adjacent to square B, which is adjacent to square C. The three squares all have their bottom sides along a common horizontal line, as shown. The upper left vertices of the three squares are collinear. If square A has area 24 and square B has area 36, find the area of square C.

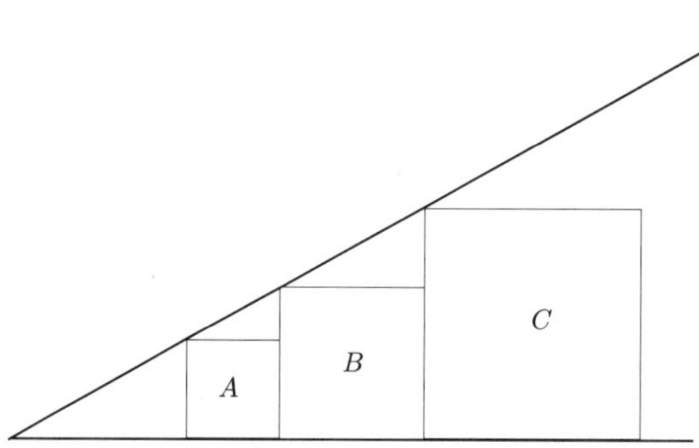

Answer: 54

Let the upper corners of square A be labeled V and W, the upper corners of square B be labeled X and Y, and the upper left corner of square C be labeled Z as shown below. Since the sides of the squares are parallel, the triangles VWX and XYZ are similar. Note that since square A has area 24, it has side length $VW = \sqrt{24} = 2\sqrt{6}$, and since square B has area 36, it has side length $XY = \sqrt{36} = 6$. The ratio of corresponding sides of similar triangles are equal, so $\frac{WX}{VW} = \frac{YZ}{XY}$ and $YZ = XY \cdot \frac{WX}{VW} = 6 \cdot \frac{6-2\sqrt{6}}{2\sqrt{6}} = 3\sqrt{6} - 6$. The side length of square C is ZY plus the length of the side of square B, so the side length of square C is $(3\sqrt{6}-6)+6 = 3\sqrt{6}$. The area of square C is $\left(3\sqrt{6}\right)^2 = 9\cdot 6 = 54$.

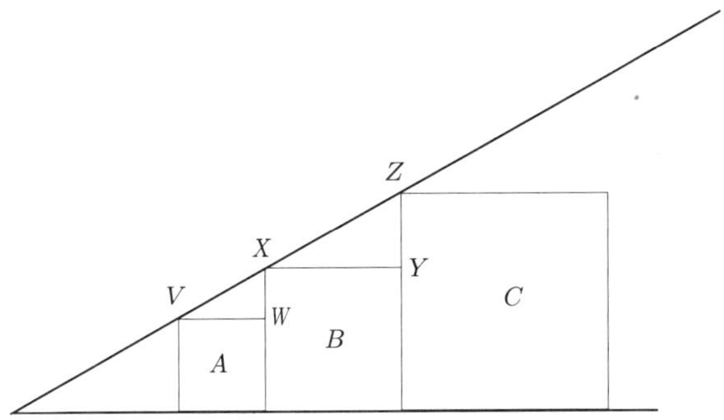

Problem 20. Suppose that f is a function such that

$$3f(x) - 5xf\left(\frac{1}{x}\right) = x - 7$$

for all non-zero real numbers x. Find $f(2010)$.

Answer: 4021

If x is not zero, replace x by $\frac{1}{x}$ in the equation to obtain

$$3f(\frac{1}{x}) - 5\frac{1}{x}f(x) = \frac{1}{x} - 7.$$

By multiplying by x, this reduces to

$$3xf(\frac{1}{x}) - 5f(x) = 1 - 7x.$$

Adding 5 times this equation to 3 times the original equation yields

$$5\left[3xf\left(\frac{1}{x}\right) - 5f(x)\right] + 3\left[3f(x) - 5xf\left(\frac{1}{x}\right)\right] = 5[1-7x]+3[x-7]$$

This simplifies to $-16f(x) = -16 - 32x$ and $f(x) = 1 + 2x$. It follows that $f(2010) = 1 + 2(2010) = 4021$.

2.16 2011 High School

Problem 1. The ratio of 3 to the positive number n is the same as the ratio of n to 192. Find n.

Answer: 24

The number n satisfies $\frac{3}{n} = \frac{n}{192}$ which is equivalent to

$$n^2 = 3 \cdot 192 = (3 \cdot 3)(64) = (3 \cdot 8)^2.$$

Thus, $n = 3 \cdot 8 = 24$.

Problem 2. The target below is made up of concentric circles with diameters 4, 8, 12, 16, and 20. The area of the dark region is $n\pi$. Find n.

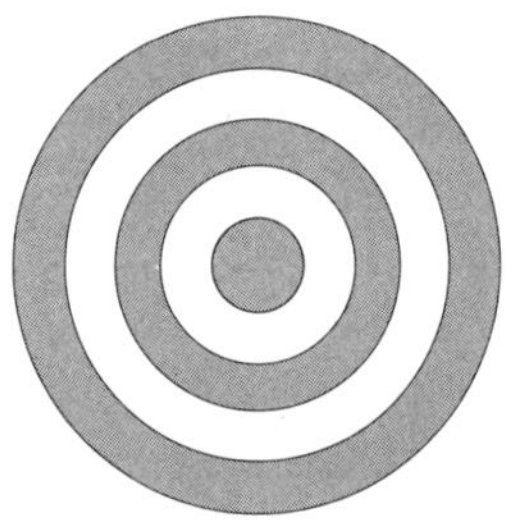

Answer: 60

The area of a circle with radius r is πr^2, so the area of the dark region is $\pi(10^2 - 8^2 + 6^2 - 4^2 + 2^2) = \pi(100 - 64 + 36 - 16 + 4) = 60\pi$. Hence $n = 60$.

Problem 3. Shirley went to the store planning to buy 120 balloons for 10 dollars. When she arrived, she was surprised to find that the balloons were on sale for 20 percent less than expected. How many balloons could Shirley buy for her 10 dollars?

Answer: 150

Shirley will spend 10 dollars to buy balloons selling at the price of $\frac{120}{10(1-0.20)}$. This lets her purchase $10 \cdot \frac{120}{8} = 150$ balloons.

Problem 4. Five non-overlapping equilateral triangles meet at a common vertex so that the angles between adjacent triangles are all congruent. What is the degree measure of the angle between two adjacent triangles?

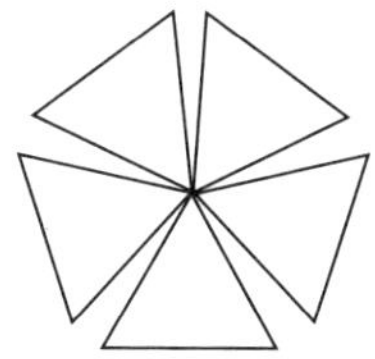

Answer: 12

There are ten angles that meet at the center point of the diagram. The angles within the five triangles are all $60°$, so the other five angles have total measure $360° - 5 \cdot 60° = 60°$. Thus, each of the other five angles has measure $\frac{60°}{5} = 12°$, and the answer is 12.

Problem 5. Let $a_1 = 2$, and for $n \geq 1$, let $a_{n+1} = 2a_n + 1$. Find the smallest value of an a_n that is not a prime number.

Answer: 95

Simply calculating, one obtains

$$a_2 = 2 \cdot 2 + 1 = 5,\ a_3 = 2 \cdot 5 + 1 = 11,\ a_4 = 2 \cdot 11 + 1 = 23,$$

$$a_5 = 2 \cdot 23 + 1 = 47,\ \text{and}\ a_6 = 2 \cdot 47 + 1 = 95.$$

Because 2, 5, 11, 23, and 47 are all prime, and 95 is not prime, the answer is 95. Note that in general, $a_n = 3 \cdot 2^{n-1} - 1$.

Problem 6. Working alone, the expert can paint a car in one day, the amateur can paint a car in two days, and the beginner can paint a car in three days. If the three painters work together at these speeds to paint three cars, it will take them $\frac{m}{n}$ days where m and n are relatively prime positive integers. Find $m + n$.

Answer: 29

The expert paints cars at the rate of 1 per day. The amateur paints cars at the rate of $\frac{1}{2}$ per day. The beginner paints cars at the rate of $\frac{1}{3}$ per day. They work together to paint cars at a rate of $1 + \frac{1}{2} + \frac{1}{3}$ per day. Thus, they can paint three cars in $\frac{3}{1+\frac{1}{2}+\frac{1}{3}} = \frac{3 \cdot 6}{6+3+2} = \frac{18}{11}$ days. The requested sum is $18 + 11 = 29$.

Problem 7. Find the prime number p such that $71p + 1$ is a perfect square.

Answer: 73

Note that if $71p + 1 = n^2$, then $71p = (n-1)(n+1)$. Since 71 is prime, either $n - 1$ or $n + 1$ must be multiples of 71. If $n + 1$

is $71k$, then $71p = (71k)(71k-2)$, and p must be $k(71k-2)$ which cannot happen for any prime p. If $n-1$ is $71k$, then $71p = (71k)(71k+2)$, and p must be $k(71k+2)$ which can only happen when $k=1$ and p is 73.

Problem 8. When 126 is added to its reversal, 621, the sum is $126+621=747$. Find the greatest integer which when added to its reversal yields 1211.

Answer: 952

Assume that the answer is a three digit number with digits, in order, given by a, b, and c. Then the number is $100a+10b+c$, and its reversal is $100c+10b+a$. The sum of the number and its reversal is $101(a+c)+20b=1211$. For the sum to end in 1, either $a+c=1$ which is not possible, or $a+c=11$. Since 1211 ends in 11, $20b$ must end in 00, so b must be 5. The largest possible value for a is 9 which would require c to be 2. The number 952 does satisfy the requirement that it plus its reversal add to 1211 as do 853, 754, 655, 556, 457, 358, and 259, but 952 is clearly the greatest of these.

Problem 9. There are integers m and n so that $9+\sqrt{11}$ is a root of the polynomial x^2+mx+n. Find $m+n$.

Answer: 52

From the quadratic formula $9+\sqrt{11} = \frac{-m+\sqrt{m^2-4n}}{2}$. Thus, $-\frac{m}{2} = 9$ and $\frac{m^2-4n}{4} = \left(\frac{m}{2}\right)^2 - n = 11$. So $m=-18$ and $n = 81-11 = 70$. The requested sum is $-18+70=52$.

Alternatively, if $9+\sqrt{11}$ is a root of the polynomial, then so is its conjugate, $9-\sqrt{11}$. Thus, the given polynomial factors as $x^2+mx+n = (x-9+\sqrt{11})(x-9-\sqrt{11})$. Substituting $x=1$ yields $1+m+n = (1-9+\sqrt{11})(1-9-\sqrt{11}) = (8+\sqrt{11})(8-\sqrt{11})$ $= 64-11 = 53$, and $m+n=52$.

Problem 10. The diagram shows a large circular dart board with four smaller shaded circles each internally tangent to the larger circle. Two of the internal circles have half the radius of the large circle, and are, therefore, tangent to each other. The other two smaller circles are tangent to these circles. If a dart is thrown so that it sticks to a point randomly chosen on the dart board, then the probability that the dart sticks to a point in the shaded area is $\frac{m}{n}$ where m and n are relatively prime positive integers. Find $m+n$.

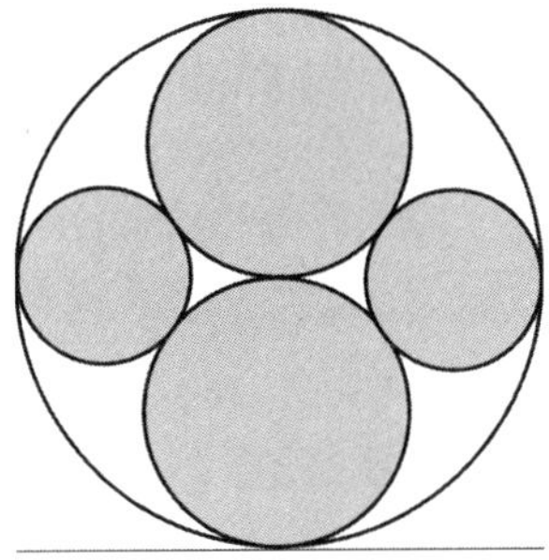

Answer: 31

Without loss of generality, assume that the dart board has radius 2. Then the larger shaded circles have radius 1, and the smaller shaded circles have radius r. The Pythagorean Theorem can be used to relate the distance between the centers of a larger and a smaller shaded circle to the distances of these two circles to the center of the dart board as shown in the diagram. Here $1^2 + (2-r)^2 = (1+r)^2$. Solving this yields $r = \frac{2}{3}$. The desired probability is the ratio of the shaded area to the area of the dart board which is $\frac{2\cdot\pi 1^2 + 2\cdot\pi\left(\frac{2}{3}\right)^2}{\pi 2^2} = \frac{13}{18}$. The requested sum is then $13 + 18 = 31$.

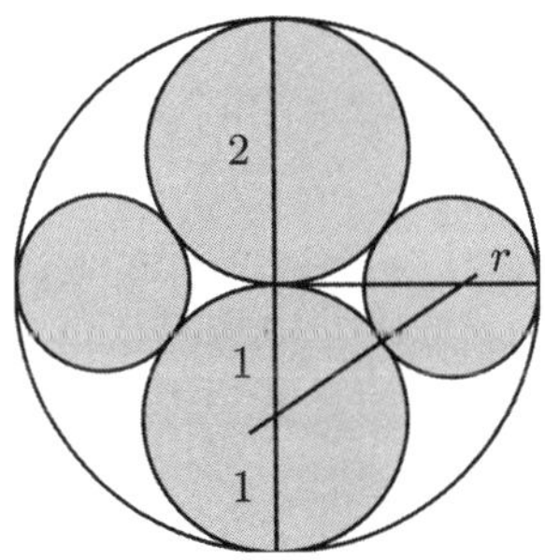

Problem 11. Six distinct positive integers are randomly chosen between 1 and 2011, inclusive. The probability that some pair of the six chosen integers has a difference that is a multiple of 5 is n percent. Find n.

Answer: 100

There are only 5 different possible remainders when a number is divided by 5. If six numbers are selected, by the Pigeonhole Principle, at least 2 of the selected numbers have the same remainder when divided by 5. The difference of these two numbers will be an integer divisible by 5. Hence the probability is equal to 1 or 100 percent.

Problem 12. Find the area of the region in the coordinate plane satisfying the three conditions

- $x \leq 2y$
- $y \leq 2x$
- $x + y \leq 60.$

Answer: 600

The line $x = 2y$ intersects the line $x + y = 60$ at the point $(40, 20)$. The line $y = 2x$ intersects the line $x + y = 60$ at the point $(20, 40)$. The region satisfying the inequalities is a triangle with vertices at $A = (0, 0)$, $B = (40, 20)$, and $C = (20, 40)$. This is an isosceles triangle with base $\overline{BC}$. The midpoint of the base is $(30, 30)$. Thus, the triangle has base length $20\sqrt{2}$ and height $30\sqrt{2}$, so its area is $\frac{20\sqrt{2} \cdot 30\sqrt{2}}{2} = 600$.

Problem 13. A 3 by 3 determinant has three entries equal to 2, three entries equal to 5, and three entries equal to 8. Find the maximum possible value of the determinant.

Answer: 405

The optimal configuration of such a determinant is

$$\begin{vmatrix} 8 & 5 & 2 \\ 2 & 8 & 5 \\ 5 & 2 & 8 \end{vmatrix}.$$

The added terms of the determinant's expansion sum to $8^3 + 5^3 + 2^3 = 645$ which is the largest possible such sum, and the subtracted terms add to $3 \cdot (2 \cdot 5 \cdot 8) = 240$ which is the smallest possible such sum. Thus, the difference $645 - 240 = 405$ yields the greatest possible value for such a determinant.

Problem 14. The lengths of the three sides of a right triangle form a geometric sequence. The sine of the smallest of the angles in the triangle is $\frac{m+\sqrt{n}}{k}$ where m, n, and k are integers, and n is not divisible by the square of any prime. Find $m + n + k$.

Answer: 6

There are real numbers $a > 0$ and $r > 1$ so that the sides of the triangle have lengths a, ar, and ar^2. By the Pythagorean Theorem, $a^2 + (ar)^2 = (ar^2)^2$ or $1 + r^2 = r^4$. The quadratic formula shows that this equation has one positive solution satisfying $r^2 = \frac{1+\sqrt{5}}{2}$. The sine of the smallest angle of the triangle is $\frac{a}{ar^2} = \frac{1}{r^2} = \frac{2}{1+\sqrt{5}} = \frac{-1+\sqrt{5}}{2}$. The requested sum is $-1+5+2 = 6$.

Problem 15. A pyramid has a base which is an equilateral triangle with side length 300 centimeters. The vertex of the pyramid is 100 centimeters above the center of the triangular base. A mouse starts at a corner of the base of the pyramid and walks up the edge of the pyramid toward the vertex at the top. When the mouse has walked a distance of 134 centimeters, how many centimeters above the base of the pyramid is the mouse?

Answer: 67

An equilateral triangle with side length a has altitude $\frac{a\sqrt{3}}{2}$, and the distance from a vertex to the center (the centroid, incenter, and circumcenter are all at the same point in an equilateral triangle) is two-thirds the altitude or $\frac{a}{\sqrt{3}}$. Thus, for the base of the pyramid, the distance from a vertex to the center is $100\sqrt{3}$ centimeters. By the Pythagorean Theorem, the length an edge of the pyramid going from a corner of the base to the top vertex is

$$\sqrt{(100\sqrt{3})^2 + 100^2} = \sqrt{30000 + 10000} = 200 \text{ centimeters.}$$

Thus, the edge is twice as long as the height of the pyramid. It follows by similar triangles that the mouse walks a distance equal to twice the height that the mouse is from the base of the pyramid. When the mouse has walked 134 centimeters, it is at a height of $\frac{134}{2} = 67$ centimeters above the base.

Problem 16. Evaluate $1^3 - 2^3 + 3^3 - 4^3 + 5^3 - \cdots + 101^3$.

Answer: 522801

Let $S = 1^3 - 2^3 + 3^3 - 4^4 + 5^3 - \cdots + 101^3$. Then

$$\begin{aligned} S &= \sum_{i=1}^{101} i^3 - 2\sum_{i=1}^{50}(2i)^3 = \sum_{i=1}^{101} i^3 - 2^4 \sum_{i=1}^{50} i^3 \\ &= \left[\frac{101 \cdot 102}{2}\right]^2 - 16\left[\frac{50 \cdot 51}{2}\right]^2 \\ &= 51^2 \cdot 101^2 - 4^2 \cdot 25^2 \cdot 51^2 = 51^2(101^2 - 100^2) \\ &= 51^2(101 - 100)(101 + 100) = 51^2 \cdot 201 = 522801. \end{aligned}$$

Alternatively,

$$\begin{aligned} S - 1^3 &= \sum_{k=1}^{50}[(2k+1)^3 - (2k)^3] = \sum_{k=1}^{50}(12k^2 + 6k + 1) \\ &= 12\left(\frac{50 \cdot 51(2 \cdot 50 + 1)}{6}\right) + 6\left(\frac{50 \cdot 51}{2}\right) + 50 \\ &= 515100 + 7650 + 50 = 522800. \end{aligned}$$

Thus, $S = 522801$.

Problem 17. In how many distinguishable rearrangements of the letters $ABCCDEEF$ does the A precede both C's, the F appears between the 2 C's, and the D appears after the F?

Answer: 336

The conditions in the problem indicate that the letters $ACFC$ must appear in that order. Into this sequence of letters, there are two positions into which the D can be placed, either before or after the final C. Into this sequence of five letters, insert three positions for the remaining three letters. Those positions can be inserted into any of the six positions surrounding the five letters. The number of ways of inserting three new positions can be counted by the well-known stars-and-bars technique to be $\binom{5+3}{3} = 56$. Finally, select a way to place the three remaining letters, B, E, F into those three positions. This can be done in $\frac{3!}{2!} = 3$ ways. Thus, the number of distinguishable rearrangements satisfying the given conditions is $2 \cdot 56 \cdot 3 = 336$.

Problem 18. Let a be a positive real number such that $\frac{a^2}{a^4-a^2+1} = \frac{4}{37}$. Then $\frac{a^3}{a^6-a^3+1} = \frac{m}{n}$, where m and n are relatively prime positive integers. Find $m+n$.

Answer: 259

Let $a+\frac{1}{a} = x$. It is given that $\frac{1}{a^2-1+\frac{1}{a^2}} = \frac{4}{37}$ hence $a^2+\frac{1}{a^2} = \frac{41}{4}$ or equivalently, $x^2-2 = \frac{41}{4}$, hence $x = \frac{7}{2}$. Then $\left(a^2+\frac{1}{a^2}\right)\left(a+\frac{1}{a}\right) = a^3+\frac{1}{a^3}+a+\frac{1}{a}$. Hence $a^3+\frac{1}{a^3} = x^3-3x = \left(\frac{7}{2}\right)^3-3\cdot\frac{7}{2} = \frac{259}{8}$. Thus, $\frac{a^3}{a^6-a^3+1} = \frac{1}{a^3-1+\frac{1}{a^3}} = \frac{1}{\frac{259}{8}-1} = \frac{8}{251} = \frac{m}{n}$. Thus $m+n = 259$.

Problem 19. The diagrams below shows a 2 by 2 grid made up of four 1 by 1 squares. Shown are two paths along the grid from the lower left corner to the upper right corner of the grid, one with length 4 and one with length 6. A path may not intersect itself by moving to a point where the path has already been. Find the sum of the lengths of all the paths from the lower left corner to the upper right corner of the grid.

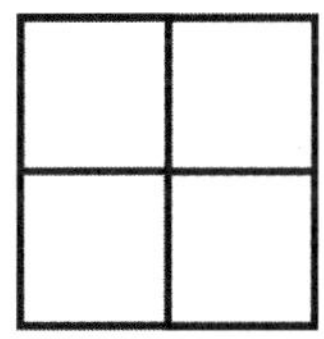

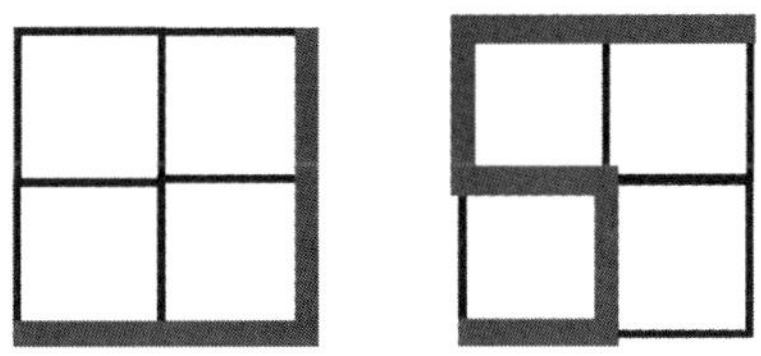

Answer: 64

By alternately coloring the corners of the squares black and white and noticing that each edge of a square connects a black corner to a white corner, it is seen that any path from the lower left corner of the grid to the upper right corner of the grid must be of even length. Because a path from the lower left corner of the grid to the upper right corner of the grid must move right along at least two sides of the 1 by 1 squares and move up along at least two sides of the 1 by 1 squares, such a path must have length at least 4. Since there are only 9 corners of the 1 by 1 squares in the grid, and a path cannot cross any corner twice, a path must have length at most 8. Thus, a path from the lower left corner of the grid to the upper right corner of the grid can have lengths 4, 6, or 8. A path of length 4 consists of movements along two horizontal edges and two vertical edges of 1 by 1 squares. The number of ways of arranging two horizontal moves and two vertical moves in a sequence of four moves is $\binom{4}{2} = 6$, so there are 6 paths of length 4. There are 4 paths of length 6, a fact easily verified by considering that a path of length 6 is determined by the first two edges of the path. There are only 2 paths of length 8, a fact easily verified by noticing that a path of length 8 must reach each of the corners which can be done only by including either all the horizontal edges or all the vertical edges of the 1 by 1 squares. Thus, the sum of the lengths of all the paths is $6 \cdot 4 + 4 \cdot 6 + 2 \cdot 8 = 64$.

Problem 20. Points A and B are the endpoints of a diameter of a circle with center C. Points D and E lie on the same diameter so that C bisects segment $\overline{DE}$. Let F be a randomly chosen point within the circle. The probability that $\triangle DEF$ has a perimeter less than the length of the diameter of the circle is $\frac{17}{128}$. There are relatively prime positive integers m and n so that the ratio of DE to AB is $\frac{m}{n}$. Find $m+n$.

Answer: 47

Without loss of generality, assume that the circle has radius 1 so its diameter is 2. Assume that points D and E are each a distance x from C. If $\triangle DEF$ has perimeter less than 2, then the sum $DF + EF + 2x < 2$ so $DF + EF < 2 - 2x$. It follows that the point F lies inside an ellipse with foci D and E where the sum of the distances from a point on the ellipse to the points D and E is the constant $2 - 2x$. It follows that the endpoints of the major axis of the ellipse are each a distance $1 - x$ from C. Let G be an endpoint of the minor axis of the ellipse. Then $DG + EG = 2 - 2x$ and $DG = EG$, so $DG = 1 - x$. Since $\triangle CDG$ is a right triangle with hypotenuse $DG = 1 - x$ and leg $CD = x$, the leg

$$CG = \sqrt{(1-x)^2 - x^2} = \sqrt{1-2x}.$$

The semiminor axis of the ellipse, therefore, has length $\sqrt{1-2x}$ while the semimajor axis has length $1 - x$. It follows that the ellipse has area $\pi(1-x)\sqrt{1-2x}$. The circle has area π, so the probability that the point F lies inside the ellipse is

$$\frac{\pi(1-x)\sqrt{1-2x}}{\pi} = (1-x)\sqrt{1-2x} = \frac{17}{128}.$$

Squaring and simplifying yields $2x^3 - 5x^2 + 4x - 1 + \frac{17^2}{2^{14}}$. Multiplying by 2^{14}, reduces this to $2^{15}x^3 - 2^{14}\cdot 5x^2 + 2^{16}x - 16,095 = 0$. It is known that this equation has a positive rational solution, so substitute $x = \frac{m}{2^5}$ to obtain $m^3 - 80m^2 + 2^{11}m - 16,095 = 0$. By the Rational Root Theorem, any rational solution to this equation must be an integer factor of $16,095 = 3 \cdot 5 \cdot 29 \cdot 37$. Testing the possibilities shows that $m = 15$ works, so $x = \frac{15}{32}$. This is the desired ratio of DE to AB, and the requested sum is $15 + 32 = 47$.

Problem 21. If a, b, and c are non-negative real numbers satisfying

$$a + b + c = 400,$$

find the maximum possible value of $\sqrt{2a+b}+\sqrt{2b+c}+\sqrt{2c+a}$.

Answer: 60

The graph of $x^2+y^2+z^2=k^2$ is a sphere radius k centered at the origin. For a given constant m, the graph of $x+y+z=m$ is a plane perpendicular to the line $x=y=z$. It follows that the largest possible value for m such that the plane intersects the sphere is the value of m for which the plane is tangent to sphere at a point where $x=y=z$. This occurs at $x=y=z=\sqrt{\frac{k}{3}}$, so the largest value of m such that the plane intersects the sphere is $m=\sqrt{3}k$. Thus, the largest value of $x+y+z$ for points satisfying $x^2+y^2+z^2=k^2$ is $\sqrt{3}k$. In other words, the largest value of $\sqrt{r}+\sqrt{s}+\sqrt{t}$ subject to $r+s+t=k^2$ is $\sqrt{3}k$, and this occurs when $r=s=t$.

Note that $(2a+b)+(2b+c)+(2c+a)=3(a+b+c)$. It follows that the maximum of $\sqrt{2a+b}+\sqrt{2b+c}+\sqrt{2c+a}$ occurs when $2a+b=2b+c=2c+a=\frac{3(a+b+c)}{3}=a+b+c=400$. This happens when $a=b=c=\frac{400}{3}$, and the maximum value is $3\sqrt{\frac{3\cdot 400}{3}}=60$.

Problem 22. Five congruent circles have centers at the vertices of a regular pentagon so that each of the circles is tangent to its two neighbors. A sixth circle (shaded in the diagram below) congruent to the other five is placed tangent to two of the five. If this sixth circle is allowed to roll without slipping around the exterior of the figure formed by the other five circles, then it will turn through an angle of k degrees before it returns to its starting position. Find k.

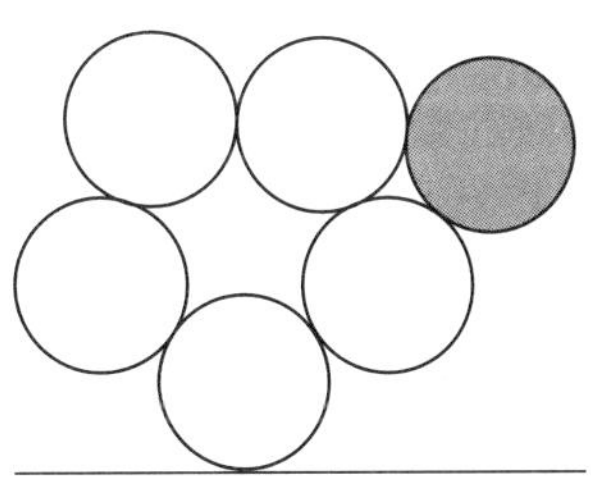

Answer: 1320

Label three adjacent vertices of the regular pentagon A, B, and C, and let the shaded circle begin tangent to the circles with centers at B and C as shown.

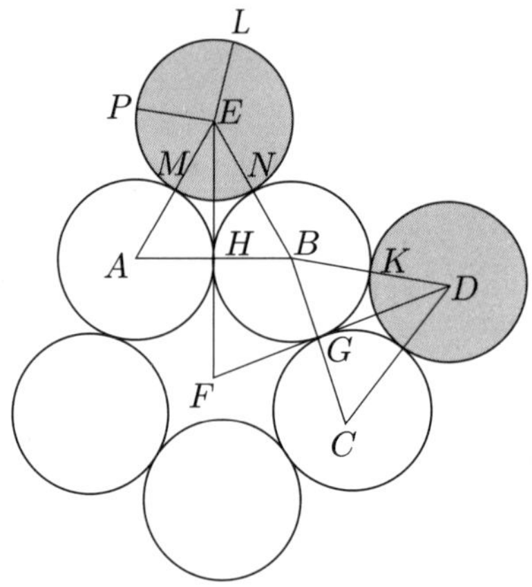

When the shaded circle rolls from its starting position centered at D to the position of the circle centered at E, the point K, where $\overline{BD}$ intersects the circle, rolls to a point L. Let F be the center of the regular pentagon. Let G be the intersection of $\overline{DF}$ and $\overline{BC}$, and H the intersection of $\overline{EF}$ and $\overline{AB}$. Let M and N be the intersections of the circles with $\overline{AE}$ and $\overline{BE}$, respectively. Let P be on the circle centered at E so that $\overline{EP}$ is parallel to $\overline{DK}$. As the circle centered at D rolls to the position of the circle centered at E, an arc of the rolling circle rolls along the arc of the circle centered at B. The arc of the circle centered at B is $\overset{\frown}{KN}$ which is congruent to $\overset{\frown}{NL}$ of the circle centered at E, and the point K on the rolling circle moves through an obtuse angle congruent to the obtuse $\angle PEL$ with the point K ending up in the position of the point L. Then $\angle PEM$ is congruent to $\angle DFE$ which is $\frac{360°}{5} = 72°$. Note that $\triangle ABE$ and $\triangle BCD$ are equilateral, so $\angle MEN$ is $60°$. Because $\angle DFE$ is $72°$, and $\angle FGB$ and $\angle FHB$ are right angles, $\angle GBH$ is $360° - 72° - 2 \cdot 90° = 108°$. (This is also seen as the measure of the internal angle of the regular pentagon.) Angle KBN is then $360° - \angle KBG - \angle GBH - \angle HBN = 360° - 60° - 108° - 60° = 132°$. From the congruence of $\angle KBN$ and $\angle NEL$, it follows that the obtuse $\angle PEL$ is $\angle PEM + \angle MEN + \angle NEL = 72° + 60° + 132° = 264°$. Thus, the shaded circle rolls through an angle of $264°$ when rolling from a position tangent to two of the five congruent circle to the next position where it is tangent to two of the five circles. It follows that the number of degrees the shaded circle must roll to return to its starting position is $5 \cdot 264° = 1320°$.

Problem 23. Let x be a real number in the interval $\left(0, \frac{\pi}{2}\right)$ such that

$$\frac{1}{\sin x \cos x} + 2 \cot 2x = \frac{1}{2}.$$

Evaluate

$$\frac{1}{\sin x \cos x} - 2 \cot 2x.$$

Answer: 8

For any x satisfying the given condition, $\frac{2}{\sin 2x} + 2 \cot 2x = \frac{1}{2}$ yielding $\csc 2x + \cot 2x = \frac{1}{4}$. Multiplying this relation by $\csc 2x - \cot 2x$ and taking into account that $\csc^2 a - \cot^2 a = 1$ implies $1 = \frac{1}{4}\left(\csc 2x - \cot 2x\right)$. It follows that $2 \csc 2x - 2 \cot 2x = 8$; hence $\frac{1}{\sin x \cos x} - 2 \cot 2x = 8$.

Problem 24. The diagram below shows a regular hexagon with an inscribed square where two sides of the square are parallel to two sides of the hexagon. There are positive integers m, n, and p such that the ratio of the area of the hexagon to the area of the square can be written as $\frac{m+\sqrt{n}}{p}$ where m and p are relatively prime. Find $m + n + p$.

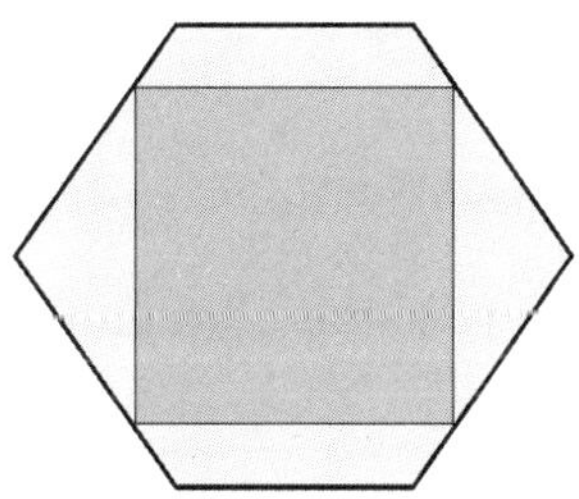

Answer: 19

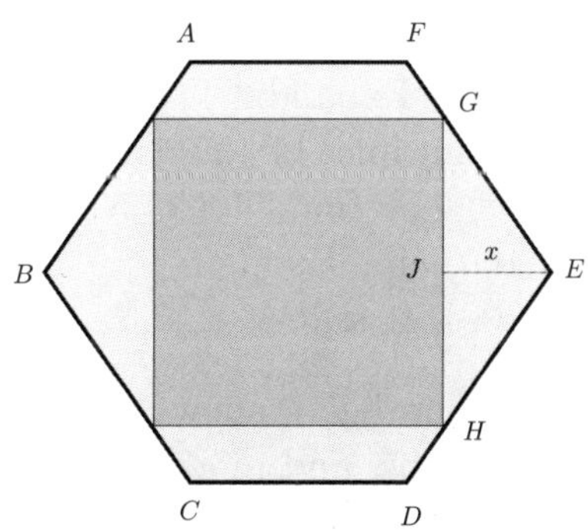

Label the vertices of the hexagon as shown. Let G and H be points where one side of the square touches the hexagon. Let J be the point on the side of the square closest to vertex E. Let the distance from E to J be given by x. Without loss of generality, assume that the hexagon has side length 1. Since $\angle DEF$ is $120°$ and side $\overline{DE}$ has length 1, the distance from D to F is twice the height of a 30-60-90 triangle with hypotenuse 1, or $2 \cdot 1 \cdot \sin 60° = \sqrt{3}$. So the hexagon is made up of a rectangle measuring 1 by $\sqrt{3}$ and four 30-60-90 triangles with legs measuring $\frac{1}{2}$ and $\frac{\sqrt{3}}{2}$. Thus, the area of the hexagon is $\sqrt{3} + 4 \cdot \frac{\sqrt{3}}{8} = \frac{3\sqrt{3}}{2}$. The width of the hexagon is 2, so the width of the square is $2 - 2x$. The height of the square is the length of $\overline{GH}$ which is $2x\sqrt{3}$. The width and height of the square are equal, so $2 - 2x = 2x\sqrt{3}$ and $x = \frac{1}{1+\sqrt{3}}$. Thus, the area of the square is

$$(2-2x)^2 = \left(2 - \frac{2}{1+\sqrt{3}}\right)^2 = \left(\frac{2\sqrt{3}}{1+\sqrt{3}}\right)^2 = \frac{6}{2+\sqrt{3}}.$$

The ratio of the areas is

$$\frac{\frac{3\sqrt{3}}{2}}{\frac{6}{2+\sqrt{3}}} = \frac{3+\sqrt{12}}{4}.$$

The requested sum is $3 + 12 + 4 = 19$.

Problem 25. Find the remainder when $3^3 \cdot 33^{33} \cdot 333^{333} \cdot 3333^{3333}$ is divided by 100.

Answer: 19

The problem seeks the positive integers k with $k < 100$ so that $3^3 \cdot 33^{33} \cdot 333^{333} \cdot 3333^{3333} \equiv k \pmod{100}$. Then $3^3 \cdot 99^{33} \cdot 999^{333} \cdot 9999^{3333} \equiv 3^{33}3^{333}3^{3333}k \equiv 3^{3699}k \pmod{100}$. So $(-1)^{33}(-1)^{333}(-1)^{3333} \equiv 3^{3696}k \equiv 9^{1848}k \equiv (10-1)^{1848}k \pmod{100}$. Expanding $(10-1)^{1848}$ yields only two terms that are not multiples of 100, that is, $(10-1)^{1848}k \equiv (1-18480)k \equiv -79k \equiv 21k \pmod{100}$. Now $21k \equiv (-1)^{33+333+3333} \equiv -1 \pmod{100}$ implies that $-19 \equiv 19 \cdot 21k \equiv (20^2-1)k \equiv -k \pmod{100}$. It follows that $k = 19$.

Problem 26. The diagram below shows two parallel rows with seven points in the upper row and nine points in the lower row. The points in each row are spaced one unit apart, and the two rows are two units apart. How many trapezoids which are not parallelograms have vertices in this set of 16 points and have area of at least six square units?

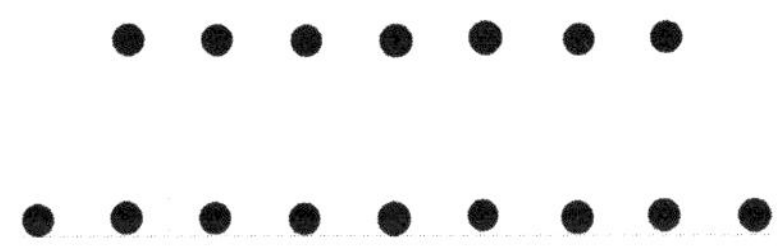

Answer: 361

Any choice of two vertices from the upper row of points and two vertices from the lower row of points results in vertices of a trapezoid. If the two vertices from the top row are k units apart, and the two vertices from the bottom row are m units apart, then the trapezoid will have area $2\left(\frac{k+m}{2}\right) = k + m$. The trapezoid will be a parallelogram if $k = m$. Also note that there are $7-k$ ways to choose two vertices in the upper row of vertices so that the two vertices are k units apart, and there are $9 - m$ ways to choose two vertices in the lower row of vertices so that the two vertices are m units apart. The possible values of k and m giving an acceptable trapezoid are listed in the following table:

k	m	Upper points choices	Lower points choices	Total choices
1	5,6,7,8	6	$4+3+2+1=10$	60
2	4,5,6,7,8	5	$5+4+3+2+1=13$	75
3	4,5,6,7,8	4	$5+4+3+2+1=13$	60
4	2,3,5,6,7,8	3	$7+6+4+3+2+1=23$	69
5	1,2,3,4,6,7,8	2	$8+7+6+5+3+2+1=32$	64
6	1,2,3,4,5,7,8	1	$8+7+6+5+4+2+1=33$	33

Thus, the number of acceptable trapezoids is

$$60 + 75 + 60 + 69 + 64 + 33 = 361.$$

Problem 27. Find the smallest prime number that does not divide

$$9 + 9^2 + 9^3 + \cdots + 9^{2010}.$$

Answer: 17

Let $N = 9 + 9^2 + 9^3 + \cdots + 9^{2010}$.

- Then N is divisible by 2 because it is the sum of 2010 odd integers.
- Clearly, N is a multiple of 9, so it is a multiple of 3.
- Because $N = 9(1+9)+9^3(1+9)+9^5(1+9)+\cdots+9^{2009}(1+9)$, it follows that N is a multiple of $1 + 9 = 10$, so it is a multiple of 5.
- Note that
$$N = 9(1+9+9^2) + 9^4(1+9+9^2)$$
$$+9^7(1+9+9^2) + \cdots + 9^{2008}(1+9+9^2)$$
shows that N is a multiple of $1 + 9 + 9^2 = 91 = 7 \cdot 13$, and thus N is a multiple of both 7 and 13.
- Since $2^5 = 32 \equiv -1 \pmod{11}$, it follows that $2^{10} \equiv 1 \pmod{11}$. Now, summing the geometric series gives $N = 9 \cdot \frac{9^{2010}-1}{9-1}$. But $9^{2010} \equiv (-2)^{2010} \equiv 2^{2010} \equiv \left(2^{10}\right)^{201} \equiv 1 \pmod{11}$. It follows that $9^{2010} - 1$ is a multiple of 11, and N is a multiple of 11.

On the other hand, N is not a multiple of 17. Indeed, note that $9^2 = 81 = 17 \cdot 5 - 4 \equiv -4 \pmod{17}$. Thus, $9^4 \equiv (-4)^2 \equiv 16 \equiv -1 \pmod{17}$, and $9^8 \equiv (-1)^2 \equiv 1 \pmod{17}$. It follows that $9^{2010} \equiv 9^{8\cdot 251+2} \equiv 9^2 \equiv 13 \pmod{17}$. Then, from above, $N = 9 \cdot \frac{9^{2010}-1}{8}$ which cannot be a multiple of 17 since $9^{2010} - 1$ has a remainder of $13 - 1 = 12$ when divided by 17.

Problem 28. Pictured below is part of a large circle with radius 30. There is a chain of three circles with radius 3, each internally tangent to the large circle and each tangent to its neighbors in the chain. There are two circles with radius 2 each tangent to two of the radius 3 circles. The distance between the centers of the two circles with radius 2 can be written as $\frac{a\sqrt{b}-c}{d}$ where a, b, c, and d are positive integers, c and d are relatively prime, and b is not divisible by the square of any prime. Find $a + b + c + d$.

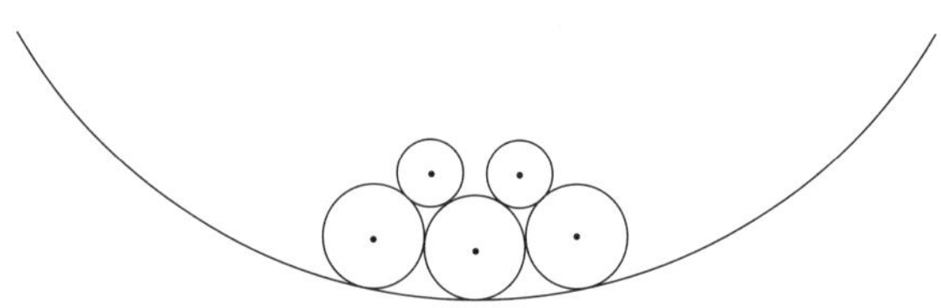

Answer: 46

Let G be the center of the circle with radius 30, A, B, and C be the centers of the circles with radius 3, and D and E be the centers of the circles with radius 2 as shown in the diagram. Also shown in the diagram are radii of the circle with radius 30 passing through the points A, D, B, and E. Line GD also passes through point F where the circles with centers at A and B are tangent because point D is equally distant from points A and B. Thus, $\overline{GD}$ is tangent to the circle centered at B, so $\triangle FDB$ is a right triangle with $FB = 3$ and $BD = 3+2 = 5$. Hence $FD = 4$. Also $GB = 30-3 = 27$, so it follows that $GF = \sqrt{GB^2 - BF^2} = \sqrt{27^2 - 3^2} = \sqrt{720} = 12\sqrt{5}$. Then $GD = GF - FD = 12\sqrt{5} - 4$. Now since $\triangle GAB$ is similar to $\triangle GDE$, it follows that $\frac{DE}{GD} = \frac{AB}{GA}$ or $DE = \frac{6}{27} \cdot (12\sqrt{5} - 4) = \frac{24\sqrt{5}-8}{9}$. The requested sum is $24 + 5 + 8 + 9 = 46$.

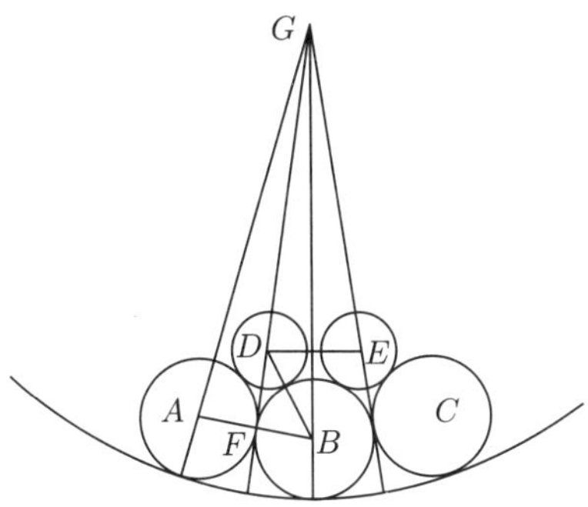

Problem 29. Let S be a randomly selected four-element subset of $\{1, 2, 3, 4, 5, 6, 7, 8\}$. Let m and n be relatively prime positive integers so that the expected value of the maximum element in S is $\frac{m}{n}$. Find $m + n$.

Answer: 41

If S has maximum element k, then it has three other elements less than k. Thus, the number of four-element subsets S with maximum element $k > 3$ is $\binom{k-1}{3}$. The expected value of the maximum element is $\frac{4\binom{3}{3}+5\binom{4}{3}+6\binom{5}{3}+7\binom{6}{3}+8\binom{7}{3}}{\binom{8}{4}}$. But since $k\binom{k-1}{3} = \frac{k\cdot(k-1)!}{3!\cdot(k-4)!} = 4 \cdot \frac{k!}{4!\cdot(k-4)!} = 4\binom{k}{4}$, the required expected value can be rewritten as $\frac{4\left[\binom{4}{4}+\binom{5}{4}+\binom{6}{4}+\binom{7}{4}+\binom{8}{4}\right]}{\binom{8}{4}}$. But $\left[\binom{4}{4} + \binom{5}{4} + \binom{6}{4} + \binom{7}{4} + \binom{8}{4}\right]$ is the number of ways of selecting a subset of size 5 from the set $\{1,2,3,4,5,6,7,8,9\}$ since each term counts the number of subsets of size five with maximum elements 5, 6, 7, 8, and 9. Thus, the required expected value is

$$\frac{4\binom{9}{5}}{\binom{8}{4}} = \frac{4\cdot\frac{9!}{5!\cdot 4!}}{\frac{8!}{4!\cdot 4!}} = \frac{36}{5}.$$

The requested sum is $36+5=41$.

Problem 30. Four congruent spheres are stacked so that each is tangent to the other three. A larger sphere, R, contains the four congruent spheres so that all four are internally tangent to R. A smaller sphere, S, sits in the space between the four congruent spheres so that all four are externally tangent to S. The ratio of the surface area of R to the surface area of S can be written $m+\sqrt{n}$ where m and n are positive integers. Find $m+n$.

Answer: 2449

The centers of the four congruent spheres are at the vertices of a regular tetrahedron. Without loss of generality, three of those vertices are located in three-dimensional coordinate space at $(2,0,0)$, $(-1,\sqrt{3},0)$, and $(-1,-\sqrt{3},0)$. Note that the distance between any two of these points is $2\sqrt{3}$. The fourth vertex can be located at $(0,0,2\sqrt{2})$ which is a distance $2\sqrt{3}$ from each of the other three vertices. Each of the four congruent spheres then has radius $\sqrt{3}$.

The center of both spheres R and S is equidistant from these three vertices and is located at $(0,0,h)$ where the square of its distance from $(2,0,0)$ is $2^2+h^2=4+h^2$, and its distance squared from $(0,0,2\sqrt{2})$ is $(2\sqrt{2}-h)^2=8-4\sqrt{2}h+h^2$. Equating these two squared distances and solving yields $h=\frac{1}{\sqrt{2}}$. The radius of S is the distance from $(0,0,\frac{1}{\sqrt{2}})$ to $(0,0,2\sqrt{2})$ minus the radius of one of the congruent spheres. Thus, S has radius $2\sqrt{2}-\frac{1}{\sqrt{2}}-\sqrt{3}=\frac{3}{\sqrt{2}}-\sqrt{3}$. The radius of R is the distance from $(0,0,\frac{1}{\sqrt{2}})$ to $(0,0,2\sqrt{2})$ plus the radius of one of the congruent spheres. Thus, R has radius $2\sqrt{2}-\frac{1}{\sqrt{2}}+\sqrt{3}=\frac{3}{\sqrt{2}}+\sqrt{3}$. The ratio of surface areas is the square of the ratio of these radii or $\left(\frac{\frac{3}{\sqrt{2}}+\sqrt{3}}{\frac{3}{\sqrt{2}}-\sqrt{3}}\right)^2=\left(5+2\sqrt{6}\right)^2=49+20\sqrt{6}=49+\sqrt{2400}$. The requested sum is $49+2400=2449$.

2.17 2011 Middle School

Problem 1. There are relatively prime positive integers m and n so that

$$\frac{\frac{1}{2}}{\frac{\frac{1}{3}}{\frac{1}{4}+\frac{1}{4}+\frac{1}{4}}+\frac{\frac{1}{3}}{\frac{1}{4}+\frac{1}{4}+\frac{1}{4}}}=\frac{m}{n}.$$

Find $m+2n$.

Answer: 41

$$\frac{\frac{1}{2}}{\frac{\frac{1}{3}}{\frac{1}{4}+\frac{1}{4}+\frac{1}{4}}+\frac{\frac{1}{3}}{\frac{1}{4}+\frac{1}{4}+\frac{1}{4}}}=\frac{\frac{1}{2}}{\frac{\frac{1}{3}}{\frac{3}{4}}+\frac{\frac{1}{3}}{\frac{3}{4}}}=\frac{\frac{1}{2}}{\frac{4}{9}+\frac{4}{9}}=\frac{\frac{1}{2}}{\frac{8}{9}}=\frac{9}{16}=\frac{m}{n}.$$

Thus, $m=9$ and $n=16$. The requested sum is $9+2\cdot 16=41$.

Problem 2. The diagram below shows a 12-sided figure made up of three congruent squares. The figure has total perimeter 60. Find its area.

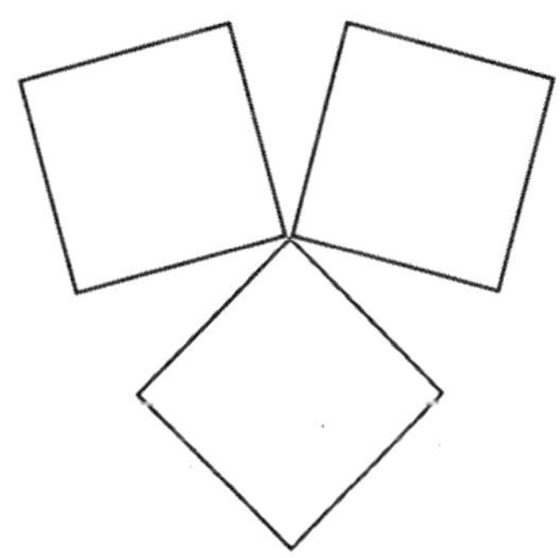

Answer: 75

If the total perimeter is 60, each square has perimeter 20. Thus, each square has side length 5. Since each square then has area $5^2=25$, the entire figure has area $3\cdot 25=75$.

Problem 3. Find the sum of all two-digit integers which are both prime and are 1 more than a multiple of 10.

Answer: 215

The two-digit integers which are both prime and are 1 more than a multiple of 10 are 11, 31, 41, 61, and 71. Their sum is $11+31+41+61+71=215$.

Problem 4. Jerry buys a bottle of 150 pills. Using a standard 12 hour clock, he sees that the clock reads exactly 12 when he takes the first pill. If he takes one pill every five hours, what hour will the clock read when he takes the last pill in the bottle?

Answer: 1

Jerry will take the last pill 5×149 hours after he takes the first pill. Since $5 \times 149 = 5 \times (144 + 5) \equiv 5 \times 5 = 25 \equiv 1 \pmod{12}$, the clock will read 1 hour past 12, or 1 when Jerry takes the last pill.

Problem 5. Given that $\frac{6}{11} - \frac{10}{19} = \frac{9}{19} - \frac{n}{11}$, find n.

Answer: 5

Add $\frac{10}{19}$ to each side of the given equation to yield

$$\frac{6}{11} = \frac{9}{19} - \frac{n}{11} + \frac{10}{19} = \frac{19}{19} - \frac{n}{11} = \frac{11}{11} - \frac{n}{11} = \frac{11-n}{11}.$$

Thus, $6 = 11 - n$ and $n = 5$.

Problem 6. The following addition problem is not correct if the numbers are interpreted as base 10 numbers. In what number base is the problem correct?

$$\begin{array}{r} 6\ 6 \\ 8\ 7 \\ 8\ 5 \\ +\ 4\ 8 \\ \hline 1\ 3\ 2 \end{array}$$

Answer: 24

Let the number base for this problem be b. The ones column in the problem adds to $6 + 7 + 5 + 8 = 26 \equiv 2 \pmod{b}$. Thus, b must be a divisor of 24. Since the problem includes the digit 8, b must be at least 9. This implies that either $b = 12$ or $b = 24$. It is straightforward to check that only $b = 24$ works.

Problem 7. When 12^{18} is divided by 18^{12}, the result is $\left(\frac{m}{n}\right)^3$, where m and n are relatively prime positive integers. Find $m - n$.

Answer: 247

$$\frac{12^{18}}{18^{12}} = \frac{4^{18}3^{18}}{2^{12}9^{12}} = \frac{2^{36}3^{18}}{2^{12}3^{24}} = \frac{2^{24}}{3^6} = \left(\frac{2^8}{3^2}\right)^3 = \left(\frac{256}{9}\right)^3.$$

The requested difference is $256 - 9 = 247$.

Problem 8. A square measuring 15 by 15 is partitioned into five rows of five congruent squares as shown below. The small squares are alternately colored black and white as shown. Find the total area of the part colored black.

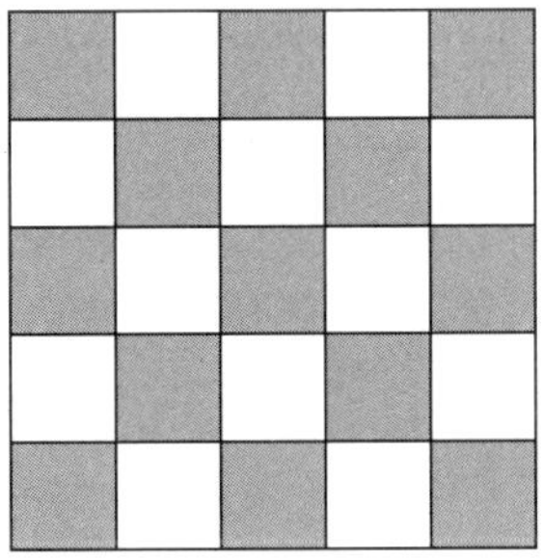

Answer: 117

Each of the small squares has side length $\frac{15}{5} = 3$, so each has area $3^2 = 9$. There are 13 shaded squares. Thus, the shaded area is $9 \cdot 13 = 117$.

Problem 9. A jar contains one quarter red jelly beans and three quarters blue jelly beans. If Chris removes three quarters of the red jelly beans and one quarter of the blue jelly beans, what percent of the jelly beans remaining in the jar will be red?

Answer: 10

At the start, $\frac{1}{4}$ of the jelly beans are red, and $\frac{3}{4}$ of them are removed, so at the end, the number of red jelly beans will be $\frac{1}{4} - \frac{3}{4} \cdot \frac{1}{4} = \frac{4}{16} - \frac{3}{16} = \frac{1}{16}$ of the original number of jelly beans. Similarly, at the end, the number of blue jelly beans will be $\frac{3}{4} - \frac{1}{4} \cdot \frac{3}{4} = \frac{12}{16} - \frac{3}{16} = \frac{9}{16}$ of the original number of jelly beans. Therefore, at the end, the fraction of red jelly beans in the jar is $\frac{\frac{1}{16}}{\frac{1}{16}+\frac{9}{16}} = \frac{1}{1+9} = \frac{1}{10} = 10$ percent.

Problem 10. Five rays $\overrightarrow{OA}$, $\overrightarrow{OB}$, $\overrightarrow{OC}$, $\overrightarrow{OD}$, and $\overrightarrow{OE}$ radiate in a clockwise order from O forming four non-overlapping angles such that $\angle EOD = 2\angle COB$, $\angle COB = 2\angle BOA$, while $\angle DOC = 3\angle BOA$. If E, O, A are collinear with O between A and E, what is the degree measure of $\angle DOB$?

Answer: 90

Let $\angle BOA = x$. Then $\angle COB = 2x, \angle DOC = 3x$, and $\angle DOE = 4x$. Thus $x+2x+3x+4x = 10x = 180° \implies x = 18°$. Then $\angle DOB = \angle DOC + \angle COB = 3x + 2x = 5x = 90°$.

Problem 11. How many numbers are there that appear both in the arithmetic sequence 10, 16, 22, 28, ... 1000 and the arithmetic sequence 10, 21, 32, 43, ..., 1000?

Answer: 16

The first sequence includes every sixth integer from 10 up to 1000. The second sequence includes every eleventh integer from 10 up to 1000. The numbers contained in both sequences are, therefore, every sixty-sixth number from 10 to 1000. That is, 10, 76, 142, ... 1000. There are $\frac{1000-10}{66} + 1 = 16$ such numbers.

Problem 12. When Troy writes his digits, his 0, 1, and 8 look the same right-side-up and upside-down as seen in the figure below. His 6 and 9 look like upside-down images of each other. None of his other digits look like digits when they are inverted. How many different five-digit numbers (which do not begin with the digit zero) can Troy write which read the same right-side-up and upside-down?

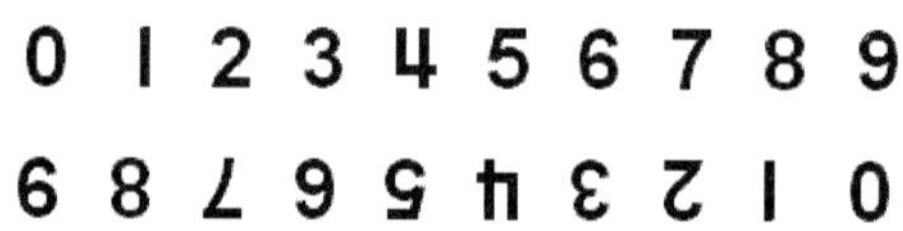

Answer: 60

Each of the digits that Troy writes in a five-digit number must be one of the five: 0, 1, 8, 6, 9. In addition, the numbers in positions 4 and 5 must be the up-side-down versions of the digits he writes in positions 2 and 1, respectively. Also, the digit in position 3 must be 0, 1, or 8 since it must look the same right-side-up and up-side-down. Finally, the digit in the first position cannot be 0 since no five-digit number can begin with the digit 0. Thus, there are 4 possible choices for the first digit, 5 choices for the second digit, three choices for the third digit, and the last two digits are determined by the choices for the first two digits. Therefore, there are $4 \cdot 5 \cdot 3 = 60$ choices for five-digit numbers.

Problem 13. The diagram shows two equilateral triangles with side length 4 mounted on two adjacent sides of a square also with side length 4. The distance between the two vertices marked A and B can be written as $\sqrt{m} + \sqrt{n}$ for two positive integers m and n. Find $m + n$.

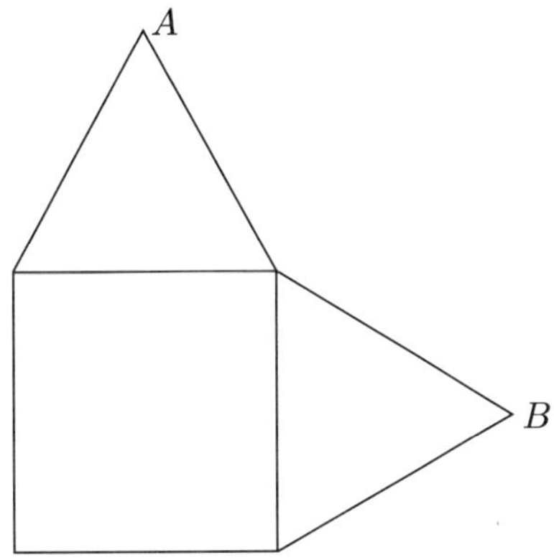

Answer: 32

Let the center of the square be point C. The distance from A to C is given by the height of the equilateral triangle, $4 \cdot \frac{\sqrt{3}}{2} = \sqrt{12}$, plus half the length of the side of the square, 2. Thus, the distance from A to C is $\sqrt{12} + 2$. Triangle ACB is an isosceles right triangle, so its hypotenuse $\overline{AB}$ has length $\sqrt{2} \cdot (\sqrt{12} + 2) = \sqrt{24} + \sqrt{8}$. The requested sum is $24 + 8 = 32$.

Problem 14. The five-digit number 12110 is divisible by the sum of its digits $1 + 2 + 1 + 1 + 0 = 5$. Find the greatest five-digit number which is divisible by the sum of its digits.

Answer: 99972

Let n be the greatest five digit number which is divisible by the sum of its digits, and suppose that n ends with the digit a. If $n = \underline{9}\,\underline{9}\,\underline{9}\,\underline{9}\,\underline{a} = 99990 + a$, then it would be divisible by $9 + 9 + 9 + 9 + a = 36 + a$. Then $36 + a$ would also divide $(99990 + a) - (36 + a) = 99954$. But $99954 = 2 \cdot 3^4 \cdot 617$. Because $36 + a$ is between 36 and 45, and no integer between 36 and 45 has factors in common with the prime 617 or divides $2 \cdot 3^4$, n is not of the form $\underline{9}\,\underline{9}\,\underline{9}\,\underline{9}\,\underline{a}$.

If $n = \underline{9}\,\underline{9}\,\underline{9}\,\underline{8}\,\underline{a} = 99980 + a$ for some digit a, then n would be divisible by $9 + 9 + 9 + 8 + a = 35 + a$. It would follow that $35 + a$ would also divide $(99980 + a) - (35 + a) = 99945$. But $99945 = 3^2 \cdot 5 \cdot 2221$. Because $35 + a$ must be between 35 and 44, there is no way for it to divide $3^2 \cdot 5 = 45$ or have any factors in common with the prime 2221, so n is not of the form $\underline{9}\,\underline{9}\,\underline{9}\,\underline{8}\,\underline{a}$.

If $n = \underline{9}\,\underline{9}\,\underline{9}\,\underline{7}\,\underline{a} = 99970 + a$ for some digit a, then by arguing as above, this would mean the number $34 + a$ would divide $(99970 + a) - (34 + a) = 99936 = 2^5 \cdot 3^2 \cdot 347$. The number $34 + a$ cannot have any factors in common with the prime 347, but it could divide $2^5 \cdot 3^2$. The only integer between 34 and 43 equal to a power of 2 times a power of 3 is 36. This happens when $a = 2$, and the number 99972 is divisible by the sum of its digits which is 36.

Problem 15. In the diagram below, $\overline{AB}$ and $\overline{CD}$ are parallel, $\angle BXY = 45°$, $\angle DZY = 25°$, and $XY = YZ$. What is the degree measure of $\angle YXZ$?

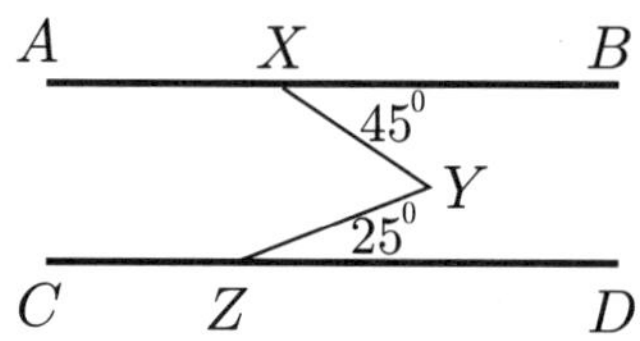

Answer: 55

Extend $\overline{XY}$ until it intersects $\overline{CD}$ at X'. Then $\angle XX'Z = 45°$. Hence $\angle ZYX' = 180° - 25° - 45° = 110°$. This implies that $\angle XYZ = 180° - 110° = 70°$. Since $\triangle XYZ$ is isosceles, $2\angle YXZ + 70° = 180°$ and $\angle YXZ = 55°$.

Problem 16. Let a and b be nonzero real numbers such that $\frac{1}{3a} + \frac{1}{b} = 2011$ and $\frac{1}{a} + \frac{1}{3b} = 1$. What is the quotient when $a + b$ is divided by ab?

Answer: 1509

Adding fractions gives $\frac{3a+b}{3ab} = 2011$ and $\frac{a+3b}{3ab} = 1$. Adding these two equations together gives $\frac{4a+4b}{3ab} = \frac{4}{3} \cdot \frac{a+b}{ab} = 2011 + 1 = 2012$. Thus, the requested ratio is $\frac{a+b}{ab} = 2012 \cdot \frac{3}{4} = 1509$.

Problem 17. Find the number of ordered quadruples (a, b, c, d) where each of a, b, c, and d are (not necessarily distinct) elements of $\{1, 2, 3, 4, 5, 6, 7\}$ and $3abc + 4abd + 5bcd$ is even. For example, $(2, 2, 5, 1)$ and $(3, 1, 4, 6)$ satisfy the conditions.

Answer: 2017

The given sum is even exactly when the products abc and bcd are either both even or both odd. This happens either when the product bc is even and or when bc is odd and a and d are either both even or both odd. Of the $7 \cdot 7 = 49$ ways to assign values to b and d, $4 \cdot 4 = 16$ ways result in bd being odd, and $49 - 16 = 33$ ways result in bc being even. Similarly, there are 49 ways to assign values to a and d. Of these ways, $4 \cdot 4 = 16$

result in both a and d being odd, and $3 \cdot 3 = 9$ result in both a and d being even. It follows that the total number of ways to assign values to a, b, c, and d so that $3abc + 4abd + 5bcd$ is even is $33 \cdot 49 + 16 \cdot (16 + 9) = 2017$.

Problem 18. Find the positive integer n so that n^2 is the perfect square closest to $8 + 16 + 24 + \cdots + 8040$.

Answer: 2011

The sum

$$8 + 16 + 24 + \cdots + 8040 = 8 \sum_{i=1}^{1005} i = 8 \left(\frac{1005 \cdot 1006}{2} \right)$$

$$= 2010 \cdot 2012 = (2011 - 1)(2011 + 1) = 2011^2 - 1.$$

Hence the requested number is 2011.

Problem 19. How many ordered pairs of sets (A, B) have the properties:

(a) $A \subseteq \{1, 2, 3, 4, 5, 6\}$

(b) $B \subseteq \{2, 3, 4, 5, 6, 7, 8\}$

(c) $A \cap B$ has exactly 3 elements.

Answer: 720

The sets $\{1, 2, 3, 4, 5, 6\}$ and $\{2, 3, 4, 5, 6, 7, 8\}$ have five elements in common, so there are $\binom{5}{3} = 10$ ways to select three elements that the sets A and B have in common. Having selected the three shared elements, there are two choices of what to do with the element 1 which can either be in or out of set A. Similarly, there are two choices for each of the elements 7 and 8 which can be in or out of set B. Of the elements in $\{2, 3, 4, 5, 6\}$, three of the elements have been chosen to appear in both A and B, but there are three choices of what to do with each of the other two elements which can be placed either in the set A, in the set B, or left out of both sets. It follows that the number of ways to choose the sets A and B is $10 \cdot 2 \cdot 2 \cdot 2 \cdot 3 \cdot 3 = 720$.

Problem 20. Let V be the set of vertices of a regular 25 sided polygon with center at point C. How many triangles have vertices in V and contain the point C in the interior of the triangle?

Answer: 650

The number of triangles with vertices among the 25 vertices is $\binom{25}{3} = 2300$. The triangles that do **not** contain the center C

are exactly the triangles that are obtuse. Number the vertices of the regular 25 sided polygon in order from 1 to 25. A triangle will have an obtuse angle at the point numbered 11 if one of its other two vertices is numbered with a number less than 11, one is numbered with a number greater than 11, and the difference between these two numbers does not exceed 11. Thus, one of these vertices could be numbered k for $1 \leq k \leq 10$ while the other could be numbered m with $12 \leq m \leq k+11$. That is, for each choice of k, there are k choices for m. It follows that the number of triangles with an obtuse angle at vertex 11 is $1+2+3+\cdots+10 = \frac{10 \cdot 11}{2} = 66$. An obtuse triangle could have its obtuse angle at any of the 25 vertices, so it follows that there are $25 \cdot 66 = 1650$ obtuse triangles. Hence, the number of triangles that contain the point C is $2300 - 1650 = 650$.

2.18 2012 High School

Problem 1. Last month a pet store sold three times as many cats as dogs. If the store had sold the same number of cats but eight more dogs, it would have sold twice as many cats as dogs. How many cats did the pet store sell last month?

Answer: 48

Let x be the number of dogs sold. Then $3x$ is the number of cats sold, and $2(x+8) = 3x$. Thus, $2x + 16 = 3x$ and $x = 16$. It follows that the number of cats sold is $3x = 3 \cdot 16 = 48$.

Problem 2. What is the greatest three-digit divisor of 111777?

Answer: 703

Factoring yields $111777 = 111 \cdot 1007 = 3 \cdot 37 \cdot 19 \cdot 53$. The product of any three of the prime factors of 111777 exceeds 1000, so consider the products taking two at a time: $3 \cdot 19 = 57$, $3 \cdot 37 = 111$, $3 \cdot 53 = 159$, $19 \cdot 37 = 703$, $19 \cdot 53 = 1007$, and $37 \cdot 53$ is clearly greater than 1000. The greatest of the 3-digit products is 703.

Problem 3. The diagram below shows a large square divided into nine congruent smaller squares. There are circles inscribed in five of the smaller squares. The total area covered by all the five circles is 20π. Find the area of the large square.

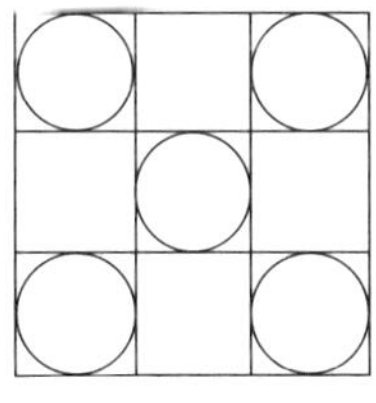

Answer: 144

Let the circles each have radius r. Then the total area covered by all the five circles is $5\pi r^2 = 20\pi$, so $r = 2$. Each small square has side length $2r = 4$, and the larger square has side length $3 \cdot 4 = 12$. It follows that the larger square has area $12^2 = 144$.

Problem 4. The following diagram shows an equilateral triangle and two squares that share common edges. The area of each square is 75. Find the distance from point A to point B.

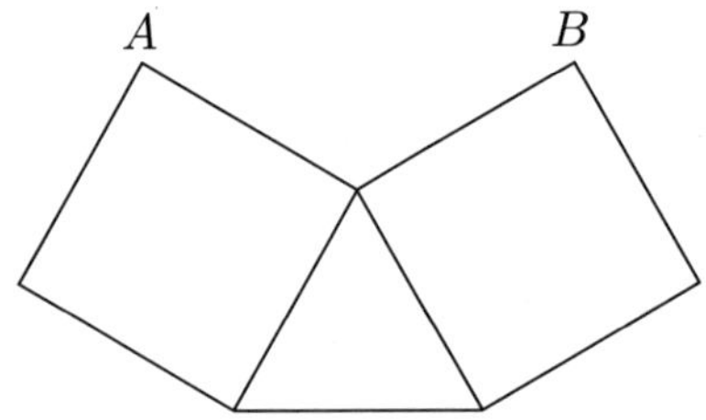

Answer: 15

Let the squares and triangle have side length s. Let the vertex common to the two squares and the triangle be labeled C. Note that $\angle ACB = 120°$. Then the Law of Cosines gives

$$AB^2 = s^2 + s^2 - 2 \cdot s \cdot s\cos(120°) = 3 \cdot s^2 = 3 \cdot 75 = 225.$$

Thus, $AB = 15$.

Problem 5. Find the sum of the squares of the values x that satisfy

$$\frac{1}{x} + \frac{2}{x+3} + \frac{3}{x+6} = 1.$$

Answer: 33

Multiply the equation by $x(x+3)(x+6)$ to get

$$(x+3)(x+6) + 2x(x+6) + 3x(x+3) = x(x+3)(x+6).$$

This simplifies to

$$x^3 + 3x^2 - 12x - 18 = (x-3)(x^2 + 6x + 6) = 0.$$

The possible values of x are 3, $-3 + \sqrt{3}$, and $-3 - \sqrt{3}$. The requested sum of squares is

$$3^2+(-3+\sqrt{3})^2+(-3-\sqrt{3})^2 = 9+(9-6\sqrt{3}+3)+(9+6\sqrt{3}+3) = 33.$$

Problem 6. Find the least positive integer n so that both n and $n+1$ have prime factorizations with exactly four (not necessarily distinct) prime factors.

Answer: 135

Note that either n or $n+1$ must be odd. The smallest odd positive integer with four prime factors is $3^4 = 81$, but neither $80 = 2^4 \cdot 5$ nor $82 = 2 \cdot 41$ have four prime factors. The second smallest odd positive integer with four prime factors is $3^3 \cdot 5 = 135$. The neighbors of 135 are $134 = 2 \cdot 67$ and $136 = 2^3 \cdot 17$, so it follows that $n = 135$ is the requested number.

Problem 7. Two convex polygons have a total of 33 sides and 243 diagonals. Find the number of diagonals in the polygon with the greater number of sides.

Answer: 189

If a convex polygon has n sides, it has

$$\frac{n(n-1)}{2} - n = \frac{n^2 - 3n}{2} \text{ diagonals.}$$

So if one convex polygon has n sides and another has $33 - n$ sides, together the number of diagonals in the two polygons is

$$\frac{n^2 - 3n}{2} + \frac{(33-n)^2 - 3(33-n)}{2} = \frac{2n^2 - 66n + 990}{2}.$$

Setting this equal to 243 yields $2n^2 - 66n + 990 = 2 \cdot 243$ which simplifies to $n^2 - 33n + 252 = 0$. The polynomial factors to give $(n-12)(n-21) = 0$. Thus, the two polygons have 12 and 21 sides. The number of diagonals in the polygon with 21 sides is, therefore,

$$\frac{21^2 - 3 \cdot 21}{2} = \frac{18 \cdot 21}{2} = 9 \cdot 21 = 189.$$

Problem 8. In the tribe of Zimmer, being able to hike long distances and knowing the roads through the forest are both extremely important, so a boy who reaches the age of manhood is not designated as a man by the tribe until he completes an interesting rite of passage. The man must go on a sequence of hikes. The first hike is a 5 kilometer hike down the main road. The second hike is a $5\frac{1}{4}$ kilometer hike down a secondary road. Each hike goes down a different road and is a quarter kilometer longer than the previous hike. The rite of passage is completed at the end of the hike where the cumulative distance walked by the man on all his hikes exceeds 1000 kilometers. So in the tribe of Zimmer, how many roads must a man walk down, before you call him a man?

Answer: 73

The total distance walked by a man in n hikes is

$$5 + 5\frac{1}{4} + 5\frac{1}{2} + \cdots + \frac{19+n}{4} = \sum_{k=1}^{n} \frac{19+k}{4} = \frac{1}{4}\left[19n + \frac{n(n+1)}{2}\right].$$

This exceeds 1000 when $n^2 + 39n > 8000$. The smallest natural number n for which this inequality is satisfied is $n = 73$.

Problem 9. Find the value of x that satisfies $\log_3(\log_9 x) = \log_9(\log_3 x)$.

Answer: 81

Using the change of base formula for logarithms, the logarithms base 9 can be changed to logarithms base 3 to get $\log_3\left(\frac{\log_3 x}{\log_3 9}\right) = \frac{\log_3 \log_3 x}{\log_3 9}$. Thus, $2\log_3\left(\frac{\log_3 x}{2}\right) = \log_3 \log_3 x$, so $\left(\frac{\log_3 x}{2}\right)^2 = \log_3 x$. Let $u = \log_3 x$. Then $\frac{u^2}{4} = u$, so $u = 0$ or $u = 4$. But $u = 0$ is not a possible solution since $\log_3 u$ must be defined. Therefore, $u = 4$ and $\log_3 x = 4$. Finally, $x = 3^4 = 81$.

Problem 10. Consider a sequence of eleven squares that have side lengths $3, 6, 9, 12, \ldots, 33$. Eleven copies of a single square each with area A have the same total area as the total area of the eleven squares of the sequence. Find A.

Answer: 414

Recall that

$$\sum_{k=1}^{n} k^2 = \frac{n(n+1)(2n+1)}{6}.$$

The total area of the squares in the sequence is

$$\begin{aligned} 3^2 + 6^2 + 9^2 + \cdots + 33^2 &= 3^2\left[1^2 + 2^2 + 3^2 + \cdots + 11^2\right] \\ &= 9 \cdot \frac{11 \cdot 12 \cdot 23}{6} = 11A. \end{aligned}$$

Thus, $A = 9 \cdot \frac{12 \cdot 23}{6} = 414$.

Problem 11. Define $f(x) = 2x + 3$ and suppose that

$$g(x+2) = f\left(f(x-1) \cdot f(x+1) + f(x)\right).$$

Find $g(6)$.

Answer: 259

One has

$$g(6) = g(4+2) = f\left(f(4-1) \cdot f(4+1) + f(4)\right)$$

$$= f\left(f(3) \cdot f(5) + f(4)\right) = f\left(9 \cdot 13 + 11\right) = f(128) = 259.$$

Problem 12. Ted flips seven fair coins. There are relatively prime positive integers m and n so that $\frac{m}{n}$ is the probability that Ted flips at least two heads given that he flips at least three tails. Find $m + n$.

Answer: 190

There are $\binom{7}{k}$ equally likely ways to flip k tails with seven flips. Thus, the probability Ted flips at most five tails given that Ted flips at least three tails is

$$\frac{\binom{7}{3}+\binom{7}{4}+\binom{7}{5}}{\binom{7}{3}+\binom{7}{4}+\binom{7}{5}+\binom{7}{6}+\binom{7}{7}} = \frac{35+35+21}{35+35+21+7+1} = \frac{91}{99}.$$

The requested sum is $91 + 99 = 190$.

Problem 13. Find the least n for which $n!(n+1)!(2n+1)! - 1$ ends in thirty digits that are all 9's.

Answer: 34

If $n!(n+1)!(2n+1)! - 1$ ends in thirty digits all equal to 9, then $n!(n+1)!(2n+1)!$ must end in thirty 0's. The number of trailing zeros in $n!$ is the number of factors of 5 in $n!$ which, for $n < 100$, is $\lfloor \frac{n}{5} \rfloor + \lfloor \frac{n}{25} \rfloor$. (Here, $\lfloor x \rfloor$ refers to the greatest integer less than or equal to x.) For example, when $n = 27$, the number of trailing zeros in $n!$ is $\lfloor \frac{27}{5} \rfloor + \lfloor \frac{27}{25} \rfloor = 5 + 1 = 6$. The least n for which $n!(n+1)!(2n+1)!$ ends with thirty 0's can be found by searching. If $n = 27$, the number of zeros is $6 + 6 + 13 = 25$. If $n = 30$, the number of zeros is $7 + 7 + 14 = 28$. If $n = 33$, the number of zeros is $7+7+15 = 29$. But if $n = 34$, the number of zeros is $7+8+15 = 30$, so the least n with the desired property is $n = 34$.

Problem 14. A circle in the first quadrant with center on the curve

$$y = 2x^2 - 27$$

is tangent to the y-axis and the line $4x = 3y$. The radius of the circle is $\frac{m}{n}$ where m and n are relatively prime positive integers. Find $m + n$.

Answer: 11

The distance from a point to the line $4x - 3y$ is given by the normal form for the equation of the line and is $\frac{3y-4x}{\sqrt{3^2+4^2}} = \frac{3y-4x}{5}$. For the circle to be tangent to the line and the y-axis, its center $(x, 2x^2 - 27)$ must be equidistant from the y-axis and the line, so $x = \frac{3\cdot(2x^2-27)-4x}{5}$ or $6x^2 - 9x - 81 = 0$. This has solutions $x = -3$ and $x = \frac{9}{2}$. Thus, the circle in the first quadrant has radius $\frac{9}{2}$, and the requested sum is $9 + 2 = 11$.

Problem 15. Let N be a positive integer whose digits add up to 23. What is the greatest possible product the digits of N can have?

Answer: 4374

Digits 0 and 1 do not increase the product of the digits, so the N with the largest product of digits has no digits of 0 or 1. A digit of 4 can be replaced by two digits of 2 which have the same sum and same product as the one digit of 4. Any digit $k \geq 5$ can be replaced by the digits 2 and $k-2$ because $2(k-2) > k$ for these digits. Hence, an N with the greatest product of digits has all of its digits being 2's and 3's. The largest product is then of the form $2^a 3^b$ where $2a+3b = 23$. Note that $2+2+2 = 3+3$, but $2 \cdot 2 \cdot 2 < 3 \cdot 3$, so two 3's give a larger product than three 2's. It follows that the largest product is $2 \cdot 3^7 = 4374$.

Problem 16. Let a, b, and c be non-zero real numbers such that $\frac{ab}{a+b} = 3$, $\frac{bc}{b+c} = 4$, and $\frac{ca}{c+a} = 5$. There are relatively prime positive integers m and n so that $\frac{abc}{ab+bc+ca} = \frac{m}{n}$. Find $m+n$.

Answer: 167

Taking the reciprocals of each side of the given equations shows $\frac{1}{3} = \frac{a+b}{ab} = \frac{1}{b} + \frac{1}{a}$, $\frac{1}{4} = \frac{b+c}{bc} = \frac{1}{c} + \frac{1}{b}$, and $\frac{1}{5} = \frac{c+a}{ca} = \frac{1}{a} + \frac{1}{c}$. Add these three equations to get $\frac{2}{a} + \frac{2}{b} + \frac{2}{c} = \frac{1}{3} + \frac{1}{4} + \frac{1}{5} = \frac{47}{60}$. Note that $\frac{n}{m} = \frac{ab+bc+ca}{abc} = \frac{1}{a} + \frac{1}{b} + \frac{1}{c} = \frac{47}{120}$. The requested sum is $m+n = 120+47 = 167$.

Problem 17. How many positive integer solutions are there to

$$w + x + y + z = 20$$

where $w + x \geq 5$ and $y + z \geq 5$?

Answer: 781

Assume a solution assigns $w + x = k$ and $y + z = 20 - k$. There are $k - 1$ ways to assign positive integer values to w and x so that their sum is k, and $19 - k$ ways to assign positive integer values to y and z so that $y + z = 20 - k$. It follows that the number of solutions satisfying the given conditions is

$$\sum_{k=5}^{15}(k-1)(19-k) = \sum_{k=1}^{11}(k+3)(15-k) = \sum_{k=1}^{11}\left(45 + 12k - k^2\right)$$

$$= 45 \cdot 11 + 12 \cdot \frac{11(12)}{2} - \frac{11(12)(23)}{6}$$

$$= 11(45 + 72 - 46) = 11(71) = 781.$$

Problem 18. Find the number of three-digit numbers such that its first two digits are each divisible by its third digit.

Answer: 138

If the third digit of a three-digit number divides the other two digits, the third digit clearly cannot be 0.

Third Digit	First Digit	Second Digit	Possibilities
1	1,2,3,4,5,6,7,8,9	0,1,2,3,4,5,6,7,8,9	90
2	2,4,6,8	0,2,4,6,8	20
3	3,6,9	0,3,6,9	12
4	4,8	0,4,8	6
5,6,7,8,9	same as third	0 or same as third	10

Thus, the number of such numbers is $90+20+12+6+10 = 138$.

Problem 19. Find the remainder when $2^{5^9} + 5^{9^2} + 9^{2^5}$ is divided by 11.

Answer: 8

Fermat's Little Theorem states that if integer n is not a multiple of prime p, then $n^{p-1} \equiv 1 \pmod{p}$. Thus, 2^{10}, 5^{10}, and 9^{10} are all congruent to $1 \pmod{11}$. Note that $5^9 \equiv 5 \pmod{10}$, $9^2 = 1 \pmod{10}$, and $2^5 \equiv 2 \pmod{10}$. It follows that

$$2^{5^9} + 5^{9^2} + 9^{2^5} \equiv 2^5 + 5^1 + 9^2 \equiv 32 + 5 + 81 \equiv 8 \pmod{11}.$$

Problem 20. Square $ABCD$ has side length 68. Let E be the midpoint of segment $\overline{CD}$, and let F be the point on segment $\overline{AB}$ a distance 17 from point A. Point G is on segment $\overline{EF}$ so that $\overline{EF}$ is perpendicular to segment $\overline{GD}$. The length of segment $\overline{BG}$ can be written as $m\sqrt{n}$ where m and n are positive integers, and n is not divisible by the square of any prime. Find $m + n$.

Answer: 46

Let vector $\overrightarrow{BC} = \mathbf{u}$ and vector $\overrightarrow{BA} = \mathbf{v}$. Then $\overrightarrow{EF} = \frac{\mathbf{v}}{4} - \mathbf{u}$. Note that $\overrightarrow{EG}$ is the projection of $\overrightarrow{ED} = \frac{\mathbf{v}}{2}$ onto $\overrightarrow{EF}$. This projection is

$$\left(\frac{\overrightarrow{ED} \cdot \overrightarrow{EF}}{\overrightarrow{EF} \cdot \overrightarrow{EF}}\right)\overrightarrow{EF} = \left(\frac{\frac{\mathbf{v}}{2} \cdot \left(\frac{\mathbf{v}}{4} - \mathbf{u}\right)}{\left(\frac{\mathbf{v}}{4} - \mathbf{u}\right) \cdot \left(\frac{\mathbf{v}}{4} - \mathbf{u}\right)}\right)\left(\frac{\mathbf{v}}{4} - \mathbf{u}\right)$$

$$= \left(\frac{\frac{1}{8}}{\frac{1}{16} + 1}\right)\left(\frac{\mathbf{v}}{4} - \mathbf{u}\right) = \frac{1}{34}\mathbf{v} - \frac{2}{17}\mathbf{u}.$$

Then

$$\overrightarrow{BG} = \mathbf{u} + \frac{\mathbf{v}}{2} + \overrightarrow{EG} = \frac{15}{17}\mathbf{u} + \frac{9}{17}\mathbf{v}.$$

This vector has length

$$\frac{68}{17}\sqrt{15^2+9^2} = 4\sqrt{306} = 12\sqrt{34}.$$

The requested sum is $12 + 34 = 46$.

Problem 21. Each time you click a toggle switch, the switch either turns from *off* to *on* or from *on* to *off*. Suppose that you start with three toggle switches with one of them *on* and two of them *off*. On each move you randomly select one of the three switches and click it. Let m and n be relatively prime positive integers so that $\frac{m}{n}$ is the probability that after four such clicks, one switch will be *on* and two of them will be *off*. Find $m + n$.

Answer: 142

Without loss of generality assume that at the beginning the first of the three toggle switches is *on*, and the second and third are *off*. There are $3^4 = 81$ equally likely ways to select a sequence of four toggle switches to click. After four clicks, there will either be one or three switches in the *on* position. To get three toggle switches in the *on* position, one must click the first switch an even number of times and the second and third switches an odd number of times each. This can be done by clicking the first switch twice and the other switches once each, or by clicking either the second or third switch three times and the other of the two switches one time. The number of ways to do one of these is

$$\binom{4}{2,1,1}+\binom{4}{0,1,3}+\binom{4}{0,3,1} = \frac{4!}{2!\cdot 1!\cdot 1!}+\frac{4!}{0!\cdot 1!\cdot 3!}+\frac{4!}{0!\cdot 3!\cdot 1!}$$

$$= 12 + 4 + 4 = 20.$$

It follows that the desired probability is

$$\frac{81-20}{81} = \frac{61}{81}.$$

The requested sum is $61 + 81 = 142$.

Alternatively, there are four states of the toggle switches indicating whether there are 0, 1, 2, or 3 switches in the *on* position. Let $a_{i,j}$ be the probability that the system will transition from state i to state j with one random toggle switch. The transition matrix with i, j entry $a_{i,j}$ is

$$A = \begin{pmatrix} 0 & 1 & 0 & 0 \\ \frac{1}{3} & 0 & \frac{2}{3} & 0 \\ 0 & \frac{2}{3} & 0 & \frac{1}{3} \\ 0 & 0 & 1 & 0 \end{pmatrix}.$$

Then

$$A^2 = \begin{pmatrix} \frac{1}{3} & 0 & \frac{2}{3} & 0 \\ 0 & \frac{7}{9} & 0 & \frac{2}{9} \\ \frac{2}{9} & 0 & \frac{7}{9} & 0 \\ 0 & \frac{2}{3} & 0 & \frac{1}{3} \end{pmatrix} \text{ and } A^4 = \begin{pmatrix} \frac{7}{27} & 0 & \frac{20}{27} & 0 \\ 0 & \frac{61}{81} & 0 & \frac{20}{81} \\ \frac{20}{81} & 0 & \frac{61}{81} & 0 \\ 0 & \frac{20}{27} & 0 & \frac{61}{81} \end{pmatrix}.$$

Since A^2 gives the probabilities of transitioning from one state to another with two clicks, and A^4 gives the probabilities of transitioning from one state to another with four clicks, it is seen that the probability of transitioning from one toggle *on* to one toggle *on* in four clicks is $\frac{61}{81}$.

Problem 22. The diagram below shows circles radius 1 and 2 externally tangent to each other and internally tangent to a circle radius 3. There are relatively prime positive integers m and n so that a circle radius $\frac{m}{n}$ is internally tangent to the circle radius 3 and externally tangent to the other two circles as shown. Find $m+n$.

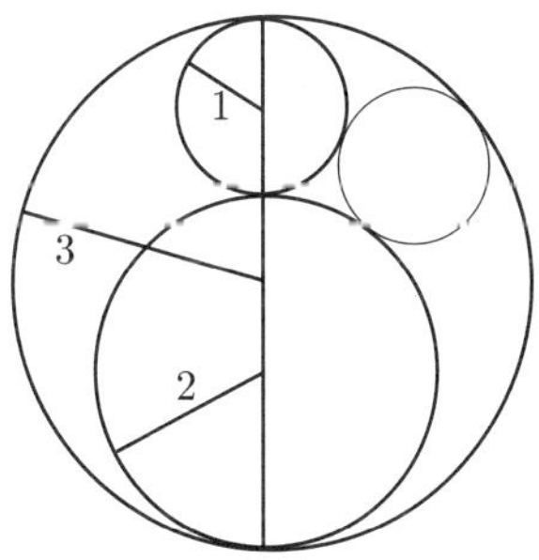

Answer: 13

Let the unknown radius be r. Let the centers of the circles radius 3, 2, 1, and r be A, B, C, and D, respectively, as shown. Applying the Law of Cosines to $\triangle ABD$ gives

$$1^2 + (2+r)^2 - 2 \cdot 1 \cdot (2+r)(\cos ABD) = (3-r)^2.$$

Applying the Law of Cosines to $\triangle CBD$ gives

$$3^2 + (2+r)^2 - 2 \cdot 3 \cdot (2+r)(\cos ABD) = (1+r)^2.$$

Subtracting three times the first equation from the second yields

$$3(3-r)^2 - (1+r)^2 = -6 + 2(2+r)^2$$

which reduces to $24 = 28r$ implying $r = \frac{6}{7}$. The requested sum is $6+7 = 13$. Note that this can also be solved as an application of the Descartes' *Kissing Circles Theorem.*

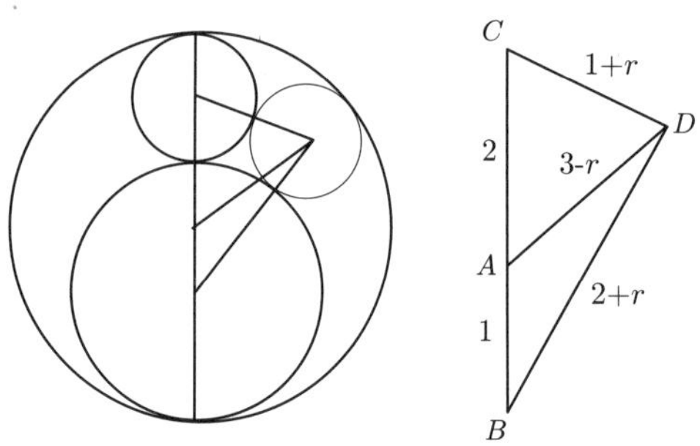

Problem 23. Find the greatest seven-digit integer divisible by 132 whose digits, in order, are 2, 0, x, y, 1, 2, z where x, y, and z are single digits.

Answer: 2093124

From the given information it follows that

- Since 132 is divisible by 3, $5 + x + y + z \equiv 0 \pmod{3}$.
- Since 132 is divisible by 11, $1 + x - y + z \equiv 0 \pmod{11}$.
- Since 132 is divisible by 4, $z \equiv 0 \pmod{4}$.

Thus, z must be 0, 4, or 8. The number would be maximized if x could be 9. In this case, it follows that $y \equiv 1 - z \pmod{3}$ and $y \equiv z - 1 \pmod{11}$. If $z = 0$, then the digit y would need to be equivalent to 10 (mod 11) which is not possible. If $z = 8$, the digit y would need to be both equivalent to $1 - 8 \equiv 2 \pmod{3}$ and equivalent to $8 - 1 \equiv 7 \pmod{11}$ which is also not possible. But when $z = 4$, $y = 3$ satisfies both $y \equiv 1 - 4 \equiv 0 \pmod{3}$ and $y \equiv 4 - 1 \equiv 3 \pmod{11}$. Thus, the number 2093124 satisfies all the conditions and is the greatest such value that does.

Problem 24. There are positive integers m and n so that $x = m + \sqrt{n}$ is a solution to the equation $x^2 - 10x + 1 = \sqrt{x}(x+1)$. Find $m+n$.

Answer: 55

Divide both sides of the given equation by x to get $x - 10 + \frac{1}{x} = \sqrt{x} + \frac{1}{\sqrt{x}}$. Let $y = \sqrt{x} + \frac{1}{\sqrt{x}}$. Then $y^2 = x + 2 + \frac{1}{x}$. This shows

$y^2 - 12 = y$ which is equivalent to $(y-4)(y+3) = 0$. The solution $y = -3$ will not work, but $y = 4$ means $\sqrt{x} + \frac{1}{\sqrt{x}} = 4$ or $x - 4\sqrt{x} + 1 = 0$. This quadratic equation in $\sqrt{x}$ has solutions $\sqrt{x} = 2 \pm \sqrt{3}$. The desired solution is, therefore, $x = (2+\sqrt{3})^2 = 7 + 4\sqrt{3} = 7 + \sqrt{48}$. The requested sum is $7 + 48 = 55$.

Problem 25. Find the largest prime that divides

$$1 \cdot 2 \cdot 3 \; + \; 2 \cdot 3 \cdot 4 \; + \; ... \; + \; 44 \cdot 45 \cdot 46.$$

Answer: 47

The sum has terms of the form $(k-1)k(k+1) = k^3 - k$. Then

$$\sum_{k=1}^{n}(k-1)k(k+1) = \sum_{k=1}^{n} k^3 - k = \left(\frac{n(n+1)}{2}\right)^2 - \frac{n(n+1)}{2}$$

$$= \frac{n(n+1)\,[n(n+1)-2]}{4} = \frac{n(n+1)(n^2+n-2)}{4}$$

$$= \frac{n(n+1)(n-1)(n+2)}{4}.$$

The given sum is the special case when $n = 45$, so the sum equals $\dfrac{(45)(46)(44)(47)}{4}$ whose largest prime factor is 47.

Problem 26. A paper cup has a base that is a circle with radius r, a top that is a circle with radius $2r$, and sides that connect the two circles with straight line segments as shown below. This cup has height h and volume V. A second cup that is exactly the same shape as the first is held upright inside the first cup so that its base is a distance of $\frac{h}{2}$ from the base of the first cup. The volume of liquid that will fit inside the first cup and outside the second cup can be written $\frac{m}{n} \cdot V$ where m and n are relatively prime positive integers. Find $m + n$.

Answer: 93

The volume between the cups does not change if the sides of each cup are extended so that the cups form circular cones with radius $2r$ and height $2h$. The volume of the first cone is then

$$\frac{1}{3}\pi(2r)^2(2h) = \frac{8}{3}\pi r^2 h.$$

The part of the second cone inside the first is a cone that is $\frac{3}{4}$ the height of the first cone, so the volume that it takes up within the first cone is

$$\left(\frac{3}{4}\right)^3 \frac{8}{3}\pi r^2 h = \frac{9}{8}\pi r^2 h.$$

Thus, the volume inside the first cup and outside the second cup is

$$\left[\frac{8}{3} - \frac{9}{8}\right]\pi r^2 h = \frac{37}{24}\pi r^2 h.$$

The volume of the first cup is

$$\left[1 - \left(\frac{1}{2}\right)^3\right]\frac{4}{3}\pi r^2(2h) = \frac{7}{3}\pi r^2 h.$$

The desired fraction is

$$\frac{\frac{37}{24}}{\frac{7}{3}} = \frac{37}{56}.$$

The requested sum is $37 + 56 = 93$.

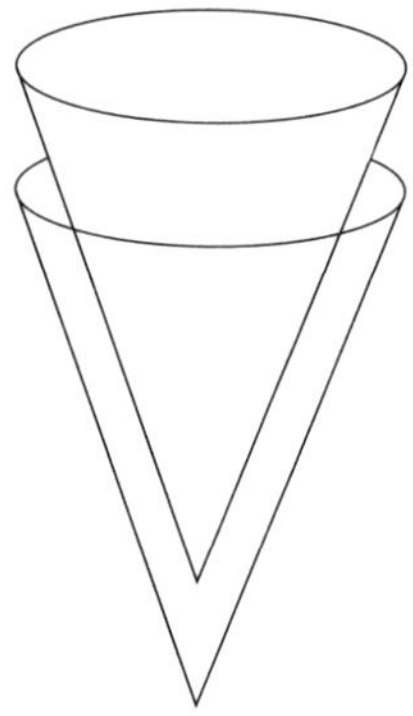

Problem 27. You have some white 1×1 tiles and some black and white 2×1 tiles as shown below. There are four different color patterns that can be generated when using these tiles to cover a 3×1 rectangle

by laying these tiles side by side (WWW, BWW, WBW, WWB). How many different color patterns can be generated when using these tiles to cover a 10×1 rectangle?

Answer: 286

Let A_j refer to the number of color patterns one can obtain by laying these tiles together to tile a $j \times 1$ rectangle. Clearly, $A_0 = A_1 = 1$ and $A_2 = 3$. When $j > 2$, categorize the coloring patterns by the position of the first black square among the j squares. If the coloring pattern begins with a black square in its first position, the pattern must have a 2×1 tile followed by one of the A_{j-2} patterns of length $j - 2$. Similarly, if the coloring pattern has a white square in the first position and a black square in the second position, the first two squares can be followed by any of the A_{j-2} patterns of length $j - 2$. If the first black square appears in the kth position where $k > 2$, it can be followed by any of the A_{j-k} patterns of length $j - k$. Finally, there is one pattern consisting of all white squares. Thus

$$A_j = A_{j-2} + A_{j-2} + A_{j-3} + A_{j-4} + A_{j-5} + \cdots A_1 + A_0 + 1.$$

From this one can generate $A_3 = A_1 + A_1 + A_0 + 1 = 4$ and $A_4 = A_2 + A_2 + A_1 + A_0 + 1 = 9$. The following table is then easy to complete.

0	1	2	3	4	5	6	7	8	9	10
1	1	3	4	9	14	28	47	89	155	286

Problem 28. A bag contains 8 green candies and 4 red candies. You randomly select one candy at a time to eat. If you eat five candies, there are relatively prime positive integers m and n so that $\frac{m}{n}$ is the probability that you do not eat a green candy after you eat a red candy. Find $m + n$.

Answer: 6

The five candies can be eaten in the order GRRRR, GGRRR, GGGRR, GGGGR, or GGGGG. The probabilities of eating the candies in these orders are, respectively,

$$\frac{8}{12} \cdot \frac{4}{11} \cdot \frac{3}{10} \cdot \frac{2}{9} \cdot \frac{1}{8}, \frac{8}{12} \cdot \frac{7}{11} \cdot \frac{4}{10} \cdot \frac{3}{9} \cdot \frac{2}{8}, \frac{8}{12} \cdot \frac{7}{11} \cdot \frac{6}{10} \cdot \frac{4}{9} \cdot \frac{3}{8},$$

$$\frac{8}{12}\cdot\frac{7}{11}\cdot\frac{6}{10}\cdot\frac{5}{9}\cdot\frac{4}{8}, \text{ and } \frac{8}{12}\cdot\frac{7}{11}\cdot\frac{6}{10}\cdot\frac{5}{9}\cdot\frac{4}{8}.$$

The sum of these is

$$\frac{4\cdot3\cdot2\cdot1+7\cdot4\cdot3\cdot2+7\cdot6\cdot4\cdot3+7\cdot6\cdot5\cdot4+7\cdot6\cdot5\cdot4}{12\cdot11\cdot10\cdot9}$$

$$=\frac{12(2+14+42+70+70)}{12\cdot11\cdot10\cdot9}=\frac{198}{11\cdot10\cdot9}=\frac{1}{5}.$$

The requested sum is $1+5=6$.

Problem 29. Let $A=\{1,3,5,7,9\}$ and $B=\{2,4,6,8,10\}$. Let f be a randomly chosen function from the set $A\cup B$ into itself. There are relatively prime positive integers m and n such that $\frac{m}{n}$ is the probability that f is a one-to-one function on $A\cup B$ given that it maps A one-to-one into $A\cup B$ and it maps B one-to-one into $A\cup B$. Find $m+n$.

Answer: 253

A one-to-one function f that maps $A\cup B$ into itself is just a permutation of the set $A\cup B$. Thus, there are 10! equally likely such functions. The number of one-to-one functions that map A into $A\cup B$ is ${}_{10}P_5=10\cdot9\cdot8\cdot7\cdot6$. Similarly, there are that many one-to-one functions that map B into $A\cup B$. It follows that the desired probability is

$$\frac{10\cdot9\cdot8\cdot7\cdot6\cdot5\cdot4\cdot3\cdot2\cdot1}{(10\cdot9\cdot8\cdot7\cdot6)(10\cdot9\cdot8\cdot7\cdot6)}=\frac{5\cdot4\cdot3\cdot2\cdot1}{10\cdot9\cdot8\cdot7\cdot6}=\frac{1}{\binom{10}{5}}=\frac{1}{252}.$$

The requested sum is $1+252=253$.

Problem 30. The diagram below shows four regular hexagons each with side length 1 meter attached to the sides of a square. This figure is drawn onto a thin sheet of metal and cut out. The hexagons are then bent upward along the sides of the square so that A_1 meets A_2, B_1 meets B_2, C_1 meets C_2, and D_1 meets D_2. The resulting dish is set on a table with the square lying flat on the table. If this dish is filled with water, the water will rise to the height of the corner where the A_1 and A_2 meet. There are relatively prime positive integers m and n so that the number of cubic meters of water the dish will hold is $\sqrt{\frac{m}{n}}$. Find $m+n$.

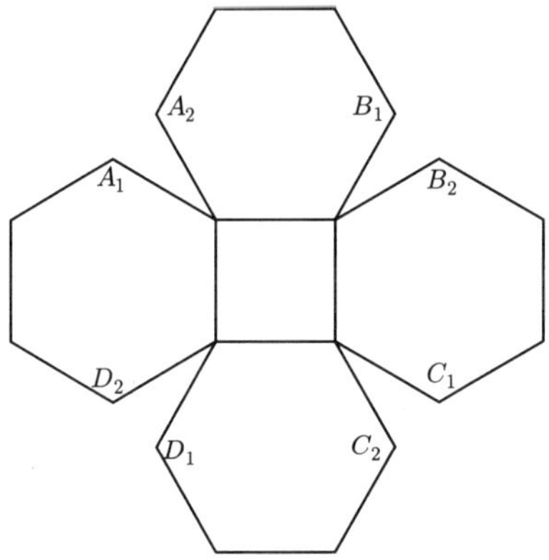

Answer: 67

Let the point where A_1 and A_2 meet be labeled A, and the point where B_1 and B_2 meet be labeled B. Let the center of the square be labeled F. Let the corner of the square nearest point A be labeled H, the corner nearest B be labeled G, and the corner diagonally opposite G be labeled E as shown in the figure below. Let J be the projection of the point A in the plane of the square. Because $\angle AHE = 120°$, the Law of Cosines gives

$$AE = \sqrt{AH^2 + HE^2 - 2 \cdot AH \cdot HE \cos 120°}$$
$$= \sqrt{1+1+1} = \sqrt{3}.$$

Note that

$$EF = HF = \frac{1}{\sqrt{2}}.$$

Then the Pythagorean Theorem gives

$$AF = \sqrt{AE^2 - EF^2} = \sqrt{3 - \frac{1}{2}} = \sqrt{\frac{5}{2}}.$$

The Law of Cosines then gives

$$\cos AHF = \frac{AF^2 - AH^2 - HF^2}{2 \cdot AH \cdot HF} = \frac{\frac{5}{2} - 1 - \frac{1}{2}}{-\sqrt{2}} = -\frac{1}{\sqrt{2}}.$$

It follows that $\angle AHF = 135°$, and $\angle AHJ = 45°$. Thus, $AJ = \frac{1}{\sqrt{2}}$ is the height of A from the plane of the square. Since the diameter of a regular hexagon is twice the length of one of its sides, it follows that $AB = 2$. If the vertical edges of the dish $\overline{AH}$ and $\overline{BG}$ are extended, they will meet at a point K as shown in the diagram. Note that K is the vertex of a downward pointing square pyramid with base having side length 2 and with height $\sqrt{2}$. The volume of this pyramid is given by $\frac{1}{3}$ the area of its base times its height $= \frac{1}{3} \cdot 4 \cdot \sqrt{2}$. The lower half of the pyramid is another square pyramid with base having side length 1 and with height $\frac{\sqrt{2}}{2}$ so its volume is $\frac{1}{3} \cdot 1 \cdot \frac{\sqrt{2}}{2}$. The desired volume is then

$$\frac{4}{3} \cdot \sqrt{2} - \frac{1}{3} \cdot \frac{\sqrt{2}}{2} = \frac{7\sqrt{2}}{6} = \sqrt{\frac{49}{18}}.$$

The requested sum is $49 + 18 = 67$.

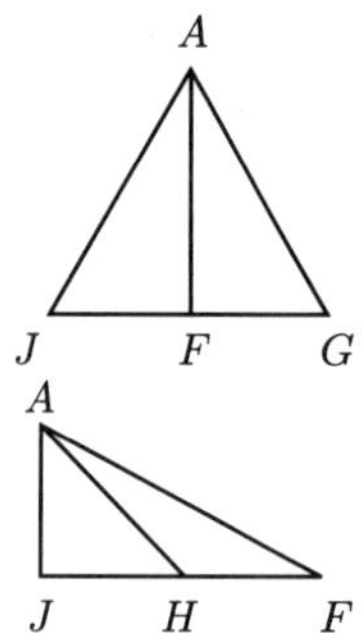

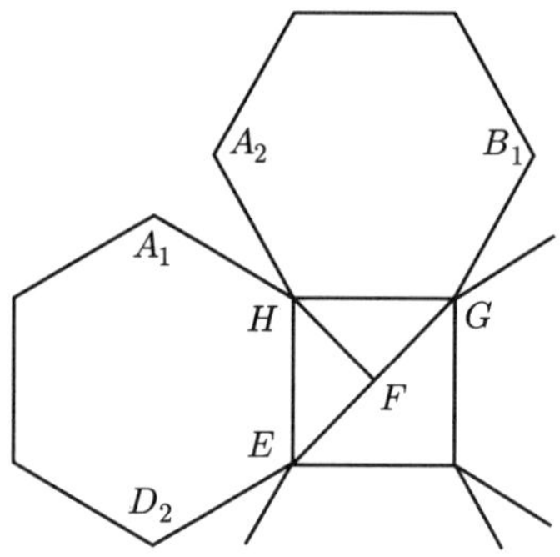

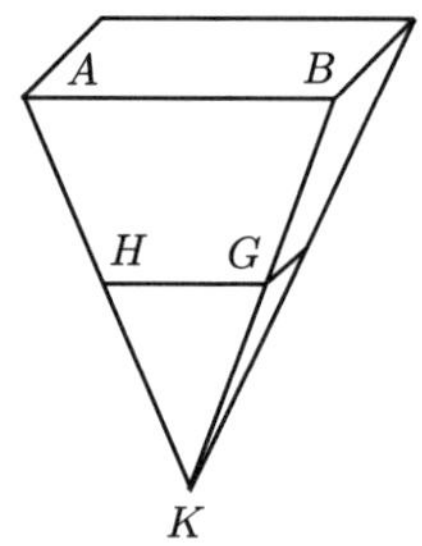

2.19 2012 Middle School

Problem 1. Evaluate $5^4 - 4^3 - 3^2 - 2^1 - 1^0$.

Answer: 549

The expression equals $625 - 64 - 9 - 2 - 1 = 549$.

Problem 2. The diagram below shows rectangle $ABDE$ where C is the midpoint of side $\overline{BD}$, and F is the midpoint of side $\overline{AE}$. If $AB = 10$ and $BD = 24$, find the area of the shaded region.

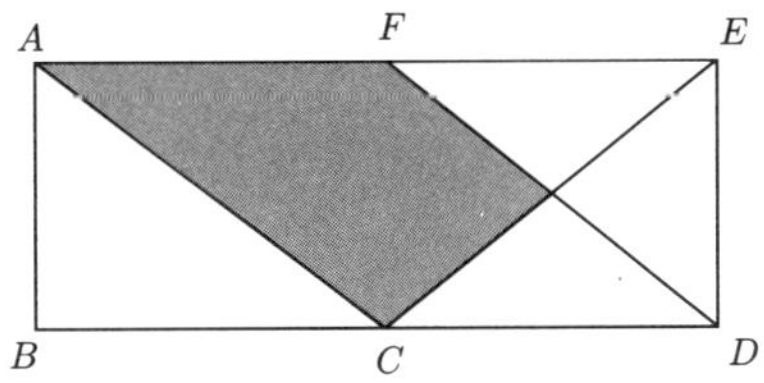

Answer: 90

Parallelogram $ACDF$ has side length $AF = 12$ and height $AB = 10$, so it has area $12 \cdot 10 = 120$. Let G be the intersection of $\overline{CE}$ and $\overline{DF}$. Then $\triangle CDG$ has base 12 and height 5, so it has area $\frac{12 \cdot 5}{2} = 30$. It follows that the shaded region has area $120 - 30 = 90$.

Alternatively, notice that $\triangle ACF$ is half the area of rectangle $ABCF$, and $\triangle FCG$ is a quarter of the area of rectangle $CDEF$. Then the shaded region has area

$$\frac{1}{2}[ABCF] + \frac{1}{4}[CDEF] = \left(\frac{1}{2} + \frac{1}{4}\right)[ABCF]$$

$$= \frac{3}{8}[ABCE] = \frac{3}{8} \cdot 240 = 90.$$

Problem 3. While Peter was driving from home to work, he noticed that after driving 21 miles, the distance he had left to drive was 30 percent of the total distance from home to work. How many miles was his complete trip from home to work?

Answer: 30

If the total distance Peter had to drive from home to work was x miles, then $(1 - 0.30)x = 21$. Thus, $x = \frac{21}{0.7} = \frac{210}{7} = 30$.

Problem 4. How many two-digit positive integers contain at least one digit equal to 5?

Answer: 18

There are nine two-digit integers which end in 5, ten two-digit integers which begin with 5, and one two-digit integer that both begins and ends with 5. Thus, there are $9+10-1 = 18$ two-digit integers that contain at least one digit equal to 5.

Problem 5. Meredith drives 5 miles to the northeast, then 15 miles to the southeast, then 25 miles to the southwest, then 35 miles to the northwest, and finally 20 miles to the northeast. How many miles is Meredith from where she started?

Answer: 20

Rotate Meredith's trip by 45 degrees. Then she drives 5 miles east, 15 miles south, 25 miles west, 35 miles north, and 20 miles east. The total distance she travels in the east-west direction is $5-25+20 = 0$. The total distance she travels in the north-south direction is $-15+35 = 20$. Thus, she ends up 20 miles from where she started.

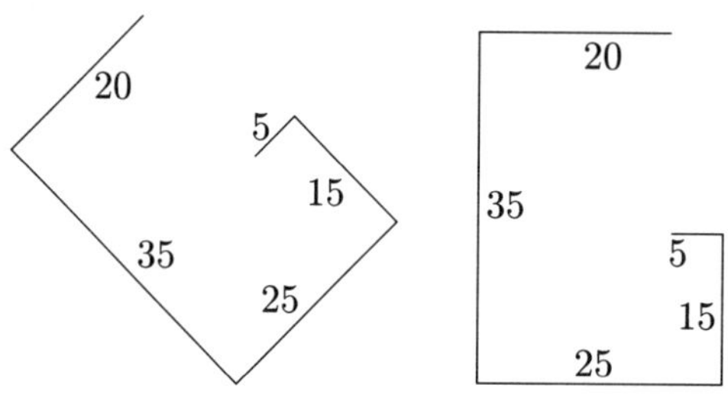

Problem 6. Volume A equals one fourth of the sum of the volumes B and C, while volume B equals one sixth of the sum of the volumes A and C. There are relatively prime positive integers m and n so that the ratio of volume C to the sum of the other two volumes is $\frac{m}{n}$. Find $m+n$.

Answer: 35

The relations between the volumes give

$$A = \frac{B+C}{4} \implies 4A - B = C$$
$$B = \frac{C+A}{6} \implies -A + 6B = C.$$

Then $4A - B = -A + 6B$ and $B = \frac{5}{7}A$, so $C = -A + 6B = -A + 6 \cdot \frac{5}{7}A = \frac{23}{7}A$. Finally, $\frac{C}{A+B} = \frac{\frac{23}{7}A}{A+\frac{5}{7}A} = \frac{23}{7+5} = \frac{23}{12}$. The requested sum is $m + n = 23 + 12 = 35$.

Problem 7. A snail crawls $2\frac{1}{2}$ centimeters in $4\frac{1}{4}$ minutes. At that rate, how many centimeters can the snail crawl in 85 minutes?

Answer: 50

The snail is moving at a rate of $\frac{2\frac{1}{2}}{4\frac{1}{4}} = \frac{\frac{5}{2}}{\frac{17}{4}} = \frac{10}{17}$ centimeters per minute. So, in 85 minutes, the snail can crawl $\frac{10}{17} \cdot 85 = 50$ centimeters.

Problem 8. Seven boys and three girls are playing basketball. In how many different ways can they make two teams of five players so that both teams have at least one girl?

Answer: 105

If there must be at least one girl on each team, exactly one team will have two girls. There are $\binom{3}{2} = 3$ ways to select the two girls who will be on that team and $\binom{7}{3} = 35$ ways to select the three boys who will be on their team. Thus, there are $3 \cdot 35 = 105$ ways to select the team with two girls which determines the two basketball teams.

Problem 9. Points E and F lie inside rectangle $ABCD$ with $AE = DE = BF = CF = EF$. If $AB = 11$ and $BC = 8$, find the area of the quadrilateral $AEFB$.

Answer: 32

Note that since E is equidistant from A and D, and F is equidistant from B and C, the segment $\overline{EF}$ is parallel to $\overline{AB}$.

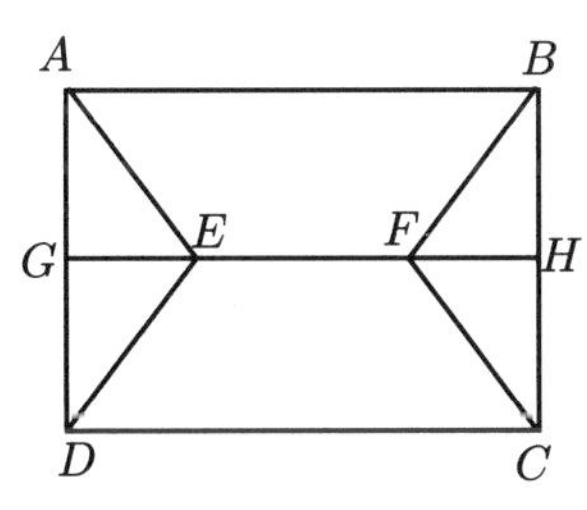

Let $\overline{EF}$ intersect the rectangle at points G and H as shown. Let $AE = EF = FB = x$, and since $\triangle AEG$ is a right triangle with $AG = 4$, the Pythagorean Theorem gives

$$EG = \sqrt{x^2 - 4^2} = \sqrt{x^2 - 16}.$$

Then $11 = GH = GE + EF + FH = 2\sqrt{x^2 - 16} + x$. Thus, $4(x^2 - 16) = (11 - x)^2 = 121 - 22x + x^2$, so $3x^2 + 22x - 185 = 0$. This factors as $(3x + 37)(x - 5) = 0$. The positive solution is $x = 5$. It follows that quadrilateral $AEFB$ is a trapezoid with bases with lengths 11 and 5 and height 4. Thus, its area is $4 \cdot \frac{11+5}{2} = 32$.

Problem 10. Find the least positive multiple of 999 that does not have 9 as a digit.

Answer: 111888

For any positive integer k, one has

$$999k = k(1000 - 1) = 1000k - k$$

which is k less than a multiple of 1000. Note that each of the integers from 889 through 999 contains 9 as a digit. Thus, in order for $999k$ not to have 9 as a digit, k must be at least $1000 - 888 = 112$. The least positive multiple of 999 that does not have 9 as a digit is $112 \cdot 999 = 112000 - 112 = 111888$.

Problem 11. For some integers a and b the function $f(x) = ax + b$ has the properties that $f(f(0)) = 0$ and $f(f(f(4))) = 9$. Find $f(f(f(f(10))))$.

Answer: 10

The given information says

$$f(f(0)) = ab + b = (a + 1)b = 0,$$

so either $b = 0$ or $a = -1$. Furthermore,

$$f(f(f(4))) = a\left(4a^2 + ab + b\right) + b = 4a^3 + a(ab + b) + b = 4a^3 + b = 9.$$

If $b = 0$, then $a = \sqrt[3]{\frac{9}{4}}$ which is not an integer. If $a = -1$, then $-4 + b = 9$ and $b = 13$. Conclude that $f(x) = -x + 13$, so that

$$f(f(f(f(10)))) = f(f(f(3))) = f(f(10)) = f(3) = 10.$$

Problem 12. Pentagon $ABCDE$ consists of a square $ACDE$ and an equilateral triangle ABC that share the side $\overline{AC}$. A circle centered at C has area 24. The intersection of the circle and the pentagon has half the area of the pentagon. Find the area of the pentagon.

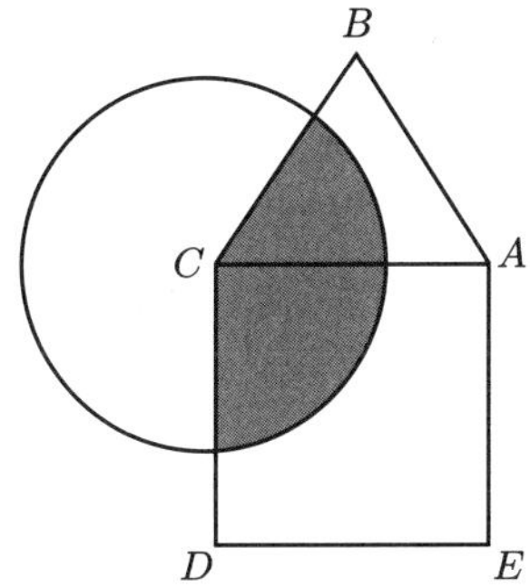

Answer: 20

Note that if the circle intersects line $\overline{AB}$, its radius would need to be at least $\frac{\sqrt{3}}{2}$ times the side length of the pentagon. In this case the circle would contain more than half the area of the both the triangle and the square. The intersection of the square $ACDE$ and the circle is a quarter of the circle. The intersection of the equilateral triangle ABC and the circle is a sixth of the circle. It follows that the area of the pentagon is $2 \cdot 24\left(\frac{1}{4}+\frac{1}{6}\right) = 2(6+4) = 20$.

Problem 13. Find the least positive integer N which is both a multiple of 19 and whose digits add to 23.

Answer: 779

Any such N must have at least 3 digits in order for the sum of its digits to be 23. If the decimal representation of N is $\overline{abc}$, then $a+b+c=23$ and $100a+10b+c = 100a+10b+(23-a-b) = 99a+9b+23 \equiv 0 \pmod{19}$. Thus, $4a+9b \equiv 15 \pmod{19}$ and multiplying this by 17 yields $11a+b \equiv 8 \pmod{19}$ and $b \equiv 8-11a \pmod{19}$. Since b and c are each at most 9, it follows that $a \geq 5$. If $a=5$, then $b \equiv 8-11\cdot 5 \equiv 10 \pmod{19}$ which is not possible. If $a=6$, then $b \equiv 8-11\cdot 6 \equiv 18 \pmod{19}$ which is not possible. If $a=7$, then $b \equiv 8-11\cdot 7 \equiv 7 \pmod{19}$. This gives the solution $a=b=7$ and $c=23-7-7=9$ and $N=779$.

Problem 14. At the 4 PM show, all the seats in the theater were taken, and 65 percent of the audience was children. At the 6 PM show, again, all the seats were taken, but this time only 50 percent of the audience was children. Of all the people who attended either of these shows, 57 percent were children although there were 12 adults and 28 children who attended both shows. How many people does the theater seat?

Answer: 520

Let s be the number of seats in the theater. Then at the first show, there were $0.65s$ children, and at the second show, there were $0.5s$ children. The number of children attending the two shows was $0.65s + 0.5s - 28$, and the number of people attending the two shows was $s + s - 28 - 12$. It follows that

$$\frac{0.65s + 0.5s - 28}{s + s - 40} = 0.57.$$

Thus, $115s - 2800 = 57(2s - 40)$, and $s = 2800 - 57 \cdot 40 = 520$.

Problem 15. The top and bottom of a rectangular tank each have area 48 square inches. The four vertical sides of the tank have a total area of 96 square inches. The longest diagonal of the tank is 13 inches. Find the sum of the height, the width, and the length of this tank in inches.

Answer: 19

Let the tank have length x, width y, and height z. Then the conditions of the problem give $xy = 48$, $2(xz + yz) = 96$, and

$$\sqrt{x^2 + y^2 + z^2} = 13.$$

It follows that $2xz + 2yz + 2xy = 96 + 2 \cdot 48 = 192$. Then

$$(x + y + z)^2 = x^2 + y^2 + z^2 + 2xz + 2yz + 2xy =$$

$$\left(x^2 + y^2 + z^2\right) + 2(xz + yz + xy) = 13^2 + 192 = 361 = 19^2.$$

Thus, the requested sum $x + y + z = 19$.

Problem 16. The following sequence lists all the positive rational numbers that do not exceed $\frac{1}{2}$ by first listing the fraction with denominator 2, followed by the one with denominator 3, followed by the two fractions with denominator 4 in increasing order, and so forth so that the sequence is

$$\frac{1}{2}, \frac{1}{3}, \frac{1}{4}, \frac{2}{4}, \frac{1}{5}, \frac{2}{5}, \frac{1}{6}, \frac{2}{6}, \frac{3}{6}, \frac{1}{7}, \frac{2}{7}, \frac{3}{7}, \dots.$$

Let m and n be relatively prime positive integers so that the 2012th fraction in the list is equal to $\frac{m}{n}$. Find $m + n$.

Answer: 61

In the sequence there are two denominators that each appear once (2 and 3), two denominators that each appear twice (4 and

5), and so forth. The number of terms of the sequence through the fraction $\frac{n}{2n+1}$ is

$$1+1+2+2+3+3+\cdots+n+n = 2\sum_{k=1}^{n} k = n(n+1).$$

The least n for which $n(n+1) > 2012$ is $n = 45$ since $44(44+1) = 1980$ and $45(45+1) = 2070$. Because $n = 44$ corresponds to fractions with denominators 88 and 89, the 2012th term of the sequence is the thirty-second fraction with denominator 90. The fraction is $\frac{32}{90} = \frac{16}{45}$. The requested sum is $16 + 45 = 61$.

Problem 17. The diagram below shows nine points on a circle where $AB = BC = CD = DE = EF = FG = GH$. Given that $\angle GHJ = 117°$ and $\overline{BH}$ is perpendicular to $\overline{EJ}$, there are relatively prime positive integers m and n so that the degree measure of $\angle AJB$ is $\frac{m}{n}$. Find $m + n$.

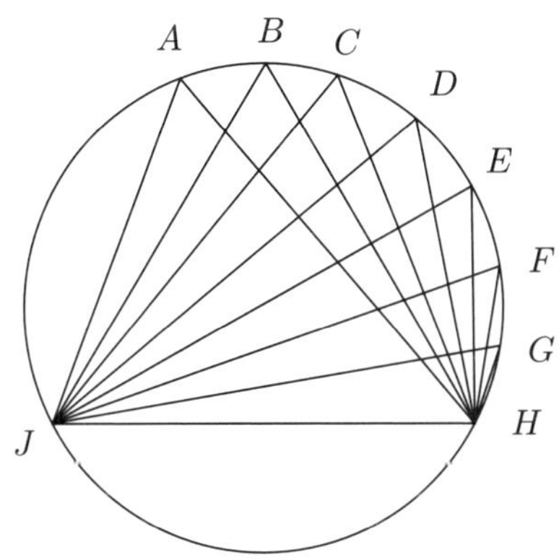

Answer: 29

Note that $\angle AJB = \angle BJC = \angle CJD = \angle DJE = \angle EJF = \angle FJG = \angle GJH$ since all the angles subtend equal sized arcs. Let x represent the degree measure of those angles. Since they subtend equal arcs, $\angle GHB = \angle GJB = 5x$. Because $\overline{BH}$ is perpendicular to $\overline{EJ}$, it follows that $\triangle JBH$ is isosceles. Thus, the sum of the angles in $\triangle JBH$ is $2\angle BHJ + \angle BJH = 2(\angle GHJ - \angle GHB) + \angle BJH = 2(117° - 5x) + 6x = 180°$. Solving yields $117° - 2x = 90°$ and $x = \frac{27}{2}$. The requested sum is $27 + 2 = 29$.

Problem 18. Find the smallest positive integer whose remainder when divided by 5 is 3, when divided by 7 is 5, and when divided by 9 is 1.

Answer: 208

Let the answer be n. Because n is 3 more than a multiple of 5, when n is divided by 35, it must have a remainder in

$$\{3, 8, 13, 18, 23, 28, 33\}.$$

Similarly, because n is 5 more than a multiple of 7, when divided by 35, it must have a remainder in

$$\{5, 12, 19, 26, 33\}.$$

Thus, n must have a remainder of 33 when divided by 35. Thus, when divided by $35 \cdot 9 = 315$, it must have a remainder in

$$\{33, 68, 103, 138, 173, 208, 243, 278, 313\}.$$

The only one of these numbers which is 1 more than a multiple of 9 is 208. Thus, the smallest possible value of n is 208.

Alternatively, note that $n + 2$ must be divisible by 5 and 7, so it is divisible by 35. Because $35 \equiv -1 \pmod 9$, it follows that $n+2$ which is $3 \pmod 9$ must satisfy $n+2 \equiv 3 \equiv 6(-1) \equiv 6 \cdot 35 \pmod 9$. So $n + 2 = 6 \cdot 35 = 210$ and $n = 208$.

Problem 19. A teacher suggests four possible books for students to read. Each of six students selects one of the four books. How many ways can these selections be made if each of the books is read by at least one student?

Answer: 1560

If each of six students can select any one of the four choices, there are $4^6 = 4096$ ways for them to make choices. For $j = 1, 2, 3, 4$ let A_j be the set of book choices where no student selects book number j. The Inclusion/Exclusion Principle says that the size of $A_1 \cup A_2 \cup A_3 \cup A_4$ is

$$4 \times \#(A_1) - \binom{4}{2} \times \#(A_1 \cap A_2) + \binom{4}{3} \times \#(A_1 \cap A_2 \cap A_3)$$

$$-\#(A_1 \cap A_2 \cap A_3 \cap A_4) = 4 \cdot 3^6 - 6 \cdot 2^6 + 4 \cdot 1 - 0 = 2536.$$

Thus, the number of ways for the students to make their selections so that each book is selected at least once is

$$4096 - 2536 = 1560.$$

Problem 20. In the following addition, different letters represent different non-zero digits. What is the 5-digit number $ABCDE$?

$$\begin{array}{rccccccc}
 & A & B & C & D & E & D & B \\
 & & B & C & D & E & D & B \\
 & & & C & D & E & D & B \\
 & & & & D & E & D & B \\
 & & & & & E & D & B \\
 & & & & & & D & B \\
+ & & & & & & & B \\
\hline
 & A & A & A & A & A & A & A
\end{array}$$

Answer: 84269

Note that since each letter is the first digit of at least one number, none of the letters represents the digit zero. From the ones column of the addition, it follows that the number $7B$ ends in the digit A which implies that A and B are either both even or both odd. Since the top row of the addition is clearly smaller than the bottom row, the digit A must be greater than the digit B. Because there is no carry from the hundred-thousands place to the millions place in the addition, B must be less than 5, and A must be at least $2B$. Since the carry into the hundred-thousands place is at most 3, A cannot exceed $2B + 3$. The possible values for (B, A) that satisfy $2B \leq A \leq 2B + 3$, $B < 5$, with A and B having the same parity are $(1, 3), (1, 5), (2, 4), (2, 6), (3, 7), (3, 9), (4, 8)$. But the ones column shows that $7B$ must end in the digit A. This limits the possibilities to $(2, 4)$ and $(4, 8)$. If $B = 2$, then there is a carry of 1 from the ones column to the tens column. But then $6D + 1$ must end with $A = 4$ which is not possible. Thus, $A = 8$ and $B = 4$. Then $6D + 2$ must exceed 10 and end in the digit 8 which requires D to be 6. Thus, there will be a carry of 3 from the tens column to the hundreds column, so $5E + 3$ ends in the digit 8 implying E is odd. Moreover, the thousands column contains four 6's and thus, with the carry, must add to 28. The carry from the hundreds column must be 4 implying that $E = 9$. Finally, the ten-thousands column has $3C + 2$ adding to 8, so $C = 2$. This determines that $ABCDE = 84269$.

Appendix

Difficulty:
1 Easy
2 Medium
3 Hard
4 Challenging
5 Extremely Difficult

Year	Level	Problem	Topic	Difficulty
2003	HS	1	Algebra	1
2003	HS	2	Calendar	1
2003	HS	3	Number Theory	1
2003	HS	4	Geometry/Algebra	1
2003	HS	5	Algebra	2
2003	HS	6	Algebra/Logarithms	2
2003	HS	7	Combinatorics/Number Theory	3
2003	HS	8	Geometry	2
2003	HS	9	Algebra	3
2003	HS	10	Algebra	2
2003	HS	11	Algebra	3
2003	HS	12	Combinatorics/Geometry	2
2003	HS	13	Algebra	3
2003	HS	14	Algebra	3
2003	HS	15	Algebra	3
2003	HS	16	Algebra	3
2003	HS	17	Trigonometry	2
2003	HS	18	Geometry	2
2003	HS	19	Algebra	3
2003	HS	20	Combinatorics	3
2003	HS	21	Algebra	4
2003	HS	22	Geometry	4
2003	HS	23	Algebra	5
2003	HS	24	Geometry	4
2003	HS	25	Trigonometry/Algebra	4
2004	MS	1	Calendar	1
2004	MS	2	Algebra	1
2004	MS	3	Geometry	2
2004	MS	4	Algebra	1
2004	MS	5	Number Theory/Number Bases	1
2004	MS	6	Number Theory	2
2004	MS	7	Geometry/Algebra	2

Year	Level	Problem	Topic	Difficulty
2004	MS	8	Algebra	2
2004	MS	9	Combinatorics/Number Theory	2
2004	MS	10	Algebra	2
2004	MS	11	Algebra	2
2004	MS	12	Algebra	2
2004	MS	13	Combinatorics	2
2004	MS	14	Geometry/Algebra	2
2004	MS	15	Number Theory	3
2004	MS	16	Algebra	3
2004	MS	17	Algebra	2
2004	MS	18	Combinatorics	3
2004	MS	19	Algebra	3
2004	MS	20	Geometry	2
2004	MS	21	Algebra	3
2004	MS	22	Geometry	3
2004	MS	23	Geometry/Algebra	3
2004	MS	24	Algebra	3
2004	MS	25	Algebra	4
2004	HS	1	Number Theory	2
2004	HS	2	Geometry	2
2004	HS	3	Algebra	3
2004	HS	4	Algebra	2
2004	HS	5	Algebra	2
2004	HS	6	Algebra	2
2004	HS	7	Combinatorics/Number Theory	2
2004	HS	8	Algebra	2
2004	HS	9	Algebra	2
2004	HS	10	Geometry	2
2004	HS	11	Algebra	3
2004	HS	12	Geometry	3
2004	HS	13	Geometry/Algebra	3
2004	HS	14	Geometry	3
2004	HS	15	Algebra	3
2004	HS	16	Algebra	3
2004	HS	17	Combinatorics	3
2004	HS	18	Algebra	4
2004	HS	19	Probability	4
2004	HS	20	Geometry 3D	3
2004	HS	21	Algebra	4
2004	HS	22	Geometry	5
2004	HS	23	Algebra	4

Year	Level	Problem	Topic	Difficulty
2004	HS	24	Algebra	4
2004	HS	25	Algebra	4
2005	MS	1	Algebra	1
2005	MS	2	Algebra	1
2005	MS	3	Geometry	1
2005	MS	4	Algebra	2
2005	MS	5	Number Theory	1
2005	MS	6	Geometry 3D	1
2005	MS	7	Algebra	1
2005	MS	8	Number Theory	2
2005	MS	9	Number Theory	2
2005	MS	10	Number Theory	2
2005	MS	11	Algebra	2
2005	MS	12	Combinatorics	2
2005	MS	13	Algebra	2
2005	MS	14	Combinatorics	3
2005	MS	15	Algebra	3
2005	HS	1	Algebra	2
2005	HS	2	Geometry 3D	1
2005	HS	3	Geometry	1
2005	HS	4	Number Theory	1
2005	HS	5	Algebra	2
2005	HS	6	Geometry	2
2005	HS	7	Algebraic Geometry	2
2005	HS	8	Algebra	2
2005	HS	9	Geometry	2
2005	HS	10	Probability	3
2005	HS	11	Algebra	2
2005	HS	12	Number Theory	2
2005	HS	13	Algebra	3
2005	HS	14	Combinatorics	2
2005	HS	15	Algebra	3
2005	HS	16	Algebra	2
2005	HS	17	Algebra	3
2005	HS	18	Geometry	3
2005	HS	19	Algebra	3
2005	HS	20	Algebra	4
2005	HS	21	Geometry	3
2005	HS	22	Number Theory	4
2005	HS	23	Algebra	5
2005	HS	24	Geometry	4

Year	Level	Problem	Topic	Difficulty
2005	HS	25	Algebra	5
2006	MS	1	Calendar	1
2006	MS	2	Algebra	1
2006	MS	3	Number Theory	1
2006	MS	4	Algebra	1
2006	MS	5	Algebra	2
2006	MS	6	Geometry	1
2006	MS	7	Algebra	2
2006	MS	8	Algebra	2
2006	MS	9	Combinatorics	2
2006	MS	10	Combinatorics	3
2006	MS	11	Algebra	3
2006	MS	12	Number Theory	2
2006	MS	13	Geometry	2
2006	MS	14	Algebra	3
2006	HS	15	Geometry 3D	3
2006	HS	1	Algebra	2
2006	HS	2	Algebra	2
2006	HS	3	Geometry	2
2006	HS	4	Algebra	2
2006	HS	5	Algebra/Number Theory	2
2006	HS	6	Algebra	2
2006	HS	7	Algebra	3
2006	HS	8	Algebra/Complex Numbers	3
2006	HS	9	Combinatorics	3
2006	HS	10	Geometry	2
2006	HS	11	Number Theory	2
2006	HS	12	Geometry	3
2006	HS	13	Combinatorics	3
2006	HS	14	Number Theory	3
2006	HS	15	Geometry 3D	3
2006	HS	16	Algebra	3
2006	HS	17	Geometry 3D	3
2006	HS	18	Combinatorics	3
2006	HS	19	Probability	4
2006	HS	20	Number Theory	3
2006	HS	21	Geometry	4
2006	HS	22	Algebra	4
2006	HS	23	Number Theory	4
2006	HS	24	Geometry	4
2006	HS	25	Trigonometry	4

Year	Level	Problem	Topic	Difficulty
2007	MS	1	Calendar	1
2007	MS	2	Algebra	2
2007	MS	3	Geometry	2
2007	MS	4	Algebra	2
2007	MS	5	Number Theory	2
2007	MS	6	Algebra	1
2007	MS	7	Algebra	2
2007	MS	8	Algebra	2
2007	MS	9	Algebra	3
2007	MS	10	Algebra	2
2007	MS	11	Number Theory	4
2007	MS	12	Combinatorics	3
2007	MS	13	Algebra	3
2007	MS	14	Geometry 3D	3
2007	HS	15	Combinatorics	4
2007	HS	1	Algebra	2
2007	HS	2	Algebra	2
2007	HS	3	Algebra	2
2007	HS	4	Geometry	2
2007	HS	5	Algebra	2
2007	HS	6	Combinatorics/Number Theory	2
2007	HS	7	Trigonometry	2
2007	HS	8	Probability	3
2007	HS	9	Combinatorics	3
2007	HS	10	Algebra/Trigonometry/ Complex Numbers	2
2007	HS	11	Probability	3
2007	HS	12	Algebra/Logarithms	3
2007	HS	13	Geometry	3
2007	HS	14	Geometry	3
2007	HS	15	Number Theory	4
2007	HS	16	Combinatorics	4
2007	HS	17	Geometry	3
2007	HS	18	Analytic Geometry	3
2007	HS	19	Probability	4
2007	HS	20	Geometry	4
2007	HS	21	Number Theory	3
2007	HS	22	Algebra	5
2007	HS	23	Geometry	5
2007	HS	24	Number Theory	5
2007	HS	25	Number Theory	5

Year	Level	Problem	Topic	Difficulty
2008	MS	1	Number Theory	1
2008	MS	2	Calendar	1
2008	MS	3	Algebra	1
2008	MS	4	Algebra	1
2008	MS	5	Geometry	2
2008	MS	6	Algebra	2
2008	MS	7	Geometry	2
2008	MS	8	Combinatorics	2
2008	MS	9	Geometry	2
2008	MS	10	Geometry	3
2008	MS	11	Algebra	3
2008	MS	12	Combinatorics	2
2008	MS	13	Combinatorics/Geometry	3
2008	MS	14	Geometry/Algebra	3
2008	HS	15	Algebra	3
2008	HS	1	Number Theory	2
2008	HS	2	Algebra	1
2008	HS	3	Number Theory	2
2008	HS	4	Algebra	2
2008	HS	5	Algebra	2
2008	HS	6	Algebra	2
2008	HS	7	Analytic Geometry	2
2008	HS	8	Geometry 3D	3
2008	HS	9	Number Theory	2
2008	HS	10	Combinatorics	3
2008	HS	11	Geometry	3
2008	HS	12	Algebra	3
2008	HS	13	Probability	3
2008	HS	14	Geometry/Algebra	3
2008	HS	15	Algebra	3
2008	HS	16	Geometry	3
2008	HS	17	Algebra	4
2008	HS	18	Geometry	3
2008	HS	19	Geometry	3
2008	HS	20	Number Theory	3
2008	HS	21	Geometry 3D	4
2008	HS	22	Algebra	4
2008	HS	23	Geometry	5
2008	HS	24	Algebra	4
2008	HS	25	Probability/Geometry	5
2009	MS	1	Geometry	1

Year	Level	Problem	Topic	Difficulty
2009	MS	2	Number Theory	1
2009	MS	3	Calendar	1
2009	MS	4	Algebra	1
2009	MS	5	Algebra	2
2009	MS	6	Algebra	2
2009	MS	7	Combinatorics	3
2009	MS	8	Combinatorics/Geometry	2
2009	MS	9	Algebra	2
2009	MS	10	Geometry	2
2009	MS	11	Algebra	2
2009	MS	12	Geometry	3
2009	MS	13	Combinatorics	3
2009	MS	14	Geometry	3
2009	HS	15	Algebra	3
2009	HS	1	Algebra	2
2009	HS	2	Number Theory	1
2009	HS	3	Geometry	1
2009	HS	4	Algebra	2
2009	HS	5	Algebra	1
2009	HS	6	Combinatorics	2
2009	HS	7	Geometry	2
2009	HS	8	Number Theory	2
2009	HS	9	Algebra	2
2009	HS	10	Algebra	3
2009	HS	11	Geometry	3
2009	HS	12	Algebra/Number Theory	2
2009	HS	13	Probability	3
2009	HS	14	Geometry	3
2009	HS	15	Number Theory	3
2009	HS	16	Algebra/Complex Numbers	3
2009	HS	17	Combinatorics	3
2009	HS	18	Geometry	3
2009	HS	19	Algebra/Complex Numbers	4
2009	HS	20	Probability	4
2009	HS	21	Geometry	4
2009	HS	22	Analytic Geometry	4
2009	HS	23	Combinatorics	4
2009	HS	24	Geometry 3D	5
2009	HS	25	Algebra	5
2010	MS	1	Algebra	1
2010	MS	2	Algebra	1

Year	Level	Problem	Topic	Difficulty
2010	MS	3	Algebra	1
2010	MS	4	Geometry	1
2010	MS	5	Number Theory	1
2010	MS	6	Number Theory	2
2010	MS	7	Algebra	2
2010	MS	8	Number Theory	2
2010	MS	9	Combinatorics	2
2010	MS	10	Algebra	2
2010	MS	11	Combinatorics	2
2010	MS	12	Geometry	3
2010	MS	13	Combinatorics	3
2010	MS	14	Geometry	3
2010	MS	15	Combinatorics	2
2010	MS	16	Algebra	2
2010	MS	17	Geometry	2
2010	MS	18	Combinatorics	3
2010	MS	19	Geometry	3
2010	MS	20	Algebra	4
2010	HS	1	Algebra	1
2010	HS	2	Number Theory	1
2010	HS	3	Geometry	2
2010	HS	4	Algebra	1
2010	HS	5	Algebra	2
2010	HS	6	Algebra	3
2010	HS	7	Geometry	3
2010	HS	8	Algebra	2
2010	HS	9	Algebra	2
2010	HS	10	Algebra	2
2010	HS	11	Probability	3
2010	HS	12	Geometry	2
2010	HS	13	Algebra	3
2010	HS	14	Algebra	3
2010	HS	15	Number Theory	3
2010	HS	16	Geometry	3
2010	HS	17	Probability	4
2010	HS	18	Trigonometry	3
2010	HS	19	Geometry	3
2010	HS	20	Combinatorics	4
2010	HS	21	Algebra	4
2010	HS	22	Combinatorics	4
2010	HS	23	Geometry	4

Year	Level	Problem	Topic	Difficulty
2010	HS	24	Number Theory	4
2010	HS	25	Algebra	5
2010	HS	26	Analytic Geometry	4
2010	HS	27	Trigonometry	4
2010	HS	28	Algebra	5
2010	HS	29	Geometry	4
2010	HS	30	Algebra	5
2011	MS	1	Algebra	1
2011	MS	2	Geometry	1
2011	MS	3	Number Theory	1
2011	MS	4	Algebra	2
2011	MS	5	Algebra	1
2011	MS	6	Number Theory/Number Bases	2
2011	MS	7	Algebra	1
2011	MS	8	Geometry	1
2011	MS	9	Algebra	2
2011	MS	10	Geometry	2
2011	MS	11	Algebra	2
2011	MS	12	Combinatorics	3
2011	MS	13	Geometry	2
2011	MS	14	Number Theory	3
2011	MS	15	Geometry	2
2011	MS	16	Algebra	4
2011	MS	17	Combinatorics	4
2011	MS	18	Algebra	3
2011	MS	19	Combinatorics	4
2011	MS	20	Combinatorics/Geometry	4
2011	HS	1	Algebra	1
2011	HS	2	Geometry	1
2011	HS	3	Algebra	2
2011	HS	4	Geometry	1
2011	HS	5	Number Theory	2
2011	HS	6	Algebra	2
2011	HS	7	Number Theory	2
2011	HS	8	Algebra	2
2011	HS	9	Algebra	2
2011	HS	10	Probability/Geometry	3
2011	HS	11	Probability	3
2011	HS	12	Analytic Geometry	3
2011	HS	13	Algebra	3
2011	HS	14	Geometry/Algebra	4

Year	Level	Problem	Topic	Difficulty
2011	HS	15	Geometry 3D	3
2011	HS	16	Algebra	3
2011	HS	17	Combinatorics	3
2011	HS	18	Algebra	4
2011	HS	19	Combinatorics	5
2011	HS	20	Probability/Geometry	4
2011	HS	21	Algebra	4
2011	HS	22	Geometry	3
2011	HS	23	Trigonometry	4
2011	HS	24	Geometry	3
2011	HS	25	Number Theory	4
2011	HS	26	Combinatorics/Geometry	4
2011	HS	27	Number Theory	5
2011	HS	28	Geometry	5
2011	HS	29	Probability	4
2011	HS	30	Geometry 3D	5
2012	MS	1	Algebra	1
2012	MS	2	Geometry	2
2012	MS	3	Algebra	1
2012	MS	4	Combinatorics	2
2012	MS	5	Geometry	2
2012	MS	6	Geometry 3D	2
2012	MS	7	Algebra	1
2012	MS	8	Combinatorics	3
2012	MS	9	Geometry	2
2012	MS	10	Number Theory	3
2012	MS	11	Algebra	3
2012	MS	12	Geometry	2
2012	MS	13	Number Theory	3
2012	MS	14	Algebra	3
2012	MS	15	Geometry 3D	3
2012	MS	16	Algebra	3
2012	MS	17	Geometry	3
2012	MS	18	Number Theory	2
2012	MS	19	Combinatorics	3
2012	MS	20	Algebra	4
2012	HS	1	Algebra	2
2012	HS	2	Number Theory	2
2012	HS	3	Geometry	2
2012	HS	4	Geometry	2
2012	HS	5	Algebra	2

Year	Level	Problem	Topic	Difficulty
2012	HS	6	Number Theory	2
2012	HS	7	Geometry/Algebra	3
2012	HS	8	Algebra	3
2012	HS	9	Algebra/Logarithms	3
2012	HS	10	Geometry/Algebra	3
2012	HS	11	Algebra	3
2012	HS	12	Probability	3
2012	HS	13	Number Theory	3
2012	HS	14	Analytic Geometry	4
2012	HS	15	Number Theory	3
2012	HS	16	Algebra	4
2012	HS	17	Combinatorics	4
2012	HS	18	Combinatorics	3
2012	HS	19	Number Theory	3
2012	HS	20	Geometry	3
2012	HS	21	Probability	3
2012	HS	22	Geometry	4
2012	HS	23	Number Theory	4
2012	HS	24	Algebra	4
2012	HS	25	Number Theory	4
2012	HS	26	Geometry 3D	5
2012	HS	27	Combinatorics	4
2012	HS	28	Probability	4
2012	HS	29	Probability	4
2012	HS	30	Geometry 3D	5

Printed by “Combinatul Poligrafic”
Com. nr. 30173